Beginning Partial Differential Equations

Third Edition

Peter V. O'Neil
The University of Alabama
at Birmingham

WILEY

Published by John Wiley & Sons, Inc., Hoboken, New Jersey. All rights reserved.
Published simultaneously in Canada.

Library of Congress Cataloging-in-Publication data:

O'Neil, Peter V.
Beginning partial differential equations / Peter V. O'Neil. — Third edition.
pages cm
Includes bibliographical references and index.
ISBN 978-1-118-62994-9 (hardback)
1. Differential equations, Partial. I. Title.
QA377.O54 2014
515'.353—dc23 2013034307

Printed in the United States of America.

10 9 8 7 6 5 4 3 2 1

Contents

Preface

This edition is based on four themes: methods of solution of initial-boundary value problems, properties and existence of solutions, applications of partial differential equations, and use of software to carry out computations and graphics.

The focus is on equations of diffusion processes and wave motion, and on Dirichlet and Neumann problems. Following an introductory chapter, we look at methods applied to these equations in bounded and unbounded media, and in one and several space dimensions. The topics are organized to make it easy to match problems in specific settings to methods for writing solutions. Methods include Fourier series and integrals, the use of characteristics, integral solutions, integral transforms, and special functions and eigenfunction expansions.

Properties of solutions that are considered include existence and uniqueness issues, maximum and mean value principles, integral representations, and sensitivity of solutions to initial and boundary conditions.

In addition to standard material for an introductory course, topics include traveling-wave solutions of Burger's equation, damped wave motion, heat and wave equations with forcing terms, a general treatment of eigenfunction expansions, a complete solution of the telegraph equation using the Fourier transform, the use of characteristics to solve Cauchy problems and vibrating string problems with moving ends, double Fourier series solutions, and the Poisson-Kirchhoff integral solution of the wave equation in two dimensions. There are also proofs of important theorems, including an existence theorem for a Dirichlet problem and a convergence theorem for Fourier series.

Finally, there is a section on the use of MAPLE™ to carry out computations and experiment with graphics. MATLAB ®, MATHEMATICA ®, and other packages may also be used for these numerical aspects of partial differential equations.

Chapter 1

First Ideas

We will begin a study of partial differential equations by deriving equations modeling diffusion processes and wave motion. These are widely applicable in the physical and life sciences, engineering, economics, and other areas. Following this, we will lay the foundations for the Fourier method, which is used to write solutions for many kinds of problems, and then solve two eigenvalue/eigenfunction problems that occur frequently when this method is used.

The chapter concludes with a proof of a theorem on the convergence of Fourier series.

1.1 Two Partial Differential Equations

1.1.1 The Heat, or Diffusion, Equation

We will derive a partial differential equation modeling heat flow in a medium. Although we will speak in terms of heat flow because it is familiar to us, the heat equation applies to general diffusion processes, which might be a flow of energy, a dispersion of insect or bacterial populations in controlled environments, changes in the concentration of a chemical dissolving in a fluid, or many other phenomena of interest. For this reason the heat equation is also called the diffusion equation.

Consider a bar of material of constant density, ρ, having uniform cross sections with area A. The lateral surface of the bar is insulated, so there is no heat loss across this surface.

Place an x-axis along the length, L, of the bar and assume that at a given time, the temperature is the same along any cross section perpendicular to this axis, although it may vary from one cross section to another. We will derive an equation for $u(x,t)$, the temperature in the cross section of the bar at x, at time t. In the context of diffusion, $u(x,t)$ is called a *density distribution function*.

Let c be the specific heat of the material of the bar. This is the amount of heat energy that must be supplied to a unit mass of the material to raise

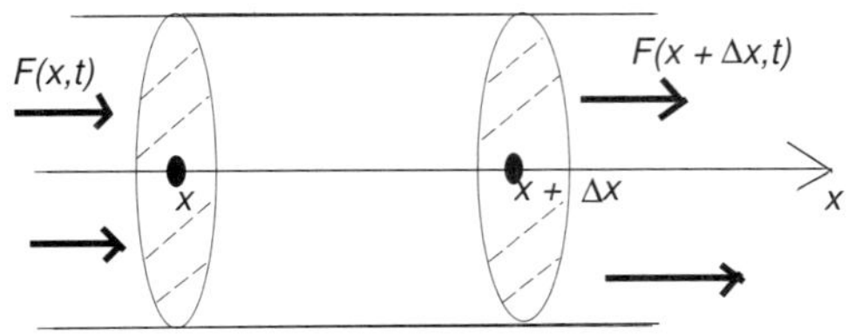

Figure 1.1: Flux in segment = rate in minus rate out.

its temperature one degree. The segment of bar between x and $x + \Delta x$ has mass $\rho A \Delta x$, and it will take approximately $\rho c A u(x,t)\Delta x$ units of heat energy to change the temperature of this segment from zero to $u(x,t)$, its temperature at time t.

The total heat energy in this segment at any time $t > 0$ is

$$E(x, \Delta x, t) = \int_x^{x+\Delta x} \rho c A u(\xi, t)\, d\xi.$$

This amount of heat energy within the segment at time t can increase in two ways: heat energy may flow into the segment across its ends (this change is the flux of the energy), and/or there may be a source or loss of heat energy within the segment. This can occur if there is, say, a chemical reaction or if the material is radioactive.

The rate of change of the temperature within the segment, with respect to time, is therefore

$$\begin{aligned}\frac{\partial E}{\partial t} &= \text{flux plus source or sink} \\ &= \int_x^{x+\Delta x} \rho c A \frac{\partial u}{\partial t}(\xi, t)\, d\xi.\end{aligned}$$

Assume for now that there is no source or loss of energy within the bar. Then

$$\text{flux} = \int_x^{x+\Delta x} \rho c A \frac{\partial u}{\partial t}(\xi, t)\, d\xi. \tag{1.1}$$

Now let $F(x,t)$ be the amount of heat energy per unit area flowing across the cross section at x at time t, in the direction of increasing x. Then the flux of the energy into the segment between x and $x + \Delta x$ at time t is the rate of flow into the segment across the section at x, minus the rate of flow out of the segment across the section at $x + \Delta x$ (Figure 1.1):

$$\text{flux} = AF(x,t) - AF(x + \Delta x, t).$$

Write this as

$$\text{flux} = -A(F(x + \Delta x, t) - F(x,t)). \tag{1.2}$$

Now recall Newton's law of cooling, which states that heat energy flows from the warmer to the cooler region, and the amount of heat energy is proportional to the temperature difference (gradient). This means that

$$F(x,t) = -K\frac{\partial u}{\partial x}(x,t).$$

The positive constant of proportionality, K, is called the *heat conductivity* of the bar. The negative sign in this equation is due to the fact that energy flows from the warmer to the cooler segment. Substitute this expression for $F(x,t)$ into equation 1.2 to obtain

$$\text{flux} = -A\left(-K\frac{\partial u}{\partial x}(x+\Delta x,t) + K\frac{\partial u}{\partial x}(x,t)\right).$$

Write this as

$$\text{flux} = \int_x^{x+\Delta x} \frac{\partial}{\partial x}\left(KA\frac{\partial u}{\partial x}(\xi,t)\right)\,d\xi. \tag{1.3}$$

From equations 1.1 and 1.3 for the flux, we have

$$\int_x^{x+\Delta x} \rho c A\frac{\partial u}{\partial t}(\xi,t)\,d\xi = \int_x^{x+\Delta x} \frac{\partial}{\partial x}\left(KA\frac{\partial u}{\partial x}(\xi,t)\right)\,d\xi.$$

Divide out the common factor A and write this equation as

$$\int_x^{x+\Delta x}\left[\rho c\frac{\partial u}{\partial t}(\xi,t) - \frac{\partial}{\partial x}\left(K\frac{\partial u}{\partial x}(\xi,t)\right)\right]\,d\xi = 0.$$

This equation must be valid for any choices of x and $x+\Delta x$, as long as

$$0 < x < x+\Delta x < L.$$

If the integrand were nonzero at some x, then, assuming continuity of this integrand (which is reasonable on physical grounds), it would be nonzero, therefore strictly positive or strictly negative on some interval $(x, x+\Delta x)$. This would force this integral to be positive or negative, not zero, for this x and Δx, and this is a contradiction. We conclude that the integrand must be identically zero, hence

$$\rho c\frac{\partial u}{\partial t} = K\frac{\partial^2 u}{\partial x^2}.$$

It is convenient to denote partial derivatives using subscripts. In this notation,

$$u_t = k u_{xx}, \tag{1.4}$$

where $k = K/c\rho$ is called the *diffusivity* of the material of the bar. Equation 1.4 is the *one-dimensional heat, or diffusion, equation.* This equation, with appropriate boundary and initial conditions, models a wide range of diffusion phenomena, providing a setting for a mathematical analysis to draw conclusions about the behavior of the process under study.

If we allow for a source term $Q(x,t)$, then the heat equation is

$$u_t = ku_{xx} + Q(x,t). \tag{1.5}$$

We say that equation 1.4 is *homogeneous*. Because of the $Q(x,t)$ term, equation 1.5 is *nonhomogeneous*. Both equations are *second-order* partial differential equations because they contain at least one second derivative term, but no higher derivative. Both equations are also *linear*, which means they are linear in the unknown function and its derivatives. By contrast, the second-order partial differential equation

$$u_t = ku_{xx} + uu_x$$

is nonlinear because of the uu_x term, which allows for an interaction between the density function, u, and its rate of change with respect to x.

The linear, homogeneous heat equation $u_t = ku_{xx}$ has the important features that a finite sum of solutions and a product of a solution by a constant are again solutions. That is, if $u_1(x,y)$ and $u_2(x,y)$ are solutions, then $au_1(x,y)+bu_2(x,y)$ is also a solution for any numbers a and b. This can be verified by substituting $au_1 + bu_2$ into equation 1.4. This is not the case with the nonhomogeneous equation 1.5, as can also be seen by substitution.

Everyday experience suggests that to know the temperature in a bar of material at any time we have to have some information, such as the temperature throughout the bar at some particular time (this is an *initial condition*), together with information about the temperatures at the ends of the bar (these are *boundary conditions*). A typical initial condition has the form

$$u(x,0) = f(x) \text{ for } 0 < x < L,$$

in which $f(x)$ is a given function. *Initial* is taken as time zero as a convenience.

Boundary conditions specify conditions at end points of the space variable (or perhaps on a surface in higher dimensional models). These can take different forms. One commonly seen set of boundary conditions is

$$u(0,t) = \alpha(t), u(L,t) = \beta(t) \text{ for } 0 < x < L,$$

where $\alpha(t)$ and $\beta(t)$ are given functions. These specify conditions at the left and right ends of the material at all times.

Boundary conditions may also reflect other physical conditions at the boundary. We will see some of these when we solve specific problems in different settings.

A problem consisting of the heat equation, together with initial and boundary conditions, is called a *initial-boundary value problem for the heat equation.*

1.1.2 The Wave Equation

Imagine a string (guitar string, wire, telephone line, power line, or the like) suspended between two points. We want to describe the motion of the string

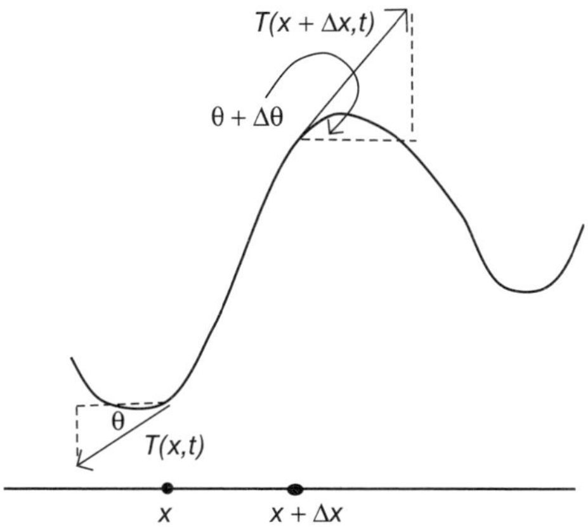

Figure 1.2: Segment of string between x and $x + \Delta x$.

if it is fixed at its ends, displaced in a specified way and released with a given velocity.

Place an x-axis along the straightened string from 0 to L, and assume that each particle of string moves only vertically in a plane. We seek a function $u(x,t)$ so that, at any time $t \geq 0$, the graph of the function $u = u(x,t)$ gives the position or shape of the string at that time. This enables us to view snapshots of the string in motion.

Begin with a simple case by neglecting damping effects, such as air resistance and the weight of the string. Let $\mathbf{T}(x,t)$ be the tension in the string at point x and time t, and assume that this acts tangentially to the string. The magnitude of this vector is $T(x,t) = \| \mathbf{T}(x,t) \|$. Also assume that the mass, ρ, per unit length is constant.

Apply Newton's second law of motion to the segment of string between x and $x+\Delta x$. This states that the net force on the segment due to the tension is equal to the acceleration of the center of mass of the segment times the mass of the segment. This is a vector equation, meaning that we can match the horizontal components and the vertical components of both sides. Looking at the vertical components in Figure 1.2 gives us approximately

$$T(x + \Delta x, t)\sin(\theta + \Delta\theta) - T(x,t)\sin(\theta) = \rho(\Delta x)u_{tt}(\overline{x}, t),$$

in which $\overline{x}$ is the center of mass of this segment of string. Then

$$\frac{T(x + \Delta x, t)\sin(\theta + \Delta\theta) - T(x,t)\sin(\theta)}{\Delta x} = \rho u_{tt}(\overline{x}, t).$$

The vertical component, $v(x,t)$, of the tension is

$$v(x,t) = T(x,t)\sin(\theta).$$

Then

$$\frac{v(x+\Delta x,t)-v(x,t)}{\Delta x}=\rho u_{tt}(\overline{x},t).$$

Let $\Delta x\to 0$. Then $\overline{x}\to x$, and this equation yields

$$v_x=\rho u_{tt}.$$

The horizontal component of the tension is

$$h(x,t)=T(x,t)\cos(\theta).$$

But

$$v(x,t)=h(x,t)\tan(\theta)=h(x,t)u_x,$$

so

$$v_x=(hu_x)_x=\rho u_{tt}.$$

By assumption, the horizontal component of the tension on the entire segment of string is zero:

$$h(x+\Delta x,t)-h(x,t)=0.$$

Therefore, $h(x,t)$ is independent of x, and

$$(hu_x)_x=hu_{xx}.$$

Then

$$hu_{xx}=\rho u_{tt}.$$

Or, in its more traditional form,

$$u_{tt}=c^2u_{xx}, \tag{1.6}$$

where $c^2=h/\rho$. Equation 1.6 is the *one-dimensional wave equation* (one space dimension).

If a *forcing term* is included to allow other forces acting on the string, then the wave equation may take the form

$$u_{tt}=c^2u_{xx}+P(x,t).$$

As with the heat equation, we attempt to solve the wave equation subject to initial and boundary conditions specifying the position of the string at time $t=0$, and the forces that set the string in motion.

The boundary conditions if the ends of the string are fixed are

$$u(0,t)=u(L,t)=0 \text{ for } t>0.$$

We will also see variations on these boundary conditions. For example, if the ends are in motion, with their positions at time t given as functions of t, then

$$u(0,t)=\alpha(t),u(L,t)=\beta(t) \text{ for } t>0,$$

for some given functions $\alpha(t)$ and $\beta(t)$.

Initial conditions take the form

$$u(x,0) = \varphi(x) \text{ and } u_t(x,0) = \psi(x) \text{ for } 0 < x < L,$$

specifying the initial position and velocity of the string. Equation 1.6, together with boundary and initial conditions, is called an *initial-boundary value problem for the wave equation.*

As we develop methods of solving these and other partial differential equations, under a variety of initial and boundary conditions, we will also explore properties of solutions and questions such as the sensitivity of solutions to small perturbations of initial and boundary conditions.

Problems for Section 1.1

1. Show that
$$u(x,t) = \cos(\alpha\pi x)e^{-\alpha^2\pi^2 t}$$
is a solution of the heat equation with $k = 1$, on any interval $[0, L]$.

2. Show that
$$u(x,t) = t^{-3/2}e^{-x^2/4kt}$$
is a solution of $u_t = ku_{xx}$ for $x > 0, t > 0$. Show also that this solution is unbounded.

3. Show that
$$u(x,t) = a\sin\left(\frac{n\pi x}{L}\right)\cos\left(\frac{n\pi ct}{L}\right)$$
satisfies wave equation 1.6, with a any constant, c and L positive constants, and n any positive integer.

4. Let f be a differentiable function of a single variable, defined on the entire real line. Show that
$$u(x,t) = \frac{1}{2}(f(x-ct) + f(x+ct))$$
is a solution of the wave equation $u_{tt} = c^2u_{xx}$ for all x and t, and that $u(x,0) = f(x)$.

5. Let $\psi(x)$ be continuous on the real line. Let
$$u(x,t) = \frac{1}{2c}\int_{x-ct}^{x+ct}\psi(s)\,ds.$$
Show that $u(x,t)$ satisfies the wave equation and that
$$u_t(x,0) = \psi(x) \text{ for } 0 < x < L.$$

6. Let φ and ψ be continuous on $[0, L]$. Let

$$u(x,t) = \frac{1}{2}(\varphi(x-ct) + \varphi(x+ct)) + \frac{1}{2c}\int_{x-ct}^{x+ct} \psi(s)\,ds.$$

Show that $u(x,t)$ satisfies the wave equation and also the initial conditions $u(x,0) = \varphi(x)$ and $u_t(x,0) = \psi(x)$.

Problems 7–12 deal with a classification of second-order partial differential equations that are linear with constant coefficients in the second derivative terms. Such an equation has the form

$$Au_{xx} + Bu_{xt} + Cu_{tt} + H(x,t,u,u_x,u_t) = 0. \tag{1.7}$$

A, B, and C are constants; A and B are not both zero; and $H(x,t,u,u_x,u_t)$ is any function of x, t, u, u_x, and u_t. Thus the equation may not be linear in the first derivative terms or terms involving u. It is always possible to transform equation 1.7 to one of three standard, or canonical, forms. These problems explore how to do this.

7. Start with a change of variables

$$\xi = x + at, \eta = x + bt.$$

Show that this transformation from the x, t-plane to a ξ, η-plane is invertible if $a \neq b$, and that

$$x = \frac{1}{b-a}(b\xi - a\eta), t = \frac{1}{b-a}(\eta - \xi).$$

8. Let $u(x(\xi,\eta), t(\xi,\eta)) = V(\xi,\eta)$, obtained by substituting for x and y in terms of ξ and η in equation 1.7. Show that the resulting partial differential equation for V is

$$\begin{aligned}&(A + aB + a^2C)V_{\xi\xi} + (2A + (a+b)B + 2abC)V_{\xi\eta}\\&+ (A + bB + b^2C)V_{\eta\eta} + K(\xi,\eta,V,V_\xi,V_\eta) = 0.\end{aligned} \tag{1.8}$$

Hint: Use the chain rule to compute u_{xx}, u_{xt}, and u_{tt} in terms of partial derivatives of $V(\xi,\eta)$.

9. Suppose $B^2 - 4AC > 0$. Try to choose a and b to make the coefficients of $V_{\xi\xi}$ and $V_{\eta\eta}$ vanish. This requires that we solve for a and b so that

$$Ca^2 + Ba + A = 0 \text{ and } Cb^2 + Bb + A = 0.$$

Notice that a and b both satisfy the same quadratic equation, having coefficients A, B, and C. Show that, if $C \neq 0$, then equation 1.7 transforms to

$$V_{\xi\eta} + K(\xi,\eta,V,V_\xi,V_\eta) = 0.$$

by choosing

$$a = \frac{-B + \sqrt{B^2 - 4AC}}{2C} \text{ and } b = \frac{-B - \sqrt{B^2 - 4AC}}{2C}.$$

In this case we say that equation 1.7 is *hyperbolic.* The transformed equation is the *canonical form of the hyperbolic equation.*

If $C = 0$, show that we can choose

$$\xi = t, \eta = -\frac{B}{A}x + t$$

to obtain the hyperbolic canonical form.

10. Show that, if $B^2 - 4AC = 0$, then by choosing $a = 0$ and $b = -B/2C$, equation 1.7 transforms to

$$V_{\xi\xi} + K(\xi, \eta, V, V_\xi, V_\eta) = 0.$$

In this case equation 1.7 is called *parabolic* and the transformed equation is called the *canonical form of the parabolic equation.*

11. Finally, suppose $B^2 - 4AC < 0$. Now the roots of $Ca^2 + Ba + A = 0$ are complex, say $p \pm iq$. Define the transformation

$$\xi = x + pt, \eta = qt.$$

and show that this transforms equation 1.7 to

$$V_{\xi\xi} + V_{\eta\eta} + K(\xi, \eta, V, V_\xi, V_\eta) = 0.$$

In this case, equation 1.7 is said to be *elliptic* and the transformed equation is the *canonical form of the elliptic equation.*

12. Classify the diffusion equation and the wave equation as being elliptic, parabolic, or hyperbolic.

In each of problems 13–17, classify the partial differential equation and determine its canonical form.

13. $4u_{xx} - 2u_{xt} + u_{tt} + 2u_x - xu = 0.$

14. $2u_{xx} + u_{xt} - 4u_{tt} + x + t = 0.$

15. $u_{xx} - 3u_{xt} - xu = 0.$

16. $u_{xx} + 9u_{tt} + x^2 - tu = 0.$

17. $u_{xx} - 2u_{xt} + 3u_{tt} + 12u^2 = 0.$

1.2 Fourier Series

In attempting to solve problems involving the heat equation, French mathematician Joseph Fourier (1768–1830) announced that he could write solutions by expanding the initial temperature function in an infinite series of sines and/or cosines of different frequencies. Because nearly any function (for example, a differentiable function) could be an initial temperature function, this led to the astounding assertion that almost any function one could think of had such a trigonometric series representation. This was too much for the rest of the scientific community to accept.

Nevertheless, Fourier's method did appear to solve significant problems. Intensive research, carried out in the eighteenth and nineteenth centuries, justified Fourier's claims, and Fourier series now have many applications. In this section we outline the fundamental idea of a Fourier series, enabling us to use these series to solve initial-boundary value problems.

1.2.1 The Fourier Series of a Function

Given $f(x)$ defined on $[-L, L]$, we want to choose numbers $a_0, a_1, \cdots$ and $b_1, b_2, \cdots$ such that

$$f(x) = \frac{1}{2}a_0 + \sum_{n=1}^{\infty}\left[a_n \cos\left(\frac{n\pi x}{L}\right) + b_n \sin\left(\frac{n\pi x}{L}\right)\right] \tag{1.9}$$

on this interval. This is not always possible, but we will explore the idea to see when it might work.

Fourier was not the first to imagine such a thing. The great Swiss mathematician Leonhard Euler (1707–1783) devised a way of calculating the $a_n's$ and $b_n's$ in the series 1.9. While lacking in rigor, Euler's approach is interesting and actually leads to the correct choice of the coefficients. We follow Euler's reasoning here, with a proof given in section 1.4.

Euler's approach was based on some easily derived trigonometric integrals. If n and k are positive integers, then

$$\int_{-L}^{L} \cos\left(\frac{n\pi x}{L}\right) \sin\left(\frac{k\pi x}{L}\right) dx = 0,$$

$$\int_{-L}^{L} \cos\left(\frac{n\pi x}{L}\right) \cos\left(\frac{k\pi x}{L}\right) dx = 0 \text{ if } n \neq k,$$

and

$$\int_{-L}^{L} \sin\left(\frac{n\pi x}{L}\right) \sin\left(\frac{k\pi x}{L}\right) dx = 0 \text{ if } n \neq k.$$

These are called *orthogonality relations* for reasons that will be clarified when we treat eigenfunction expansions in Chapter 7.

In addition,

$$\int_{-L}^{L} \cos^2\left(\frac{n\pi x}{L}\right) dx = \int_{-L}^{L} \sin^2\left(\frac{n\pi x}{L}\right) dx = L \text{ for } n = 1, 2, \cdots .$$

Now assume equation 1.9, and suppose that we can interchange the summation and an integration. In this case,

$$\begin{aligned} &\int_{-L}^{L} f(x)\, dx \\ &= \int_{-L}^{L} \frac{1}{2}a_0\, dx + \sum_{n=1}^{\infty}\left[a_n \int_{-L}^{L} \cos\left(\frac{n\pi x}{L}\right) dx + b_n \int_{-L}^{L} \sin\left(\frac{n\pi x}{L}\right) dx\right] \\ &= La_0, \end{aligned}$$

because the integrals of $\cos(n\pi x/L)$ and $\sin(n\pi x/L)$ over $[-L, L]$ are all zero. The integrated equation therefore reduces to

$$\int_{-L}^{L} f(x)\, dx = La_0,$$

from which we conclude that

$$a_0 = \frac{1}{L}\int_{-L}^{L} f(x)\, dx. \tag{1.10}$$

This is a formula for a_0. Next we want to obtain formulas for a_k with $k = 1, 2, \cdots$. Let k be any positive integer. Multiply equation 1.9 by $\cos(k\pi x/L)$ and integrate to obtain

$$\begin{aligned} &\int_{-L}^{L} f(x)\cos\left(\frac{k\pi x}{L}\right) dx \\ &= \int_{-L}^{L} \frac{1}{2}a_0 \cos\left(\frac{k\pi x}{L}\right) dx \\ &+ \sum_{n=1}^{\infty}\left[a_n \int_{-L}^{L} \cos\left(\frac{n\pi x}{L}\right)\cos\left(\frac{k\pi x}{L}\right) dx + b_n \int_{-L}^{L} \sin\left(\frac{n\pi x}{L}\right)\cos\left(\frac{k\pi x}{L}\right) dx\right]. \end{aligned}$$

Now,

$$\int_{-L}^{L} \cos\left(\frac{n\pi x}{L}\right) dx = 0.$$

Further, by the orthogonality relations, all of the integrals in the summation are zero except for the integral

$$\int_{-L}^{L} \cos\left(\frac{k\pi x}{L}\right)\cos\left(\frac{k\pi x}{L}\right) dx,$$

which occurs when $n = k$. This integral equals L. We therefore have

$$\int_{-L}^{L} f(x)\cos\left(\frac{k\pi x}{L}\right) dx = a_k L,$$

from which

$$a_k = \frac{1}{L}\int_{-L}^{L} f(x)\cos\left(\frac{k\pi x}{L}\right) \text{ for } k = 1, 2, 3, \cdots. \tag{1.11}$$

Notice that this reproduces the formula for a_0 when $k = 0$.

Similarly, if we multiply equation 1.9 by $\sin(k\pi x/L)$ and integrate term by term, all terms vanish except the $n = k$ term in the integrals of the sine terms, and we obtain

$$b_k = \frac{1}{L}\int_{-L}^{L} f(x)\sin\left(\frac{k\pi x}{L}\right) dx. \tag{1.12}$$

Equations 1.10–1.12 are the *Fourier coefficients* of $f(x)$ on $[-L, L]$. When these Fourier coefficients are used, the series on the right side of equation 1.9 is called the *Fourier series* of $f(x)$ on $[-L, L]$.

Now we must be careful not to overreach. Although we have a plausible rationale for the selection of the Fourier coefficients of a function, we have no reason to believe that this Fourier series actually converges to the function at all (or any!) points of the interval. The following two examples are revealing in this regard.

Example 1.1 Let

$$f(x) = \begin{cases} 0 & \text{for } -3 \le x < 0, \\ 2 + x & \text{for } 0 \le x \le 3. \end{cases}$$

We will write the Fourier series of $f(x)$ on $[-3, 3]$. Compute the coefficients:

$$\begin{aligned} a_0 &= \frac{1}{3}\int_{-3}^{3} f(x)\,dx \\ &= \frac{1}{3}\int_{0}^{3} (2 + x)\,dx = \frac{7}{2}, \end{aligned}$$

$$\begin{aligned} a_n &= \frac{1}{3}\int_{-3}^{3} f(x)\cos\left(\frac{n\pi x}{L}\right) dx \\ &= \frac{1}{3}\int_{0}^{3} (2 + x)\cos\left(\frac{n\pi x}{L}\right) dx \\ &= \frac{3((-1)^n - 1)}{n^2\pi^2}, \end{aligned}$$

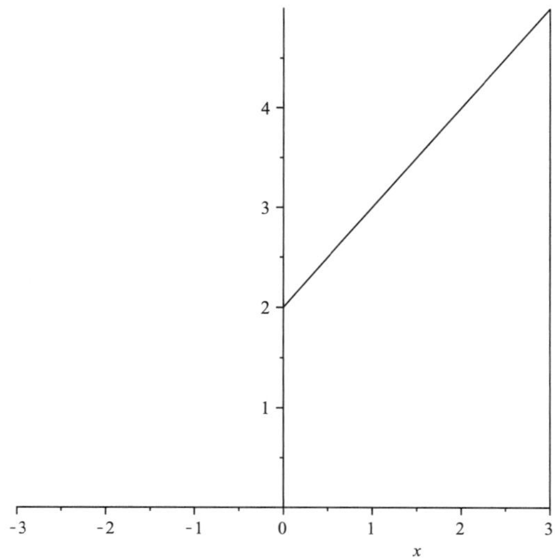

Figure 1.3: Graph of $f(x)$ from example 1.1.

and

$$\begin{aligned} b_n &= \frac{1}{3}\int_{-3}^{3} f(x)\sin\left(\frac{n\pi x}{L}\right)dx \\ &= \frac{1}{3}\int_{0}^{3} (2+x)\sin\left(\frac{n\pi x}{L}\right)dx \\ &= \frac{2-5(-1)^n}{n\pi}. \end{aligned}$$

The Fourier series of $f(x)$ on $[-3,3]$ is

$$\frac{7}{4}+\sum\left[\frac{3((-1)^n-1)}{n^2\pi^2}\cos\left(\frac{n\pi x}{3}\right)+\frac{2-5(-1)^n}{n\pi}\sin\left(\frac{n\pi x}{3}\right)\right].$$

Figure 1.3 is a graph of the function, and Figures 1.4 and 1.5 compare the function with the 10th and 50th partial sums, respectively, of its Fourier series. These graphs suggest that the series converges to $f(x)$ for $-3 < x < 0$ and for $0 < x < 3$. However, at $x = 0$, the series does not appear to converge to $f(0)$, which is 2. And at both 3 and -3, the Fourier series is the same:

$$\frac{7}{4}+\sum_{n=1}^{\infty}\left(\frac{3((-1)^n-1)}{n^2\pi^2}(-1)^n\right).$$

This series cannot converge to both $f(-3) = 0$ and to $f(3) = 5$.

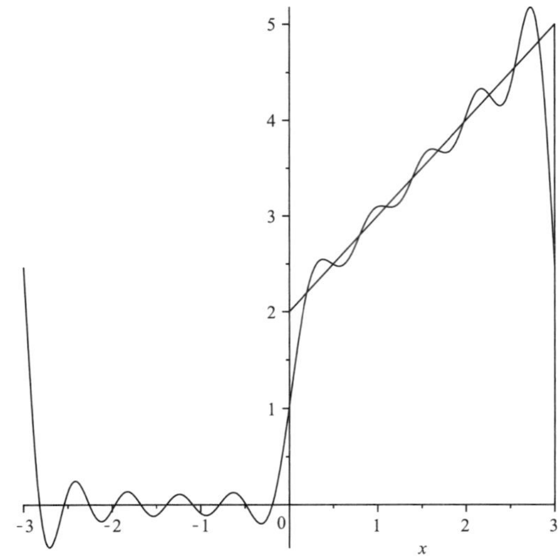

Figure 1.4: Comparison of $f(x)$ with the 10th partial sum in example 1.1.

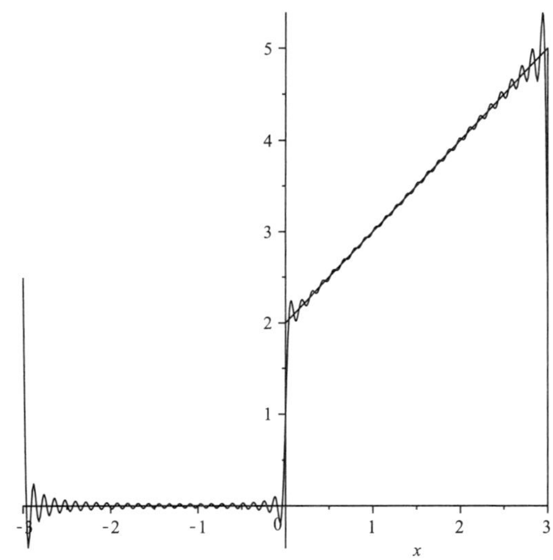

Figure 1.5: Comparison of $f(x)$ with the 50th partial sum in example 1.1.

Example 1.2 Let

$$g(x) = \begin{cases} -1 & \text{for } -2 \leq x < 1, \\ 3 & \text{for } 1 \leq x < 3/2, \\ -5 & \text{for } 3/2 \leq x \leq 2. \end{cases}$$

The Fourier coefficients of $g(x)$ on $[-2, 2]$ are

$$\begin{aligned} a_0 &= -2, \\ a_n &= \frac{1}{n\pi}(-4\sin(n\pi/2) + 8\sin(3n\pi/4)), \\ b_n &= \frac{4}{n\pi}((-1)^n + \cos(n\pi/2) - 2\cos(3n\pi/4)). \end{aligned}$$

The Fourier series is

$$-1 + \sum_{n=1}^{\infty} \left[\frac{1}{n\pi} (-4\sin(n\pi/2) + 8\sin(3n\pi/4)) \cos\left(\frac{n\pi x}{2}\right) \right.$$
$$\left. + \frac{4}{n\pi} ((-1)^n + \cos(n\pi/2) - 2\cos(3n\pi/4)) \sin\left(\frac{n\pi x}{2}\right)\right).$$

Figure 1.6 is a graph of this function, and Figures 1.7 and 1.8 are graphs of the tenth and fiftieth partial sums of the Fourier series, respectively. It does appear that the series converges to $g(x)$ for $-2 < x < 1$, $1 < x < 3/2$, and $3/2 < x < 2$. However, it is not clear what the series converges to at $x = -2, 1, 3/2$ or 2. And, as we saw in example 1.1, this Fourier series is the same at both end points of the interval, even though $g(-2) \neq g(2)$.

Because of examples like these, we need something to tell us the sum of a Fourier at points on the interval. One criterion for convergence is in terms of the familiar notions of continuity and differentiability. We say that $f(x)$ is *piecewise continuous* on $[a, b]$ if the following three conditions are satisfied:

1. $f(x)$ is continuous at all but possibly finitely many points of $[a, b]$.
2. If there is a point c with $a < c < b$ at which $f(x)$ is discontinuous, then $\lim_{x \to c-} f(x)$ and $\lim_{x \to c+} f(x)$ are both finite. That is, $f(x)$ has finite one-sided limits at every point of discontinuity interior to the interval (if there are any such points).
3. $\lim_{x \to a+} f(x)$ and $\lim_{x \to b-} f(x)$ are both finite. This means that, at the end points of the interval, the function has finite limits as x approaches the end point from within the interval.

These conditions mean that any discontinuities the function has on the interval are *jump discontinuities*, so-called because the graph has a gap or jump at such a point. The function of example 1.1 has a jump discontinuity at $x = 0$ (Figure 1.3), while the function of example 1.2 has jump discontinuities at $x = 1$ and $x = 3/2$ (Figure 1.6). Both of these functions are piecewise continuous on their interval of definition.

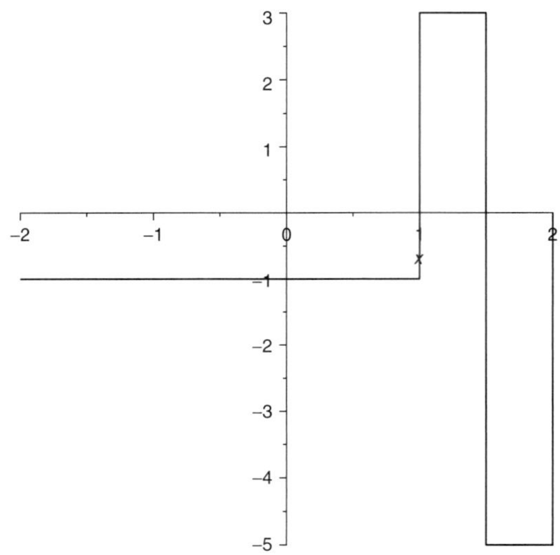

Figure 1.6: Graph of $g(x)$ of example 1.2.

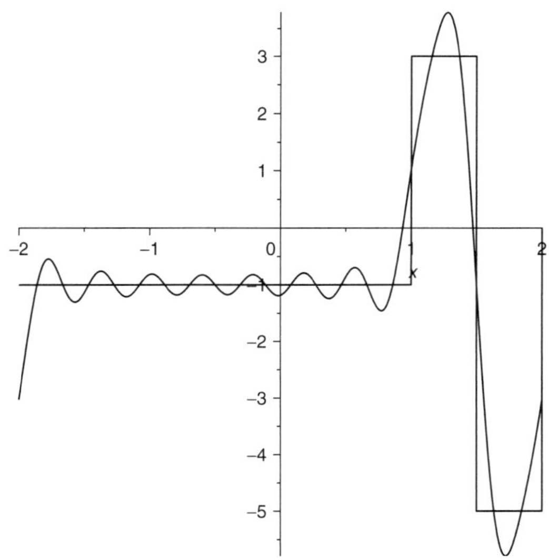

Figure 1.7: Comparison of $g(x)$ with the 10th partial sum in example 1.2.

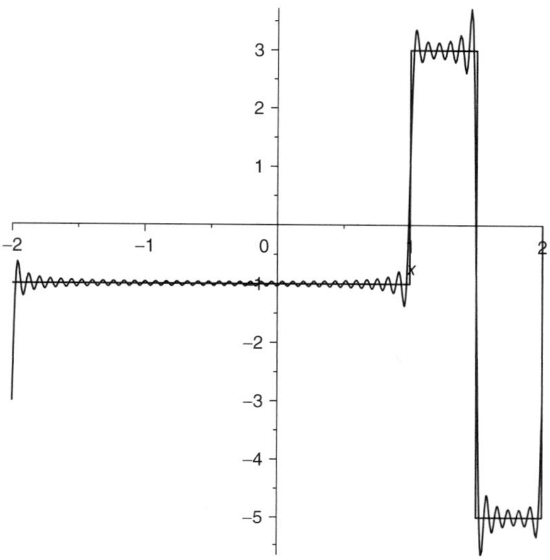

Figure 1.8: Comparison of $g(x)$ with the 50th partial sum in example 1.2.

$f(x)$ is *piecewise smooth* on $[a, b]$ if $f(x)$ and its derivative $f'(x)$ are both piecewise continuous on the interval.

Piecewise smooth means that the graph has a continuous tangent at all but finitely many points, and any discontinuities of the function exhibit themselves in finite jumps or gaps in the graph. The functions of examples 1.1 and 1.2 are piecewise smooth.

Finally, we will use the standard notation

$$f(x-) = \lim_{h \to 0+} f(x-h) \text{ and } f(x+) = \lim_{h \to 0+} f(x+h).$$

$f(x-)$ is the *left limit* of the function at x, and $f(x+)$ is the *right limit* at x. The plus and minus signs in the notation refer only to left and right limits, and x itself may be positive, negative or zero.

In example 1.1,

$$f(0-) = 0 \text{ and } f(0+) = 2,$$

while for $-3 < x < 0$ and $0 < x < 3$, $f(x-) = f(x+) = f(x)$. Further, $f(-3+) = 0$ and $f(3-) = 5$.

At any point at which the function is continuous, the left and right limits equal the function value at the point.

In example 1.2,

$$g(1-) = \lim_{h \to 0+} g(1-h) = -1$$

while

$$g(1+) = \lim_{h \to 0+} g(1+h) = 3.$$

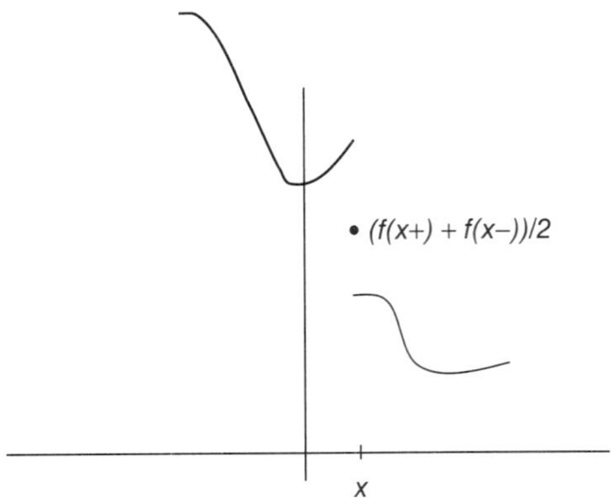

Figure 1.9: Convergence of a Fourier series at a jump discontinuity.

For $-2 < x < 1, 1 < x < 3/2$ and $3/2 < x < 2$, $g(x-) = g(x+) = g(x)$ because $g(x)$ is continuous on these intervals. And, $g(-2+) = -1$ and $g(2-) = -5$.

With these ideas and notation, we can state the following.

Theorem 1.1 (Convergence of Fourier Series) Let $f(x)$ be piecewise smooth on $[-L, L]$. If $-L < x < L$, then the Fourier series of $f(x)$ on this interval converges to

$$\frac{1}{2}(f(x-) + f(x+)).$$

Further, at both $x = L$ and $x = -L$, the Fourier series converges to

$$\frac{1}{2}(f(-L+) + f(L-)).$$

Figure 1.9 displays this behavior. If the function has a jump discontinuity at x, then the graph has a gap at x and the Fourier series converges to the average of the left and right limits of the function at x. This is the point midway between the ends of the graph at the gap. At any x where the function is continuous, the series converges to $f(x)$, because at such a point, $f(x-) = f(x+) = f(x)$.

At both end points L and $-L$, the Fourier series converges to the average of the left limit of the function at L, and the right limit at $-L$.

In example 1.1, the Fourier series of $f(x)$ converges to

$$\begin{cases} 0 & \text{for } -3 < x < 0, \\ 1 & \text{at } x = 0, \\ 2 + x & \text{for } 0 < x < 3 \\ 5/2 & \text{at } x = -3 \text{ and at } x = 3. \end{cases}$$

This conclusion at the end points 3 and -3 follows from the facts that $f(-3+) = 0$ and $f(3-) = 5$, and the average of these limits is $5/2$.

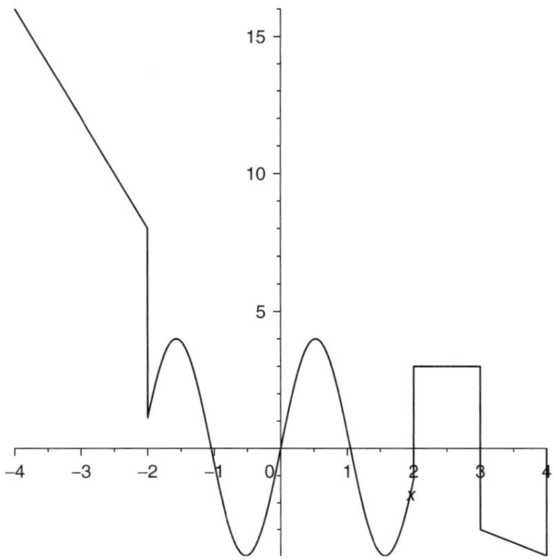

Figure 1.10: The function $w(x)$ of example 1.3.

In example 1.2, the Fourier series of $g(x)$ converges to

$$\begin{cases} -1 & \text{for } -2 < x < 1, \\ 1 & \text{at } x = 1, \\ 3 & \text{for } 1 < x < 3/2, \\ -1 & \text{at } x = 3/2, \\ -5 & \text{for } 3/2 < x < 2 \\ -3 & \text{at } x = -2 \text{ and at } x = 2. \end{cases}$$

The conclusion at the end points follows from the facts that $g(-2+) = -1$ and $g(2-) = -5$, and the average of -1 and -5 is -3.

Example 1.3 Let

$$w(x) = \begin{cases} -4x & \text{for } -4 \le x < -2, \\ 4\sin(3x) & \text{for } -2 \le x < 2, \\ 3 & \text{for } 2 \le x < 3, \\ -x & \text{for } 3 \le x \le 4. \end{cases}$$

Figure 1.10 is a graph of this function. The Fourier series of $w(x)$ on $-[4, 4]$

converges to

$$\begin{cases} -4x & \text{for } -4 < x < -2, \\ \frac{1}{2}(8 - 4\sin(6)) & \text{at } x = -2, \\ 4\sin(3x) & \text{for } -2 < x < 2, \\ \frac{1}{2}(3 + 4\sin(6)) & \text{at } x = 2, \\ -x & \text{for } 3 < x < 4 \\ 6 & \text{at } x = -4 \text{ and at } x = 4. \end{cases}$$

The conclusion at the end points follows from the calculation

$$w(-4+) = -4(-4) = 16 \text{ and } w(4-) = -4.$$

We do not have to compute the Fourier coefficients of $w(x)$ to draw these conclusions.

1.2.2 Fourier Sine and Cosine Series

In solving partial differential equations on an interval $[0, L]$, we will often need to expand a function in a series of just sines, or just cosines, on this half-interval.

The key to such a sine or cosine expansion is to recall some facts about even and odd functions. A function $f(x)$ defined on $[-L, L]$ is called an *even function* if

$$f(-x) = f(x) \text{ for } 0 < x \leq L.$$

Figure 1.11 shows a typical graph of an even function. The part of the graph to the left of the vertical axis is a reflection across this axis of the part to the right. (Fold the paper along the vertical axis and trace the part of the graph for $x > 0$). Examples of even functions are $x^2, x^6, \cos(x)$, and e^{x^2}.

If $f(x)$ is even on $[-L, L]$, then

$$\int_{-L}^{L} f(x)\,dx = 2\int_{0}^{L} f(x)\,dx.$$

This is because the area under the graph to the right of the vertical axis equals the area under the graph to the left.

We call $f(x)$ an *odd function* if

$$f(-x) = -f(-x) \text{ for } 0 < x \leq L.$$

Figure 1.12 shows the graph of a typical odd function. The graph to the left of the vertical axis is the reflection through the origin of the graph to the right of the vertical axis. That is, fold the graph for $x > 0$ over the vertical axis, then fold again over the horizontal axis. Examples of odd functions are $x^3, \sin(x)$, and $x^7 - 4x$.

If $f(x)$ is odd on $[-L, L]$, then

$$\int_{-L}^{L} f(x)\,dx = 0$$

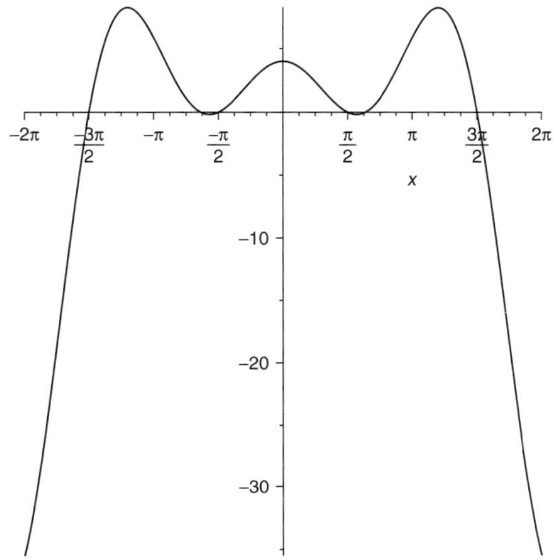

Figure 1.11: A typical even function, symmetric about the vertical axis.

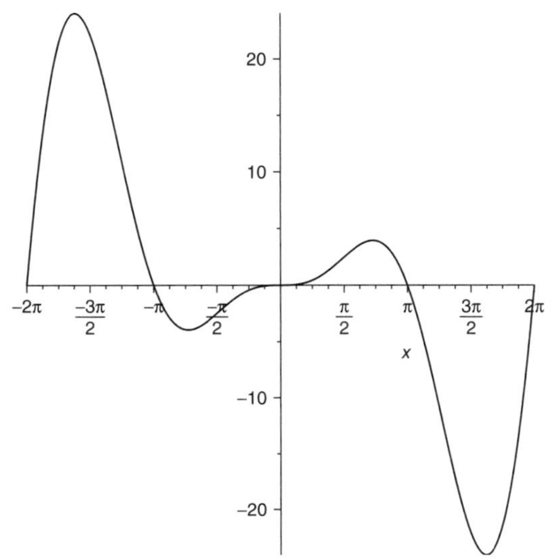

Figure 1.12: A typical odd function, symmetric through the origin.

because the area under the graph to the right of the vertical axis is the negative of the area to the left.

Notice that:

a product of odd functions is even (for example, $x^3x^5 = x^8$),

a product of even functions is even ($x^4x^2 = x^6$), and

a product of an even and an odd function is odd ($x^6x^3 = x^9$).

Now go back to the general setting of a function $g(x)$ defined on $[0, L]$. Extend $g(x)$ to an even function $G_e(x)$ on $[-L, L]$ by defining

$$G_e(x) = \begin{cases} g(x) & \text{for } 0 \le x \le L, \\ g(-x) & \text{for } -L \le x < 0. \end{cases}$$

This defines $G_e(x)$ on $[-L, L]$ by leaving $g(x)$ alone for $0 \le x \le L$, and reflecting the graph of $g(x)$ on $[0, L]$ across the vertical axis to define it for $-L \le x < 0$.

Expand $G_e(x)$ in a Fourier series on $[-L, L]$. Because $G_e(x)\cos(n\pi x/L)$ is an even function and $G_e(x) = g(x)$ for $0 \le x \le L$, the cosine coefficients in this expansion are

$$\begin{aligned} A_n &= \frac{1}{L}\int_{-L}^{L} G_e(x)\cos\left(\frac{n\pi x}{L}\right) dx \\ &= \frac{2}{L}\int_{0}^{L} G_e(x)\cos\left(\frac{n\pi x}{L}\right) dx \\ &= \frac{2}{L}\int_{0}^{L} g(x)\cos\left(\frac{n\pi x}{L}\right) dx. \end{aligned}$$

Further, $G_e(x)\sin(n\pi x/L)$ is odd on $[-L, L]$, so the coefficients of the sine terms vanish:

$$B_n = \frac{1}{L}\int_{-L}^{L} G_e(x)\sin\left(\frac{n\pi x}{L}\right) dx = 0.$$

The Fourier series of $G_e(x)$ on $[-L, L]$ is, therefore,

$$\frac{1}{2}A_0 + \sum_{n=1}^{\infty} A_n \cos\left(\frac{n\pi x}{L}\right),$$

containing only cosine terms. Because $G_e(x) = g(x)$ on $[0, L]$, this is a cosine expansion of $g(x)$ on $[0, L]$.

In summary, the *Fourier cosine series* of $g(x)$ on $[0, L]$ is

$$\frac{1}{2}A_0 + \sum_{n=1}^{\infty} A_n \cos\left(\frac{n\pi x}{L}\right), \tag{1.13}$$

in which

$$A_n = \frac{2}{L}\int_{0}^{L} g(x)\cos\left(\frac{n\pi x}{L}\right) dx \text{ for } n = 0, 1, 2, 3, \cdots. \tag{1.14}$$

Using $G_e(x)$ on $[-L, L]$ and the convergence theorem, we can determine the sum of this cosine expansion on $[0, L]$. Assuming that $g(x)$ is piecewise smooth, use the fact that $G_e(x) = g(x)$ for $0 \leq x \leq L$ to conclude that the cosine series converges to

$$\frac{1}{2}(g(x-) + g(x+)) \text{ if } 0 < x < L.$$

For $x = 0$, compute the limits:

$$\begin{aligned} G_e(0+) &= \lim_{h \to 0+} G_e(0 + h) = \lim_{h \to 0+} G_e(h) \\ &= \lim_{h \to 0+} g(x) = g(0+) \end{aligned}$$

and

$$\begin{aligned} G_e(0-) &= \lim_{h \to 0+} G_e(0 - h) = \lim_{h \to 0+} G_e(-h) \\ &= \lim_{h \to 0+} g(h) = g(0+). \end{aligned}$$

Therefore, at $x = 0$, the cosine expansion of $g(x)$ converges to

$$\frac{1}{2}(G(0-) + G(0+)) = \frac{1}{2}(g(0+) + g(0+)) = g(0+).$$

A similar argument shows that, at $x = L$, the cosine series for $g(x)$ on $[0, L]$ converges to $g(L-)$.

The even extension of $g(x)$ to $G_e(x)$ was a device used to obtain this half-interval expansion of $g(x)$ from the already known Fourier series of $G_e(x)$ on $[L, L]$. In computing the coefficients A_n in a cosine expansion, we need only $g(x)$ and do not have to explicitly construct $G_e(x)$. Just write the series 1.13, with coefficients from equation 1.14.

Example 1.4 (A Fourier Cosine Expansion) Let $g(x) = e^x$ for $0 \leq x \leq 2$. From equation 1.14, the cosine coefficients of $g(x)$ on this interval are

$$A_0 = \frac{2}{2} \int_0^2 e^x \, dx = e^2 - 1$$

and, for $n = 1, 2, \cdots$,

$$A_n = \frac{2}{2} \int_0^2 e^x \cos\left(\frac{n\pi x}{L}\right) dx = \frac{4}{4 + n^2\pi^2}((-1)^n e^2 - 1).$$

Further, using the convergence theorem, this series will converge to e^x for $0 \leq x \leq 2$:

$$e^x = \frac{1}{2}(e^2 - 1) + \sum_{n=1}^{\infty} \frac{4}{4 + n^2\pi^2}((-1)^n e^2 - 1) \cos\left(\frac{n\pi x}{2}\right).$$

Figure 1.13 shows a graph of $g(x) = e^x$ compared with the 10th partial sum of this cosine series on $[0, 2]$. This cosine expansion appears to converge very quickly to $g(x)$.

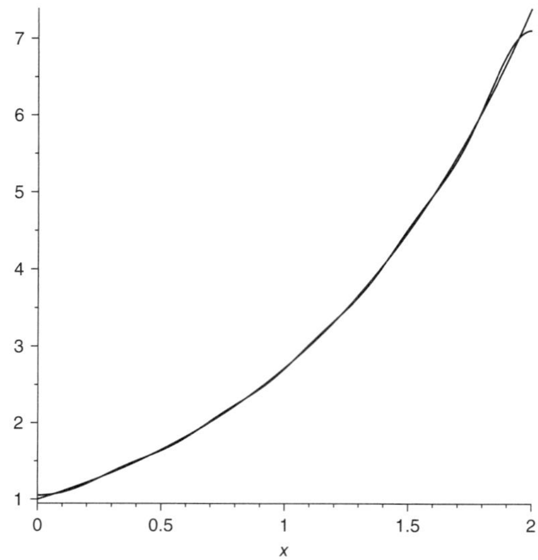

Figure 1.13: $g(x)$ compared with the 10th partial sum of the cosine expansion in example 1.4.

We can develop a Fourier sine expansion of $g(x)$ on $[0, L]$ by a similar tactic. Now, however, because the sine terms in the series are odd functions, extend $g(x)$ to an odd function $G_o(x)$ on $[-L, L]$ by setting

$$G_o(x) = \begin{cases} g(x) & \text{for } 0 \leq x \leq L, \\ -g(-x) & \text{for } -L \leq x < 0. \end{cases}$$

$G_o(x)$ is an odd function and $G_o(x) = g(x)$ on $[0, L]$. Further, the Fourier series for $G_o(x)$ on $[-L, L]$ will contain only sine terms, because the coefficients of the cosine terms involve integrals of $G_o(x)\cos(n\pi x/L)$, and these integrals are zero because this is an odd function on $[-L, L]$.

In summary, the *Fourier sine expansion* of $g(x)$ on $[0, L]$ is

$$\sum_{n=1}^{\infty} B_n \sin\left(\frac{n\pi x}{L}\right), \tag{1.15}$$

where

$$B_n = \frac{2}{L}\int_0^L g(x)\sin\left(\frac{n\pi x}{L}\right) dx \text{ for } n = 1, 2, \cdots . \tag{1.16}$$

As with cosine series, we do not actually have to write out $G_o(x)$ to compute these coefficients.

Assuming that $g(x)$ is piecewise smooth on $[0, L]$, this sine series will converge to

$$\frac{1}{2}(g(x-) + g(x+)) \text{ for } 0 < x < L.$$

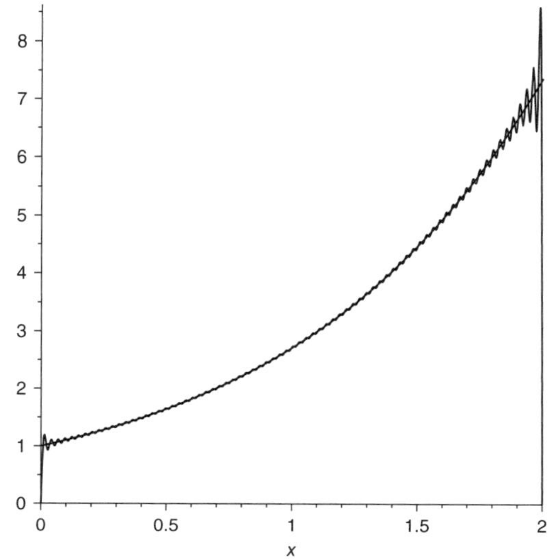

Figure 1.14: e^x compared with the 150th partial sum of the sine expansion in example 1.5.

Without any computation, a sine series converges to 0 at $x = 0$ and at $x = L$, because $\sin(0) = 0$ and $\sin(n\pi) = 0$ for every integer n.

Example 1.5 (A Fourier Sine Expansion) We will write a Fourier sine series for e^x on $[0, 2]$. The coefficients are

$$B_n = \int_0^2 e^x \sin\left(\frac{n\pi x}{L}\right) dx = \frac{2n\pi}{4 + n^2\pi^2}\left(1 - (-1)^n e^2\right).$$

This sine series converges to e^x for $0 < x < 2$ (but not at 0 or 2), so we can write

$$e^x = \sum_{n=1}^{\infty} \frac{2n\pi}{4 + n^2\pi^2}\left(1 - (-1)^n e^2\right) \sin\left(\frac{n\pi x}{L}\right) \text{ for } 0 < x < 2.$$

Figure 1.14 shows graphs of $g(x)$ and the 150th partial sum of this sine expansion. Contrast this with the much more rapid convergence of the cosine expansion of this function in example 1.4.

Problems for Section 1.2

In each of problems 1–6, write the Fourier series of the function, and determine the sum of this series on the interval. Compare graphs of some partial sums of the series with a graph of the function.

1. $f(x) = -x$ for $-1 \le x \le 1$.

2. $f(x) = \cos(3x)$ for $-\pi \leq x \leq \pi$.

3. $f(x) = \sin(2x)$ for $-2 \leq x \leq 2$.

4. $f(x) = 1 - |x|$ for $-2 \leq x \leq 2$.

5. $f(x) = \begin{cases} -4 & \text{for } -\pi \leq x < 0, \\ 4 & \text{for } 0 \leq x \leq \pi. \end{cases}$

6. $f(x) = \cos(x/2) - \sin(x)$ for $-\pi \leq x \leq \pi$

In each of problems 7–12, determine the sum of the Fourier series of the function on the interval. In doing this, it is not necessary to compute the Fourier coefficients.

7. $f(x) = \begin{cases} 2x & \text{for } -3 \leq x \leq 0, \\ 0 & \text{for } -2 < x < 1, \\ x^2 & \text{for } 1 \leq x \leq 3. \end{cases}$

8. $f(x) = \begin{cases} \cos(x) & \text{for } -2 \leq x \leq 1/2, \\ \sin(x) & \text{for } 1/2 \leq x \leq 2. \end{cases}$

9. $f(x) = \begin{cases} -x & \text{for } -4 \leq x < 2, \\ 2 & \text{for } 0 \leq x \leq 4. \end{cases}$

10. $f(x) = \begin{cases} 1 & \text{for } -2 \leq x \leq 0, \\ -1 & \text{for } 0 < x < 1/2, \\ x^2 & \text{for } 1/2 \leq x \leq 2. \end{cases}$

11. $f(x) = \begin{cases} \cos(\pi x) & \text{for } -2 \leq x < 0, \\ x & \text{for } 0 \leq x \leq 2. \end{cases}$

12. $f(x) = \begin{cases} 1 - x & \text{for } -3 <\leq x \leq -1/2, \\ 2 + x & \text{for } -1/2 < x \leq 1, \\ 4 - x^2 & \text{for } 1 < x \leq 2, \\ 1 - x - x^2 & \text{for } 2 < x \leq 3. \end{cases}$

13. Sum both of the series

$$\sum_{n=1}^{\infty} \frac{1}{n^2} \text{ and } \sum_{n=1}^{\infty} \frac{(-1)^n}{n^2}.$$

Hint: Expand $f(x) = \frac{1}{2}x^2$ in a Fourier series on $[-\pi, \pi]$. Now make choices of x to obtain these series.

14. Suppose

$$f(x) = \frac{1}{2}a_0 + \sum_{n=1}^{\infty} a_n \cos\left(\frac{n\pi x}{L}\right) + b_n \sin\left(\frac{n\pi x}{L}\right)$$

for $-L \leq x \leq L$. Multiply this equation by $f(x)$ and assume that the resulting expression can be integrated term by term to derive *Parseval's equation*:

$$\frac{1}{2}a_0^2 + \sum_{n=1}^{\infty}(a_n^2 + b_n^2) = \frac{1}{L}\int_{-L}^{L}(f(x))^2\,dx.$$

In each of problems 15–22, find the Fourier cosine series and the Fourier sine series for $f(x)$ on the interval. Determine what each series converges to on this interval. Compare graphs of some partial sums of this series with a graph of the function.

15. $f(x) = 4$ for $0 \leq x \leq 3$.

16. $f(x) = \begin{cases} 1 & \text{for } 0 \leq x \leq 1, \\ -1 & \text{for } 1 < x \leq 2. \end{cases}$

17. $f(x) = \begin{cases} 0 & \text{for } 0 \leq x \leq \pi/2, \\ \sin(x) & \text{for } \pi/2 < x \leq \pi. \end{cases}$

18. $f(x) = 2x$ for $0 \leq x \leq 1$.

19. $f(x) = x^2$ for $0 \leq x \leq 2$.

20. $f(x) = e^{-x}$ for $0 \leq x \leq 1$.

21. $f(x) = \sin(3x)$ for $0 \leq x \leq \pi$.

22. $f(x) = \begin{cases} x & \text{for } 0 \leq x \leq 1, \\ 2 - x & \text{for } 1 \leq x \leq 2. \end{cases}$

23. Sum the series

$$\sum_{n=1}^{\infty} \frac{(-1)^n}{4n^2 - 1}.$$

Hint: Expand $\sin(x)$ in a cosine series on $[0, \pi]$ and evaluate this series at an appropriately chosen point.

1.3 Two Eigenvalue Problems

In solving initial-boundary value problems, we will often encounter the problem:

$$X'' + \lambda X = 0; X(0) = X(L) = 0. \tag{1.17}$$

We want to find numerical values of the constant λ such that there are nontrivial (not identically zero) solutions $X(x)$ of problem 1.17. Such values of λ are called *eigenvalues* of this problem, and corresponding nontrivial solutions $X(x)$ are *eigenfunctions*.

To find the eigenvalues and eigenfunctions of problem 1.17, consider separately the cases that λ is zero, negative, or positive.

If $\lambda = 0$, then $X'' = 0$, so $X(x) = cx + d$ for some numbers c and d. Now

$$X(0) = d = 0$$

and

$$X(L) = cL = 0 \text{ implies that } c = 0.$$

This means that $X(x) = 0$ for all x. The only solution for $X(x)$ in this case is the trivial solution, so 0 is not an eigenvalue of this problem.

If $\lambda < 0$, then we may write $\lambda = -\alpha^2$, where $\alpha > 0$. Now

$$X'' - \alpha^2 X = 0,$$

with general solution

$$X(x) = ce^{\alpha x} + de^{-\alpha x}.$$

Then

$$X(0) = c + d = 0 \text{ implies that } c = -d,$$

so

$$X(x) = c(e^{\alpha x} - e^{-\alpha x}) = 2c\sinh(\alpha x).$$

And,

$$X(L) = 2c\sinh(\alpha L) = 0.$$

Since $\alpha L > 0$, $\sinh(\alpha L) > 0$, so $c = 0$ and $X(x)$ is the trivial function. This problem has no negative eigenvalue.

Finally, if $\lambda > 0$, write $\lambda = \alpha^2$, with $\alpha > 0$. Then

$$X'' + \alpha^2 X = 0,$$

with solutions of the form

$$X(x) = a\cos(\alpha x) + b\sin(\alpha x).$$

Now

$$X(0) = a = 0$$

so $X(x) = b\sin(\alpha x)$. Then

$$X(L) = b\sin(\alpha L) = 0.$$

If $b = 0$ we again have only the trivial solution and have run out of cases. Thus attempt to find solutions with $b \neq 0$. This will require that we choose α so that

$$\sin(\alpha L) = 0.$$

We can manage this if αL is an integer multiple of π, leading us to choose

$$\alpha = \frac{n\pi}{L} \text{ for } n = 1, 2, \cdots .$$

Denote

$$\lambda_n = \frac{n^2\pi^2}{L^2} \text{ for } n = 1, 2, 3, \cdots .$$

These are the eigenvalues of the problem 1.17, indexed by n. Corresponding to each eigenvalue λ_n, we have an eigenfunction

$$X_n(x) = \sin\left(\frac{n\pi x}{L}\right)$$

or any nonzero constant multiple of this function.

We will also frequently encounter the problem:

$$X'' + \lambda X = 0; X'(0) = X'(L) = 0. \tag{1.18}$$

As with problem 1.17, consider cases on λ.

If $\lambda = 0$, then $X(x) = cx + d$ for some constants c and d. Now $X'(0) = c = 0$, and $X(x) = d$ is a solution. Unlike problem 1.17, 0 is an eigenvalue of this problem, with nonzero constant eigenfunctions.

If $\lambda < 0$, write $\lambda = -\alpha^2$, with $\alpha > 0$, to obtain the general solution

$$X(x) = ce^{\alpha x} + de^{-\alpha x}$$

of the differential equation. Now

$$X'(0) = \alpha c - \alpha d = 0$$

so $c = d$ and

$$X(x) = c(e^{\alpha x} + e^{-\alpha x}) = 2c\cosh(\alpha x).$$

Then

$$X'(L) = 2c\alpha\sinh(\alpha L) = 0.$$

But $\alpha L > 0$, and $\sinh(\alpha L) > 0$, so $c = 0$ and this case has only the trivial solution. Problem 1.18 has no negative eigenvalue.

If $\lambda > 0$, set $\lambda = \alpha^2$, with $\alpha > 0$. Now $X'' + \alpha^2 X = 0$, with the general solution

$$X = c\cos(\alpha x) + d\sin(\alpha x).$$

Now $X'(0) = d\alpha = 0$, so

$$X(x) = c\cos(\alpha x).$$

Then

$$X'(L) = -\alpha c\sin(\alpha L) = 0.$$

To obtain a nontrivial solution, we need $c \neq 0$. This forces us to choose α so that $\sin(\alpha L) = 0$. As in problem 1.17, αL must be a positive integer multiple of π, say $\alpha L = n\pi$. Then α must be chosen as

$$\alpha = \frac{n\pi}{L} \text{ for } n = 1, 2, 3, \cdots .$$

For each positive integer n,

$$\lambda_n = \frac{n^2\pi^2}{L^2}$$

is an eigenvalue of problem 1.18, with corresponding eigenfunction

$$X_n(x) = \cos\left(\frac{n\pi x}{L}\right) \text{ for } n = 1, 2, \cdots .$$

To summarize, the eigenvalues and eigenfunctions of problem 1.18 are

$$\lambda_n = \frac{n^2\pi^2}{L^2}, X_n(x) = \cos\left(\frac{n\pi x}{L}\right) \text{ for } n = 0, 1, 2, \cdots .$$

We are now prepared to solve some important problems involving partial differential equations.

Problems for Section 1.3

Find the eigenvalues and eigenfunctions of each of the following problems.

1. $X'' + \lambda X = 0; X(0) = X'(L) = 0.$
2. $X'' + \lambda X = 0; X'(0) = X(L) = 0.$
3. $X'' + \lambda X = 0; X(0) = X(L), X'(0) = X'(L).$
4. $X'' + \lambda X = 0; X(0) = 0, X(L) + 2X'(L) = 0.$

1.4 A Proof of the Convergence Theorem

We will prove the Fourier convergence theorem, providing an understanding of why the series converges to the left and right limit of the function at interior points of the interval.

1.4.1 The Role of Periodicity

Recall that $f(x)$ is *periodic* if $f(x)$ is defined for all real numbers x and, for some positive number p,

$$f(x) = f(x + p)$$

for all x. We call p a *period* of f. If p is a period, so are $2p$, $3p$, and so on. When we speak of *the* period, or *fundamental period* of a function, we mean its smallest period. For example, $\sin(x)$ has period $2n\pi$ for any positive integer n, because

$$\sin(x + 2n\pi) = \sin(x) \text{ for all } x.$$

However, 2π is the smallest period of $\sin(x)$, so we say that $\sin(x)$ has fundamental period 2π.

Of course, "most" functions (for example, polynomials and exponential functions) are not periodic.

Now consider an issue we avoided in informally developing the idea of a Fourier series. For Fourier expansions on an interval $[-L, L]$, or perhaps $[0, L]$, we considered function values $f(x)$ only for x in this interval. Indeed, this is the way we look at things when we use Fourier series to solve initial-boundary value problems involving partial differential equations. If we are studying heat conduction in a bar of metal, we place the bar along an axis and do not think about values of x outside this segment.

However, even though we think of the function as living on an interval in such applications, the Fourier series on $[-L, L]$ is

$$\frac{1}{2}a_0 + \sum_{n=1}^{\infty} \left[a_n \cos\left(\frac{n\pi x}{L}\right) + b_n \sin\left(\frac{n\pi x}{L}\right) \right],$$

which is not only defined on the entire real line but is periodic of period $2L$! How can $f(x)$, which is defined only on an interval, be equal to a periodic function (its Fourier series)?

The answer is that the Fourier series actually represents not the function, but its periodic extension to the entire real line.

We can see this with a simple example. Let $f(x) = x$ for $-1 \le x < 1$. The familiar graph is shown in Figure 1.15. Figure 1.16 shows its periodic extension $\tilde{f}$ to the entire line. This extension is done by sliding the graph forward from $[-1, 1)$ onto the intervals $[1, 3), [3, 5), [5, 7), \cdots$ and backward onto the intervals $[-3, -1), [-5, -3), [-7, -5), \cdots$, as in the diagram. Then $f(x) = \tilde{f}(x)$ for $-1 \le x < 1$, but $\tilde{f}(x)$ continues on to be defined for all x. Further, $\tilde{f}$ is periodic of period 2, just as the Fourier expansion of $f(x)$ on $[-1, 1]$ is.

In making such an extension, a subtlety appears. Because $\tilde{f}$ is periodic, we cannot assign functions values to $\tilde{f}$ arbitrarily at different points. Given any a, we necessarily have $\tilde{f}(a) = \tilde{f}(a + np)$ for any integer n.

In this example, where $\tilde{f}$ has period $2, \tilde{f}(x) = \tilde{f}(x + 2n)$ for every integer n. In particular,

$$\tilde{f}(-1) = \tilde{f}(-1 + 2) = \tilde{f}(1).$$

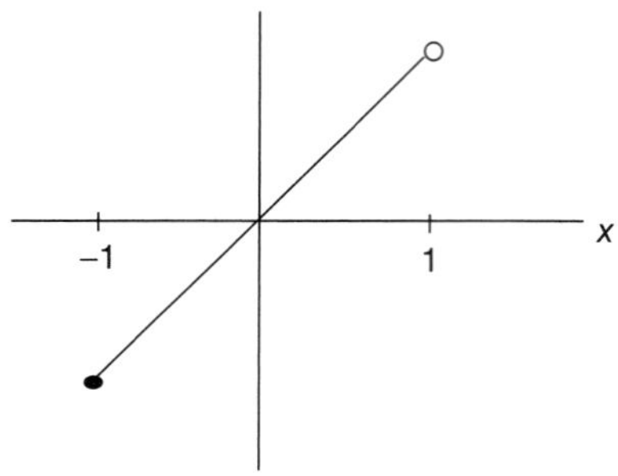

Figure 1.15: Graph of $f(x) = 1$ for $-1 \leq x < 1$.

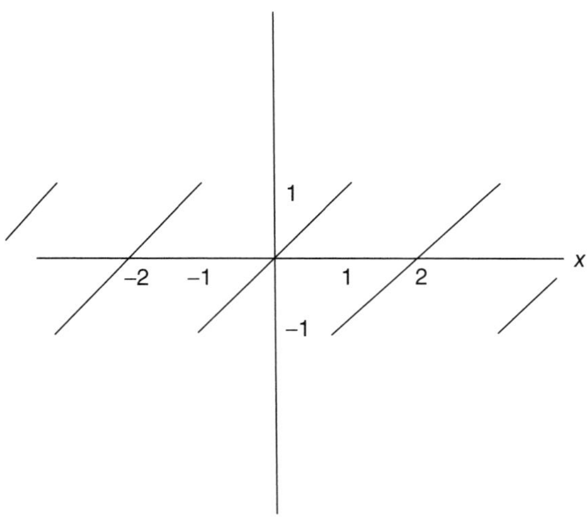

Figure 1.16: Periodic extension $\tilde{f}(x)$ of the graph of Figure 1.15.

It was in anticipation of this that we defined $f(x)$ for $-1 \leq x < 1$ instead of on $-1 \leq x \leq 1$. Once $\tilde{f}(-1)$ is specified, then $\tilde{f}(1)$ is predetermined.

Now notice that, with $\tilde{f}(x)$ defined for all x, there are no endpoints to distinguish in stating a convergence theorem. At *every* real number x, the Fourier series of $\tilde{f}(x)$ on $[-1, 1]$ converges to $(\tilde{f}(x-) + \tilde{f}(x+))/2$.

In particular, at $x = -1$, $\tilde{f}(-1+) = -1$ and $\tilde{f}(-1-) = 1$, so at -1 the series converges to

$$\frac{1}{2}(\tilde{f}(-1+) + \tilde{f}(1-)) = 0.$$

And at $x = 1$, $\tilde{f}(1-) = 1$ and $\tilde{f}(1+) = -1$, so the series also converges to

$$\frac{1}{2}(\tilde{f}(-1+) + \tilde{f}(1-)) = 0.$$

Armed with this point of view, we now develop the machinery needed to prove the convergence theorem, beginning with Dirichlet's formula. We will let $L = \pi$ and work on $[-\pi, \pi)$. This simplifies the frequently encountered expression $n\pi x/L$ to just nx.

1.4.2 Dirichlet's Formula

Convergence of any series depends on the convergence of its sequence of partial sums. Let $f(x)$ be periodic of period 2π. The Nth partial sum of the Fourier expansion of $f(x)$ on $[-\pi, \pi]$ is

$$S_N(x) = \frac{1}{2}a_0 + \sum_{n=1}^{N}(a_n \cos(nx) + b_n \sin(nx)),$$

in which the $a_n's$ and $b_n's$ are the Fourier coefficients of the function on the interval. We rewrite $S_N(x)$ in a way that will help us determine its limit as $N \to \infty$.

Insert the Fourier coefficients of $f(x)$ into $S_N(x)$ to write:

$$\begin{aligned} S_N(x) &= \frac{1}{2\pi}\int_{-\pi}^{\pi} f(\xi)\, d\xi \\ &+ \frac{1}{\pi}\sum_{n=1}^{N}\left[\int_{-\pi}^{\pi} f(\xi)\cos(n\xi)\, d\xi \cos(nx) + \int_{-\pi}^{\pi} f(\xi)\sin(n\xi)\, d\xi \sin(nx)\right] \\ &= \frac{1}{\pi}\int_{-\pi}^{\pi} f(\xi)\left[\frac{1}{2} + \sum_{n=1}^{N}(\cos(n\xi)\cos(nx) + \sin(n\xi)\sin(nx))\right] d\xi \\ &= \frac{1}{\pi}\int_{-\pi}^{\pi} f(\xi)\left[\frac{1}{2} + \sum_{n=1}^{N}\cos(n(\xi - x))\right] d\xi. \qquad (1.19) \end{aligned}$$

It is possible to derive a simple expression for the sum in square brackets. Let $y = \xi - x$ and let

$$\sigma = \frac{1}{2} + \sum_{n=1}^{N} \cos(ny).$$

Multiply σ by $2\sin(y/2)$ to obtain

$$\begin{aligned}
2\sigma \sin(y/2) &= \sin(y/2) + 2\sum_{n=1}^{N} \cos(ny)\sin(y/2) \\
&= \sin(y/2) + \sum_{n=1}^{N} [\sin((n+1/2)y) - \sin((n-1/2)y)] \\
&= \sin(y/2) + [\sin(3y/2) - \sin(y/2)] + [\sin(5y/2) - \sin(3y/2)] + \cdots \\
&+ [\sin((N-1/2)y) - \sin((N-3/2)y)] \\
&+ [\sin((N+1/2)y) - \sin((N-1/2)y)].
\end{aligned}$$

This is a telescoping sum, with all terms except one canceling on the right. We obtain

$$2\sigma \sin(y/2) = \sin((N+1/2)y).$$

Then

$$\sigma = \frac{\sin((N+1/2)(\xi - x))}{2\sin((\xi - x)/2)},$$

provided that $\sin((\xi - x)/2) \neq 0$. Inserting this result into equation 1.19, we have

$$S_N(x) = \frac{1}{\pi}\int_{-\pi}^{\pi} f(\xi)\frac{\sin((N+1/2)(\xi - x))}{2\sin((\xi - x)/2)}\,d\xi. \tag{1.20}$$

Put $t = \xi - x$ into equation 1.20 to obtain

$$S_N(x) = \frac{1}{\pi}\int_{-\pi-x}^{\pi-x} f(x+t)\frac{\sin((N+1/2)t)}{2\sin(t/2)}\,dt.$$

Because $f(x)$ is periodic of period 2π, this integrand also has period 2π and we can carry out the integration over any interval of length 2π. In particular, we can write

$$S_N(x) = \frac{1}{\pi}\int_{-\pi}^{\pi} f(x+t)\frac{\sin((N+1/2)t)}{2\sin(t/2)}\,dt. \tag{1.21}$$

This is *Dirichlet's formula*. The function

$$\frac{\sin((N+1/2)t)}{2\sin(t/2)}$$

is called the *Dirichlet kernel*. It has the following property, which we will use shortly.

Lemma 1.1

$$\frac{1}{\pi}\int_{-\pi}^{0}\frac{\sin((N+1/2)t)}{2\sin(t/2)}\,dt = \frac{1}{\pi}\int_{0}^{\pi}\frac{\sin((N+1/2)t)}{2\sin(t/2)}\,dt = \frac{1}{2}.$$

Proof Let $f(x) = 1$. The Fourier coefficients of $f(x)$ on $[-\pi, \pi]$ are

$$a_0 = \frac{1}{\pi}\int_{-\pi}^{\pi} d\xi = 2$$

and, for $n = 1, 2, \cdots$,

$$a_n = \frac{1}{\pi}\int_{-\pi}^{\pi}\cos(n\xi)\,d\xi = 0 \text{ and } b_n = \frac{1}{\pi}\int_{-\pi}^{\pi}\sin(n\xi)\,d\xi = 0.$$

Therefore, for this function, $S_N(x) = 1$ and Dirichlet's formula 1.21 becomes:

$$\frac{1}{\pi}\int_{-\pi}^{0}\frac{\sin((N+1/2)t)}{2\sin(t/2)}\,dt + \frac{1}{\pi}\int_{0}^{\pi}\frac{\sin((N+1/2)t)}{2\sin(t/2)}\,dt = 1. \tag{1.22}$$

But if we let $t = -w$ in the left integral in equation 1.22, we obtain

$$\begin{aligned}\frac{1}{\pi}\int_{-\pi}^{0}\frac{\sin((N+1/2)t)}{2\sin(t/2)}\,dt &= \frac{1}{\pi}\int_{\pi}^{0}\frac{\sin((N+1/2)w)}{2\sin(w/2)}(-1)\,dw \\ &= \frac{1}{\pi}\int_{0}^{\pi}\frac{\sin((N+1/2)t)}{2\sin(t/2)}\,dw.\end{aligned}$$

This means that the integrals on the left in equation 1.22 are equal. Because their sum is 1, each integral equals 1/2, as was to be shown.

1.4.3 The Riemann-Lebesgue Lemma

We will prove a result needed to compute the limit of $S_N(x)$ as $N \to \infty$.

Lemma 1.2 (Riemann-Lebesgue) Let g be piecewise continuous on $[a, b]$. Then

$$\lim_{\omega\to\infty}\int_{a}^{b} g(t)\sin(\omega t)\,dt = 0.$$

Proof Suppose first that g is continuous on $[a, b]$. Let

$$I = \int_{a}^{b} g(t)\sin(\omega t)\,dt.$$

Let $t = \xi + \pi/\omega$, with ω chosen large enough that $b - \pi/\omega \geq a$. Then

$$I = \int_{a-\pi/\omega}^{b-\pi/\omega} g(\xi+\pi/\omega)\sin(\omega\xi+\pi)\,d\xi = -\int_{a-\pi/\omega}^{b-\pi/\omega} g(\xi+\pi/\omega)\sin(\omega\xi)\,d\xi.$$

To maintain t as the variable of integration, replace ξ with t in the last integral:

$$I = -\int_{a-\pi/\omega}^{b-\pi/\omega} g(t+\pi/\omega)\sin(\omega t)\,dt.$$

Add this expression for I to the definition of I to write

$$\begin{aligned} 2I &= \int_a^b g(t)\sin(\omega t)\,dt - \int_{a-\pi/\omega}^{b-\pi/\omega} g(t+\pi/\omega)\sin(\omega t)\,dt \\ &= \int_a^{b-\pi/\omega} [g(t) - g(t+\pi/\omega)]\sin(\omega t)\,dt \\ &+ \int_{b-\pi/\omega}^{a-\pi/\omega} g(t)\sin(\omega t)\,dt - \int_{a-\pi/\omega}^{a} g(t+\pi/\omega)\sin(\omega t)\,dt. \end{aligned} \tag{1.23}$$

Because g is continuous on a closed interval $[a,b]$, there is some number M such that $|g(t)| \leq M$ on $[a,b]$. Then

$$\left|\int_{b-\pi/\omega}^{b} g(t)\sin(\omega t)\,dt\right| \leq M\frac{\pi}{\omega}$$

and

$$\left|\int_{a-\pi/\omega}^{a} g(t+\pi/\omega)\sin(\omega t)\,dt\right| \leq M\frac{\pi}{\omega}.$$

For the remaining integral in equation 1.23, use the fact that g is uniformly continuous on $[a,b]$. Let $\epsilon > 0$. Then there is some $\delta > 0$ such that

$$|g(x) - g(y)| < \frac{1}{3}\epsilon \text{ if } |x-y| < \delta.$$

Then

$$|g(t) - g(t+\pi/\omega)| < \frac{1}{3}\epsilon \text{ if } \frac{\pi}{\omega} < \delta.$$

Therefore, if $\omega > \pi/\delta$, and ω is also large enough that $b - \pi/\omega \geq a$ and $M\pi/\omega < \epsilon/3$, we have from equation 1.23 and the bounds just obtained that

$$|2I| < M\frac{\pi}{\omega} + M\frac{\pi}{\omega} + \frac{\epsilon}{3} < \frac{\epsilon}{3} + \frac{\epsilon}{3} + \frac{\epsilon}{3} = \epsilon.$$

But then

$$|I| < \frac{\epsilon}{2} < \epsilon \text{ if } \omega > \pi/\delta.$$

This proves that

$$\lim_{\omega\to\infty} I = 0.$$

Now we must confront the case that g has a finite number of jump discontinuities in $[a,b]$, say at $t_1 < t_2, < \cdots < t_k$. Write I as a sum of integrals, over $[a,t_1], [t_1,t_2], \cdots, [t_k,b]$. By redefining $g(t)$ at the end points of each of these intervals, if necessary, we can write I as a finite sum of integrals, each having the same form as I, but each having a continuous integrand. (This redefinition of $g(t)$ at finitely many points of $[a, b]$ does not affect the value of the integrals). By what we have just shown, each of these integrals has limit 0 as $\omega \to \infty$, completing the proof of the lemma.

1.4.4 Proof of the Convergence Theorem

We will prove that, if f is piecewise smooth on $[-\pi, \pi]$ and periodic of period 2π, then the Fourier series of $f(x)$ on this interval converges at x to

$$\frac{1}{2}\left(f(x-) + f(x+)\right).$$

The argument is essentially the one used by Dirichlet. Use Dirichlet's formula and Lemma 1.1 to write

$$\begin{aligned}
S_N(x) &= \frac{1}{\pi}\int_{-\pi}^{0} f(x+t)\frac{\sin((N+1/2)t)}{2\sin(t/2)}\,dt \\
&+ \frac{1}{\pi}\int_{0}^{\pi} f(x+t)\frac{\sin((N+1/2)t)}{2\sin(t/2)}\,dt \\
&= \frac{1}{\pi}\int_{-\pi}^{0} f(x+t)\frac{\sin((N+1/2)t)}{2\sin(t/2)}\,dt - \frac{1}{2}f(x-) + \frac{1}{2}f(x-) \\
&+ \frac{1}{\pi}\int_{0}^{\pi} f(x+t)\frac{\sin((N+1/2)t)}{2\sin(t/2)}\,dt - \frac{1}{2}f(x+) + \frac{1}{2}f(x+) \\
&= \frac{1}{\pi}\int_{-\pi}^{0} [f(x+t) - f(x-)]\frac{\sin((N+1/2)t)}{2\sin(t/2)}\,dt + \frac{1}{2}f(x-) \\
&+ \frac{1}{\pi}\int_{0}^{\pi} [f(x+t) - f(x+)]\frac{\sin((N+1/2)t)}{2\sin(t/2)}\,dt + \frac{1}{2}f(x+).
\end{aligned} \tag{1.24}$$

To complete the proof, it is enough to show that each of the last two integrals in equation 1.24 has limit zero as $N \to \infty$. In this event, we will have shown

$$S_N(x) \to \frac{1}{2}\left(f(x-) + f(x+)\right).$$

To prove that the last integral has limit zero, let

$$g(t) = \frac{f(x+t) - f(x+)}{2\sin(t/2)} \quad \text{for } 0 < t \le \pi.$$

Using the fact that $f'(x)$ is piecewise continuous on $[-\pi, \pi]$, observe that

$$\begin{aligned}
\lim_{t\to 0+} g(t) &= \lim_{t\to 0} \frac{f(x+t) - f(x+)}{2\sin(t/2)} \\
&= \lim_{t\to 0+} \frac{f(x+t) - f(x+)}{t}\,\frac{t/2}{\sin(t/2)} \\
&= \lim_{t\to 0+} \frac{f(x+t) - f(x+)}{t}\,\lim_{t\to 0+} \frac{t/2}{\sin(t/2)} \\
&= f'(x+)\cdot 1 = f'(x+).
\end{aligned}$$

Now define

$$g(0) = f'(0+).$$

Because f is piecewise smooth on $[-\pi, \pi]$ and periodic of period 2π, g is piecewise smooth on $[0, \pi]$. By the Riemann-Lebesgue lemma, with $\omega = N + 1/2$,

$$\begin{aligned}&\lim_{\omega\to\infty} \int_0^\pi g(t)\sin(\omega t)\,dt \\ &= \lim_{N\to\infty} \int_0^\pi |f(x+t) - f(x+)| \frac{\sin((N+1/2)t)}{2\sin(t/2)}\,dt = 0.\end{aligned}$$

This proves that the last integral in equation 1.24 has limit zero as $N \to \infty$. By a similar argument, we can also conclude that

$$\lim_{N\to\infty} \int_{-\pi}^0 |f(x+t) - f(x-)| \frac{\sin((N+1/2)t)}{2\sin(t/2)}\,dt = 0.$$

This proves that

$$\lim_{N\to\infty} S_N(x) = \frac{1}{2}\left(f(x+) - f(x-)\right),$$

as we wanted to show.

Problem for Section 1.4

At points of discontinuity, Fourier series exhibit the Gibbs phenomenon, named for the Yale mathematician Josiah Willard Gibbs (1839–1903), who was the first to offer an explanation. To see the Gibbs phenomenon in a specific case, let

$$f(x) = \begin{cases} -1 & \text{for } -1 \le x \le 0, \\ 1 & \text{for } 0 < x \le 1. \end{cases}$$

This function has a jump discontinuity at $x = 0$, and the Fourier series of $f(x)$ on $[-1, 1]$ converges to 0 there.

Graph $f(x)$ and partial sums $S_N(x)$ of this Fourier series, for $N = 10$, 20, 40, 60, and 100. You will see the partial sums approach closer to the graph of the function for $-1 < x < 0$ and $0 < x < 1$. However, at 0, the partial sums appear to exhibit oscillations of a height that do not decrease as x approaches 0 (even though the length of interval to the left or right of 0 on which these oscillations occur becomes shorter). This is the *Gibbs phenomenon*, and it is seen in the convergence of Fourier series at jump discontinuities.

In the late 18th century, it was thought that, if a series of functions converges to a function, then the graphs of the partial sums would vary from the limit function by less and less as $N \to \infty$. The Gibbs phenomenon showed that this is not true in the way it was then understood. The length of interval on which the oscillations occur does shorten and tends to zero as $n \to \infty$, but the oscillations remain at about the same height.

Chapter 2

Solutions of the Heat Equation

In this chapter we use Fourier series and a technique called separation of variables to solve initial-boundary value problems involving the heat equation on a bounded interval or rectangle in the plane. The chapter concludes with proofs of properties of solutions of the heat equation, including a maximum principle and a measure of how sensitive solutions are to changes in initial and boundary conditions.

2.1 Solutions on an Interval $[0, L]$

We will solve the heat equation on an interval $[0, L]$ for a variety of boundary conditions.

2.1.1 Ends Kept at Zero Temperature

The problem is

$$\begin{aligned} u_t &= ku_{xx} \text{ for } 0 < x < L, t > 0, \\ u(0,t) &= u(L,t) = 0 \text{ for } t > 0, \\ u(x,0) &= f(x) \text{ for } 0 < x < L. \end{aligned} \tag{2.1}$$

This models the temperature distribution $u(x,t)$ in a homogeneous bar of uniform cross section and length, L, if the ends are kept at temperature zero, and the initial temperature on the cross section of the bar at x is $f(x)$.

The *method of separation of variables*, or the *Fourier method*, consists of looking for solutions of the form

$$u(x,t) = X(x)T(t).$$

Substitute this into the heat equation:

$$XT' = kX''T.$$

Then

$$\frac{T'}{kT} = \frac{X''}{X}.$$

The left side depends only on t, and the right side only on x, and x and t are independent. We could, for example, fix $t = t_0$, and then $X''(x)/X(x)$ would equal the constant $T'(t_0)/kT(t_0)$ for all x in $(0, L)$. Therefore, for some constant λ, called the *separation constant*,

$$\frac{T'}{kT} = \frac{X''}{X} = -\lambda.$$

Denoting the constant $-\lambda$ instead of λ is customary. We would, however, obtain the same solution using λ.

With the variables separated, we have two ordinary differential equations

$$X'' + \lambda X = 0$$

and

$$T' + \lambda kT = 0.$$

Now use the boundary conditions. First, for all $t > 0$,

$$u(0,t) = X(0)T(t) = 0.$$

If $u(x,t)$ is not identically zero, then $T(t)$ must be nonzero for some t and we conclude that $X(0) = 0$. Similarly, $u(L,t) = X(L)T(t) = 0$ implies that $X(L) = 0$.

Then

$$X'' + \lambda X = 0; X(0) = X(L) = 0.$$

This is the eigenvalue problem 1.17, with solutions

$$\lambda_n = \frac{n^2\pi^2}{L^2}, X_n(x) = \sin\left(\frac{n\pi x}{L}\right).$$

Put these eigenvalues into the differential equation for T:

$$T' + \frac{n^2\pi^2 k}{L^2}T = 0.$$

Solutions of this equation are constant multiples of

$$T_n(t) = e^{-n^2\pi^2 kt/L^2}.$$

For each positive integer n, we now have a function

$$u_n(x,t) = X_n(t)T_n(t) = \sin\left(\frac{n\pi x}{L}\right)e^{-n^2\pi^2 kt/L^2},$$

which satisfies the heat equation and the boundary conditions. Now we must satisfy the initial condition $u(x,0) = f(x)$. Depending on $f(x)$, it may not be possible to do this with a constant multiple of some $u_n(x,t)$, or even with a finite sum of constant multiples of these functions for different choices of n. Generally, we must attempt a solution

$$u(x,t) = \sum_{n=1}^{\infty} b_n \sin\left(\frac{n\pi x}{L}\right) e^{-n^2\pi^2 kt/L^2}.$$

We need

$$u(x,0) = \sum_{n=1}^{\infty} b_n \sin\left(\frac{n\pi x}{L}\right) = f(x).$$

This is a Fourier sine expansion of $f(x)$ on $[0, L]$, so choose

$$b_n = \frac{2}{L}\int_0^L f(\xi)\sin\left(\frac{n\pi\xi}{L}\right) d\xi.$$

With these coefficients, the solution of problem 2.1 is:

$$u(x,t) = \sum_{n=1}^{\infty}\left(\frac{2}{L}\int_0^L f(\xi)\sin\left(\frac{n\pi\xi}{L}\right) d\xi\right)\sin\left(\frac{n\pi x}{L}\right) e^{-n^2\pi^2 kt/L^2}. \tag{2.2}$$

Example 2.1 Suppose we have a homogeneous bar of length L, with ends kept at temperature zero and initial temperature function

$$u(x,0) = f(x) = \begin{cases} x & \text{for } 0 \le x \le L/2, \\ L - x & \text{for } L/2 \le x \le L. \end{cases}$$

We know the solution (equation 2.2) for the temperature distribution $u(x,t)$. Compute the coefficients

$$\begin{aligned} b_n &= \frac{2}{L}\int_0^{L/2} \xi \sin\left(\frac{n\pi\xi}{L}\right) d\xi + \frac{2}{L}\int_{L/2}^{L} (L-\xi)\sin\left(\frac{n\pi\xi}{L}\right) d\xi \\ &= \frac{4L}{n^2\pi^2}\sin\left(\frac{n\pi}{2}\right). \end{aligned}$$

The solution of this initial-boundary value problem is

$$u(x,t) = \frac{4L}{\pi^2}\sum_{n=1}^{\infty}\frac{1}{n^2}\sin\left(\frac{n\pi}{2}\right)\sin\left(\frac{n\pi x}{L}\right) e^{-n^2\pi^2 kt/L^2}.$$

For example, with $k = 1$ and $L = \pi$, the solution is

$$u(x,t) = \frac{4}{\pi}\sum_{n=1}^{\infty}\frac{1}{n^2}\sin\left(\frac{n\pi}{2}\right)\sin(nx)e^{-n^2 t}.$$

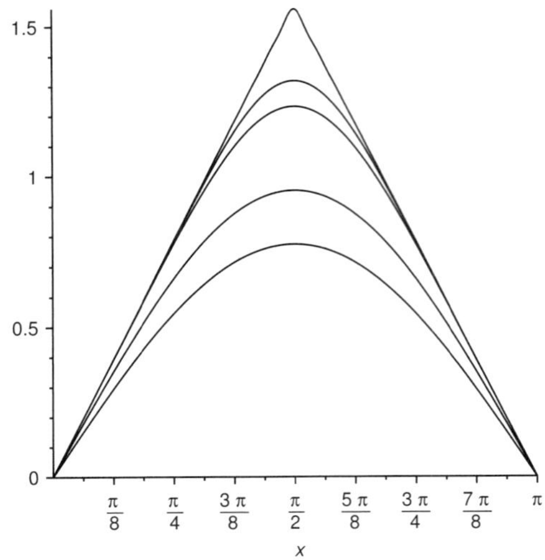

Figure 2.1: Temperature profiles in example 2.1 for $t = 0, 0.05, 0.09, 0.3$, and 0.5.

Figure 2.1 shows graphs of this temperature distribution at times $t = 0$, 0.05, 0.09, 0.3, and 0.5. The temperature decreases from the initial temperature as t increases, as we expect in a diffusion problem with no source of energy. We may also display the solution as the graph of the surface $u = u(x,t)$ in $x, t, u-$space. This is done in Figure 2.2 for $0 \le x \le \pi$ and $0 \le t \le 1.7$.

Example 2.2 Suppose the initial temperature in problem 2.1 for a bar of length 1 is

$$g(x) = x\sin(\pi x).$$

Suppose $k = 1$. For the solution, compute the coefficients

$$2\int_0^1 g(\xi)\sin(n\pi\xi)\,d\xi$$

$$= \begin{cases} 1/2 & \text{for } n = 1, \\ \frac{4n(1+(-1)^n)}{\pi^2(n^2-1)^2} & \text{for } n = 2, 3, \cdots. \end{cases}$$

The solution is

$$u(x,t) =$$

$$= -\frac{1}{2}\sin(\pi x)e^{-\pi^2 t} - \sum_{n=2}^{\infty}\frac{4n(1+(-1)^n)}{\pi^2(n^2-1)^2}\sin(n\pi x)e^{-n^2\pi^2 t}.$$

Figure 2.3 shows graphs of this solution at times $t = 0, 0.01, 0.05$, and 0.08. The temperature is close to the initial temperature at $t = 0.001$ and then decreases,

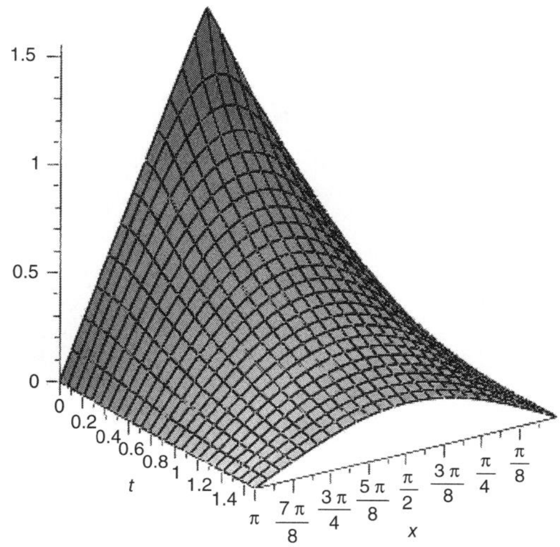

Figure 2.2: Temperature surface in example 2.1 for $0 \leq x \leq \pi$ and $0 \leq t \leq 1.7$.

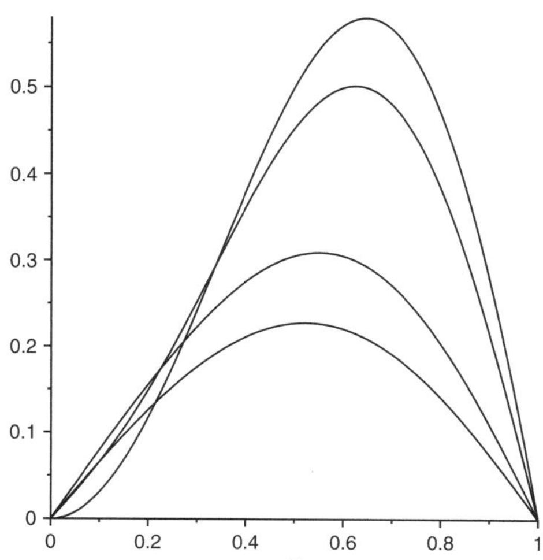

Figure 2.3: Temperature distribution in example 2.2 for $t = 0, 0.01, 0.05$, and 0.08.

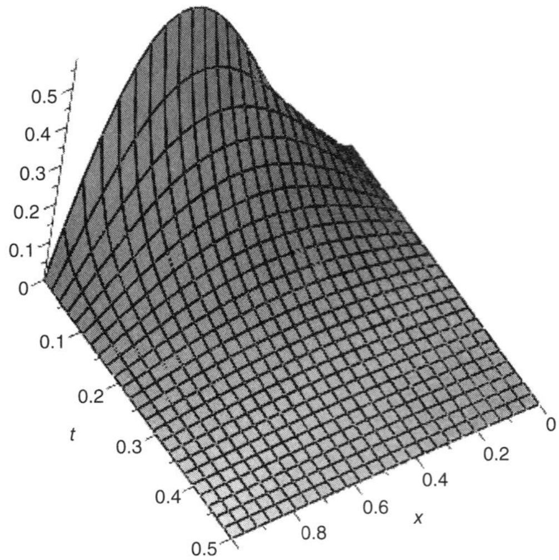

Figure 2.4: Temperature surface in example 2.2 for $0 \leq x \leq 1$, $0 \leq t \leq 1/2$.

becoming more evenly distributed throughout the bar as t increases. Figure 2.4 is a graph of the solution in $x, t, u-$space for $0 \leq x \leq 1$ and $0 \leq t \leq 1/2$.

2.1.2 Insulated Ends

If the ends of bar are insulated, then the flow of energy across the ends is zero. This model replaces the boundary conditions $u(0,t) = u(L,t) = 0$ with the zero flow conditions $u_x(0,t) = u_x(0,t) = 0$:

$$\begin{aligned} u_t &= ku_{xx} \text{ for } 0 < x < L, t > 0, \\ u_x(0,t) &= u_x(L,t) = 0 \text{ for } t > 0, \\ u(x,0) &= f(x) \text{ for } 0 < x < L. \end{aligned} \tag{2.3}$$

Although the boundary conditions have changed, the heat equation is the same, and we can set $u(x,t) = X(x)T(t)$, exactly as we did for problem 2.1, to obtain

$$X'' + \lambda X = 0 \text{ and } T' + \lambda kT = 0.$$

Now, however,

$$u_x(0,t) = X'(0)T(t) = 0 \text{ for } t > 0,$$

implying that $X'(0) = 0$. Similarly, $u_x(L,t) = 0$ for $t > 0$ implies that $X'(L) = 0$. The problem for X is

$$X'' + \lambda X = 0; X'(0) = X'(L) = 0.$$

This is the eigenvalue/eigenfunction problem 1.18, so

$$\lambda_n = \frac{n^2\pi^2}{L^2}, X_n(x) = \cos\left(\frac{n\pi x}{L}\right) \text{ for } n = 0, 1, 2, \cdots .$$

The $n = 0$ case corresponds to the constant eigenfunction $X_0(x) = 1$.

Now the problem for T is

$$T' + \frac{n^2\pi^2 k}{L^2} T = 0.$$

For $n = 0$, this has the solution

$$T_0(t) = \text{ constant}.$$

For $n = 1, 2, \cdots$,

$$T_n(t) = e^{-n^2\pi^2 kt/L^2},$$

or any constant multiple of this function.

Now we have functions

$$u_0(x,t) = \text{ constant}$$

and

$$u_n(x,t) = \cos\left(\frac{n\pi x}{L}\right) e^{-n^2\pi^2 kt/L^2},$$

which satisfy the heat equation and the boundary conditions $u_x(0,t) = u_x(L,t) = 0$. To satisfy the initial condition, we must generally try a superposition

$$u(x,t) = \frac{1}{2}a_0 + \sum_{n=1}^{\infty} a_n \cos\left(\frac{n\pi x}{L}\right) e^{-n^2\pi^2 kt/L^2}.$$

We have written the constant term as $a_0/2$ in anticipation of a Fourier cosine series. Indeed, for the initial condition, we need

$$u(x,0) = f(x) = \frac{1}{2}a_0 + \sum_{n=1}^{\infty} a_n \cos\left(\frac{n\pi x}{L}\right),$$

which is the Fourier cosine expansion of the initial temperature function on $[0, L]$. Thus choose

$$a_n = \frac{2}{L}\int_0^L f(\xi)\cos\left(\frac{n\pi\xi}{L}\right) d\xi \text{ for } n = 0, 1, 2, \cdots .$$

The solution of problem 2.3 is

$$u(x,t) = \frac{1}{L}\int_0^L f(\xi)\, d\xi + \frac{2}{L}\sum_{n=1}^{\infty}\left(\int_0^L f(\xi)\cos\left(\frac{n\pi\xi}{L}\right) d\xi\right)\cos\left(\frac{n\pi x}{L}\right) e^{-n^2\pi^2 kt/L^2}. \tag{2.4}$$

Example 2.3 Suppose a homogeneous bar of length π has insulated ends, and $k = 4$ for the material of the bar. Suppose the initial temperature is

$$f(x) = \begin{cases} 0 & \text{for } 0 \le x < \pi/2, \\ 50 & \text{for } \pi/2 \le x \le \pi. \end{cases}$$

For the solution, compute

$$\begin{aligned} a_0 &= \frac{2}{\pi} \int_0^{\pi} f(\xi)\, d\xi \\ &= \frac{2}{\pi} \int_{\pi/2}^{\pi} 50\, d\xi = 50 \end{aligned}$$

and, for $n = 1, 2, \cdots$,

$$\begin{aligned} a_n &= \frac{2}{\pi} \int_{\pi/2}^{\pi} 50 \cos(n\xi)\, d\xi \\ &= -\frac{100}{n\pi} \sin(n\pi/2). \end{aligned}$$

The solution is

$$u(x,t) = 25 - \sum_{n=1}^{\infty} \left(\frac{100}{n\pi} \sin(n\pi/2) \right) \cos(nx) e^{-4n^2 t}.$$

Figure 2.5 shows this temperature distribution at times $t = 0, 0.003, 0.025, 0.097,$ 0.038, and 0.5. The temperature starts out near the initial temperature (an approximation to the initial temperature by a partial sum of 150 terms of the cosine expansion). For $\pi/2 \le x \le \pi$, it decreases as t increases, appearing to level off and approach the steady-state value 25, which is the limit of $u(x,t)$ as $t \to \infty$. Figure 2.6 is a graph of the solution as a surface in $x, t, u-$space for $0 \le x \le \pi$, $0 \le t \le 1.5$.

2.1.3 Ends at Different Temperatures

We will solve the problem

$$\begin{aligned} u_t &= k u_{xx} \text{ for } 0 < x < L, t > 0, \\ u(0,t) &= \tau_1, u(L,t) = \tau_2 \text{ for } t > 0, \\ u(x,0) &= f(x) \text{ for } 0 < x < L. \end{aligned} \tag{2.5}$$

Here τ_1 and τ_2 are nonnegative numbers, and we assume that at least one is nonzero. The ends of the bar are maintained at different temperatures. These are nonhomogeneous boundary conditions.

Try to separate the variables. If $u(x,t) = X(x)T(t)$, then we need

$$u(0,t) = \tau_1 = X(0)T(t) \text{ and } u(L,t) = \tau_2 = X(L)T(t).$$

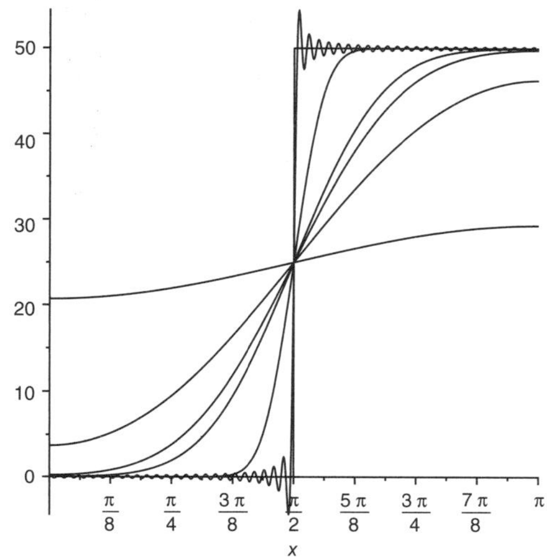

Figure 2.5: Temperatures approaching the steady-state temperature as t increases in example 2.3.

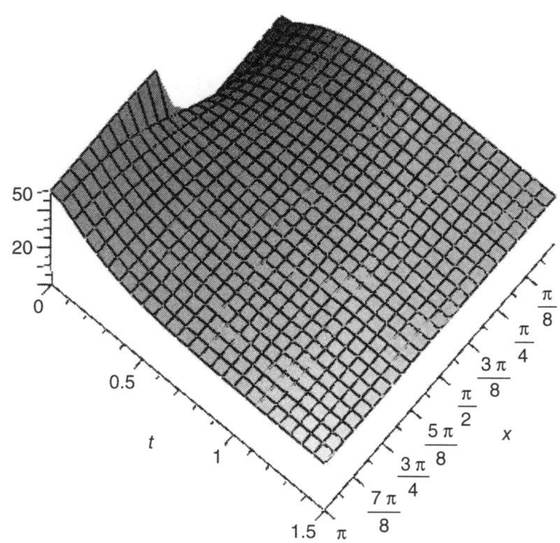

Figure 2.6: Temperature surface in example 2.3 for $0 \leq x \leq \pi$, $0 \leq t \leq 1$.

These would force $T(t)$ to be constant for all t, an unrealistic conclusion.

We can overcome this difficulty by changing the temperature function to define a new initial-boundary value problem with homogeneous boundary conditions, and that we know how to solve. The trick is to do this so that the solution of the new problem leads to the solution of the original problem.

To this end, put

$$u(x,t) = U(x,t) + \psi(x).$$

Substitute this into the heat equation to obtain

$$u_t = U_t = ku_{xx} = k(U_{xx} + \psi''(x)).$$

Then

$$U_t = kU_{xx} + k\psi''(x).$$

This will be the standard heat equation $U_t = kU_{xx}$ for $U(x,t)$ if $k\psi''(x) = 0$. For this we need

$$\psi(x) = Cx + D.$$

Now look at the boundary conditions to get a standard problem 2.1 for $U(x,t)$. First,

$$u(x,0) = \tau_1 = U(0,t) + \psi(0),$$

and this will give us $U(0,t) = 0$ if $\psi(0) = \tau_1$. Then

$$\psi(0) = D = \tau_1,$$

so

$$\psi(x) = Cx + \tau_1.$$

Next,

$$u(L,t) = \tau_2 = U(L,t) + \psi(L),$$

and we will have $U(L,t) = 0$ if

$$\psi(L) = CL + \tau_1 = \tau_2.$$

Thus choose

$$C = \frac{1}{L}(\tau_2 - \tau_1)$$

to obtain

$$\psi(x) = \frac{1}{L}(\tau_2 - \tau_1)x + \tau_1.$$

Now the problem for $U(x,t)$ is

$$\begin{aligned}
U_t &= kU_{xx} \text{ for } 0 < x < L, t > 0,\\
U(0,t) &= U(L,t) = 0 \text{ for } t > 0,\\
U(x,0) &= u(x,0) - \psi(x)\\
&= f(x) - \frac{1}{L}(\tau_2 - \tau_1)x - \tau_1.
\end{aligned}$$

This problem for U is just problem 2.1, which has the solution 2.2. Once we know $U(x,t)$, $u(x,t) = U(x,t) + \psi(x)$ is the solution of the original problem. The only price we paid for this conversion of the given problem to a simpler one is that the initial condition is changed. The initial condition for $U(x,t)$ is the original initial temperature function of the original problem, plus a new function we invented to simplify the boundary conditions.

The solution 2.2 for $U(x,t)$ is

$$U(x,t) = \sum_{n=1}^{\infty} b_n \sin\left(\frac{n\pi x}{L}\right) e^{-n^2\pi^2 kt/L^2},$$

where

$$\begin{aligned} b_n &= \frac{2}{L}\int_0^L \left(f(\xi) - \frac{1}{L}(\tau_2 - \tau_1)\xi - \tau_1\right) \sin\left(\frac{n\pi\xi}{L}\right) d\xi \\ &= \frac{2}{L}\int_0^L f(\xi)\sin\left(\frac{n\pi\xi}{L}\right) d\xi + 2\left(\frac{(-1)^n\tau_2 - \tau_1}{n\pi}\right). \end{aligned}$$

Finally,

$$u(x,t) = U(x,t) + \frac{1}{L}(\tau_2 - \tau_1)x + \tau_1.$$

We may interpret this solution as a decomposition of the temperature distribution $u(x,t)$ into a transient part $U(x,t)$, which decays to zero as $t \to \infty$, and a steady-state part $\psi(x)$, which is independent of time:

$$\lim_{t\to\infty} u(x,t) = \psi(x).$$

This type of decomposition is commonly seen in physical systems. For example, the current in an RLC circuit can be written as the sum of a transient part, which decays to zero with increasing time, and a steady-state part, which is the limit of the solution as $t \to \infty$. Solutions of mixing and other types of diffusion problems also exhibit this type of behavior.

Example 2.4 Solve the problem

$$\begin{aligned} u_t &= 7u_{xx} \text{ for } 0 < x < 5, t > 0, \\ u(0,t) &= 1, u(5,t) = 4 \text{ for } t > 0, \\ u(x,0) &= f(x) = \begin{cases} 3 - x & \text{for } 0 < x \le 3, \\ 10(x-3) & \text{for } 3 \le x < 5. \end{cases} \end{aligned}$$

Following the discussion, with $\tau_1 = 1$ and $\tau_2 = 4$, set

$$u(x,t) = U(x,t) + \frac{3}{5}x + 1,$$

where $U(x,t)$ is the solution of

$$\begin{aligned}
U_t &= 7U_{xx} \text{ for } 0 < x < 5, t > 0,\\
U(0,t) &= U(5,t) = 0 \text{ for } t < 0,\\
U(x,0) &= u(x,0) - \frac{3}{5}x - 1\\
&= \begin{cases} 2 - \frac{8}{5}x & \text{for } 0 < x \leq 3,\\ -31 + \frac{47}{5}x & \text{for } 3 \leq x < 5. \end{cases}
\end{aligned}$$

We know that the solution of this problem for $U(x,t)$ is

$$U(x,t) = \sum_{n=1}^{\infty} b_n \sin\left(\frac{n\pi x}{5}\right) e^{-7n^2\pi^2 t/25},$$

where

$$\begin{aligned}
b_n &= \frac{2}{5}\int_0^5 U(\xi,0)\sin\left(\frac{n\pi x}{5}\right)\,d\xi\\
&= \frac{2}{5}\left[\int_0^3 \left(2 - \frac{8}{5}\xi\right)\sin\left(\frac{n\pi\xi}{5}\right)\,d\xi\right]\\
&+ \frac{2}{5}\left[\int_3^5 \left(-31 + \frac{47}{5}\xi\right)\sin\left(\frac{n\pi\xi}{5}\right)\,d\xi\right]\\
&= \frac{-110\sin(3n\pi/5)}{n^2\pi^2} + \frac{4}{n\pi} - \frac{32}{n\pi}(-1)^n.
\end{aligned}$$

Then

$$\begin{aligned}
u(x,t) &= \frac{3}{5}x + 1\\
&+ \sum_{n=1}^{\infty}\left(\frac{-110\sin(3n\pi/5)}{n^2\pi^2} + \frac{4}{n\pi} - \frac{32}{n\pi}(-1)^n\right)\sin\left(\frac{n\pi x}{5}\right) e^{-7n^2\pi^2 t/25}.
\end{aligned}$$

Figure 2.7 shows the temperature distribution at times $t = 0$ and $t = 0.003$. Here the distribution at $t = 0$ is an approximation made by taking the 200th partial sum in the series solution. Figure 2.8 shows this distribution at times $t = 0.005, 0.07$, and 0.09. These show the temperature decreasing toward the graph of the straight line $u = \frac{3}{5}x + 1$, which is the steady-state limit of $u(x,t)$ as $t \to \infty$. Figure 2.9 is a plot of the surface for $0 \leq x \leq 5$ and $0 \leq t \leq 1.5$.

2.1.4 A Diffusion Equation with Additional Terms

Some models of diffusion processes include additional terms in the heat equation to allow for other effects. If we include terms proportional to u and u_x, then a

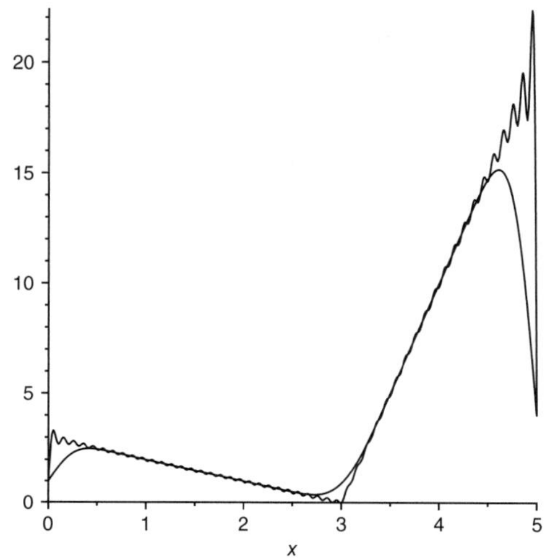

Figure 2.7: Comparison of the solution at $t = 0$ and at $t = 0.003$.

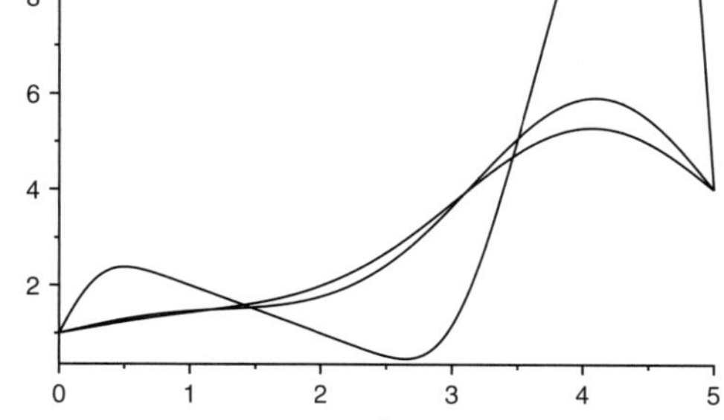

Figure 2.8: Graphs of (approximate) solutions at $t = 0.005, 0.07$, and 0.09.

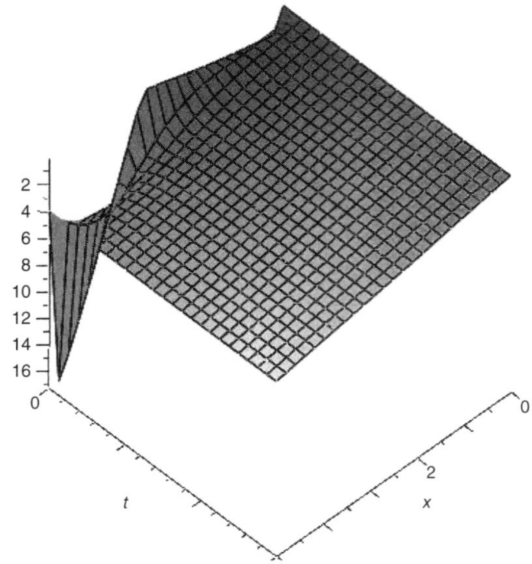

Figure 2.9: Temperature surface for $0 \leq x \leq 5$, $0 \leq t \leq 1.5$.

typical problem can have the form

$$\begin{aligned} u_t &= ku_{xx} + Au_x + Bu \text{ for } 0 < x < L, \\ u(0,t) &= u(L,t) = 0 \text{ for } t > 0, \\ u(x,0) &= f(x) \text{ for } 0 < x < L. \end{aligned}$$

If we attempt a separation of variables $u(x,t) = X(x)T(t)$, we find that the variables do separate, but the problem for X is

$$X'' + \frac{A}{k}X' + \left(\frac{B}{k} + \lambda\right) X = 0; X(0) = X(L) = 0.$$

This is an unpleasant problem to solve for the eigenvalues and eigenfunctions.

Here is another approach. In the spirit of the preceding section, we will show how to transform this problem to another one we have already solved. Let

$$u(x,t) = e^{\alpha x + \beta t} v(x,t)$$

and choose α and β to obtain a familiar problem for $v(x,t)$. Substitute $u(x,t)$ into the partial differential equation. After some routine computation, we obtain

$$v_t = kv_{xx} + (2k\alpha + A)v_x + (k\alpha^2 + A\alpha + B - \beta)v.$$

From this formulation, we can choose α and β to obtain a standard heat equation $v_t = kv_{xx}$ for v. We need

$$2k\alpha + A = 0 \text{ and } k\alpha^2 + A\alpha + B - \beta = 0.$$

Solve these to obtain

$$\alpha = -\frac{A}{2k}$$

and

$$\beta = -\frac{A^2}{4k} + B.$$

Then $v_t = kv_{xx}$. Next,

$$u(0,t) = e^{\beta t} v(0,t) = 0$$

implies that $v(0,t) = 0$. Similarly, $v(L,t) = 0$. Finally,

$$u(x,0) = e^{\alpha x} v(x,0) = f(x)$$

implies that

$$v(x,0) = e^{-\alpha x} f(x).$$

The problem for $v(x,t)$ is

$$\begin{aligned} v_t &= kv_{xx} \text{ for } 0 < x < L, t > 0, \\ v(0,t) &= v(L,t) = 0 \text{ for } t > 0, \\ v(x,0) &= e^{-\alpha x} f(x) \text{ for } 0 < x < L. \end{aligned}$$

The solution of this problem is

$$v(x,t) = \sum_{n=1}^{\infty} b_n \sin\left(\frac{n\pi x}{L}\right) e^{-n^2\pi^2 kt/L^2},$$

where

$$b_n = \frac{2}{L} \int_0^L e^{-\alpha \xi} f(\xi) \sin\left(\frac{n\pi \xi}{L}\right) d\xi.$$

The solution of the original problem is

$$u(x,t) = e^{\alpha x + \beta t} v(x,t),$$

with α and β found in terms of k, A, and B.

We will give an application of these ideas to a model for the diffusion of charges on a transistor. Let $h(x,t)$ be the concentration of positive charge carriers at time t and position x on a transistor occupying an interval $[0, L]$. Solid state scientists and engineers have shown that $h(x,t)$ is modeled by the initial-boundary value problem

$$\begin{aligned} h_t &= k\left(h_{xx} - \frac{a}{L} h_x\right) \text{ for } 0 < x < L, t > 0, \\ h(0,t) &= h(L,t) = 0 \text{ for } t > 0, \\ h(x,0) &= \frac{KL}{ka}\left(1 - e^{-a(1-x/L)}\right) \text{ for } 0 < x < L, \end{aligned}$$

with a, k, and K positive constants.

We will use the approach just discussed, with $A = -ka/L$ and $B = 0$. Set

$$h(x,t) = e^{\alpha x+\beta t}v(x,t),$$

where

$$\alpha = \frac{a}{2L}$$

and

$$\beta = -\frac{ka^2}{4L^2}.$$

The problem for v is

$$\begin{aligned} v_t &= kv_{xx} \text{ for } 0 < x < L, t > 0, \\ v(0,t) &= v(L,t) = 0 \text{ for } t > 0, \\ v(x,0) &= \frac{KL}{ka}e^{-ax/2L}\left(1 - e^{-a(1-x/L)}\right) \text{ for } 0 < x < L. \end{aligned}$$

The solution of this problem is

$$v(x,t) = \sum_{n=1}^{\infty} b_n \sin\left(\frac{n\pi x}{L}\right) e^{-n^2\pi^2 kt/L^2},$$

in which

$$\begin{aligned} b_n &= \frac{2K}{ka}\int_0^L e^{-a\xi/2L}\left(1 - e^{-a(1-\xi/L)}\right) \sin\left(\frac{n\pi\xi}{L}\right) d\xi \\ &= \frac{8KL}{ka}\frac{n\pi(1-e^{-a})}{4n^2\pi^2 + a^2}. \end{aligned}$$

Therefore

$$v(x,t) =$$
$$\frac{8KL}{ka}(1-e^{-a})\sum_{n=1}^{\infty}\frac{n\pi}{4n^2\pi^2 + a^2}\sin\left(\frac{n\pi x}{L}\right)e^{-n^2\pi^2 kt/L^2}.$$

Finally,

$$h(x,t) = e^{ax/2L}e^{-ka^2t/4L^2}v(x,t).$$

Figure 2.10 shows the diffusion function decreasing over times $t = 0.005, 0.052$, and 0.37. Here we took $L = \pi, K = 2, k = 3$ and $a = 1$. Figure 2.11 is part of the graph of the solution as a surface in 3−space.

2.1.5 One Radiating End

Suppose a homogeneous bar of length L with constant cross section has one end kept at temperature zero while the other end radiates energy into the surrounding medium. This medium is also kept at temperature zero. We want the temperature distribution $u(x,t)$ of the bar.

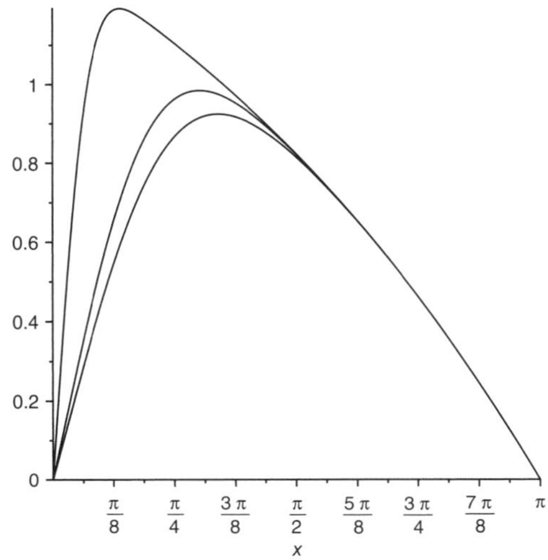

Figure 2.10: $u(x,t)$ for times $t = 0.005, 0.052$, and 0.037.

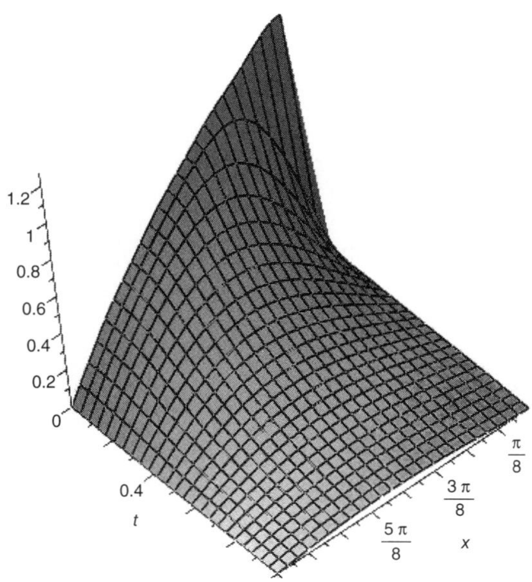

Figure 2.11: Surface graph for $0 \leq x \leq \pi$ and $0 \leq t \leq 1$.

This distribution is modeled by the initial-boundary value problem

$$u_t = ku_{xx} \text{ for } 0 < x < L, t > 0,$$
$$u(0,t) = 0, u_t(L,t) = -Au(L,t) \text{ for } t > 0,$$
$$u(x,0) = f(x) \text{ for } 0 < x < L,$$

in which A is a positive constant called the *transfer coefficient*.

Separate variables in the heat equation by putting $u(x,t) = X(x)T(t)$ to obtain

$$X'' + \lambda X = 0, T' + \lambda kT = 0.$$

Because $u(0,t) = X(0)T(t) = 0$, $X(0) = 0$. Next,

$$X'(L)T(t) = -AX(L)T(t),$$

so, assuming that $T(t)$ is not identically zero,

$$X'(L) + AX(L) = 0.$$

The problem for X is

$$X'' + \lambda X = 0; X(0) = 0, X'(L) + AX(L) = 0.$$

This is not one of the standard eigenvalue problems 1.17 or 1.18 that we have solved. The difference is due to the radiating end condition, which is unlike conditions we have seen in other problems. Consider cases on λ.

Case 1. $\lambda = 0$

Now $X(x) = cx + d$. Then $X(0) = d = 0$, and

$$X'(L) + AX(L) = c + AcL = 0.$$

If $A > 0$, this requires that $c = 0$, so this case has only the trivial solution.

Case 2. $\lambda < 0$

Write $\lambda = -\alpha^2$ with $\alpha > 0$. Then

$$X(x) = ce^{\alpha x} + de^{-\alpha x}.$$

Because

$$X(0) = c + d = 0,$$

$c = -d$, so

$$X(x) = c\left(e^{\alpha x} - e^{-\alpha x}\right).$$

The radiation condition at L now becomes

$$c\alpha\left(e^{\alpha L} + e^{-\alpha x}\right) = -Ac\left(e^{\alpha L} - e^{-\alpha L}\right).$$

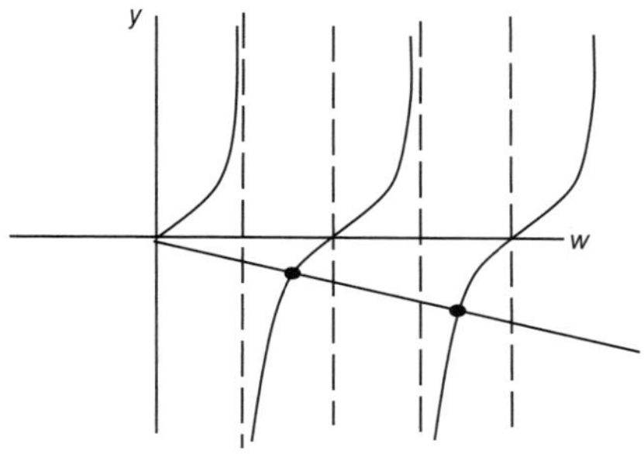

Figure 2.12: First two intersections of $y = -w/AL$ with $y = \tan(w)$ for $w > 0$.

For a nontrivial solution, we need $c \neq 0$. But in this event, we can divide by c to obtain

$$\alpha\left(e^{\alpha L} + e^{-\alpha x}\right) = -A\left(e^{\alpha L} - e^{-\alpha L}\right).$$

Since $e^{\alpha L} - e^{-\alpha L} > 0$, the left side of this equation is positive while the right side is negative. This is a contradiction, so the assumption that $c \neq 0$ is untenable and this case yields only the trivial solution.

Case 3. $\lambda > 0$

Write $\lambda = \alpha^2$ with $\alpha > 0$. In this case,

$$X(x) = c\cos(\alpha x) + d\sin(\alpha x).$$

Because $X(0) = c = 0$, we are left with $X(x) = d\sin(\alpha x)$. Now the radiation condition is

$$d\alpha\cos(\alpha L) = -dA\sin(\alpha L).$$

If $d = 0$, we are out of cases without a nontrivial solution. The other alternative is to suppose $d \neq 0$, in which case we must have

$$\frac{\sin(\alpha L)}{\cos(\alpha L)} = \tan(\alpha L) = -\frac{\alpha}{A}.$$

This equation must be solved for α. The difference between this and other eigenvalue problems we have seen (requiring, for example, zeros of the sine function), is that it may not obvious whether this equation has positive solutions for α, or what these solutions are.

Look at the problem graphically. Let $w = \alpha L$. Then

$$\tan(w) = -\frac{1}{AL}w.$$

Figure 2.12 shows part of a graph of $y = \tan(w)$ for $w > 0$ and the line $y = -w/AL$. There are infinitely many disjoint branches of the tangent function, and each of these intersects the line $y = -w/AL$ in a point with a positive first coordinate. Let these positive first coordinates be $w_1, w_2, \cdots$. Then $\alpha_n = w_n/L$ for $n = 1, 2, \cdots$. Given A and L, these numbers have to be computed numerically. The eigenvalues of this problem are

$$\lambda_n = \frac{w_n^2}{L^2} \text{ for } n = 1, 2, \cdots$$

Corresponding eigenfunctions are

$$X_n(x) = \sin(\alpha_n x) = \sin\left(\frac{w_n x}{L}\right).$$

This is a far cry from eigenvalues and eigenfunctions encountered with ends kept at temperature zero or with insulated ends. There we found the eigenvalues by solving an equation $\sin(\alpha L) = 0$, which results in the familiar expression $\alpha = n\pi/L$. With α defined implicitly by $A\tan(\alpha L) = -\alpha$, there is no such simple formula for solutions for α.

We can, however, proceed to an approximate solution for $u(x,t)$ with the following observation. Work with the eigenfunctions $X_n(x) = \sin(w_n x/L)$. With $\lambda_n = w_n^2/L^2$, the problem for T is

$$T' + \frac{kw_n^2}{L^2}T = 0,$$

with solutions that are constant multiples of

$$T_n(t) = e^{-kw_n^2 t/L^2}.$$

Formally, then, for each positive integer n, we have a function

$$u_n(x,t) = \sin\left(\frac{w_n x}{L}\right) e^{-kw_n^2 t/L^2}.$$

This is similar in form to previous problems in which we had $\alpha = n\pi/L$. Forging ahead, write

$$\begin{aligned} u(x,t) &= \sum_{n=1}^{\infty} c_n u_n(x,t) \\ &= \sum_{n=1}^{\infty} c_n \sin\left(\frac{w_n x}{L}\right) e^{-kw_n^2 t/L^2}. \end{aligned}$$

We must choose the coefficients so that

$$u(x,0) = f(x) = \sum_{n=1}^{\infty} c_n \sin\left(\frac{w_n x}{L}\right).$$

This is a series of sines, but not a Fourier sine series. However, we will attempt an informal calculation like the one carried out in Chapter 1 to find Euler's formulas for the Fourier coefficients. Multiply both sides of this proposed series by $\sin(w_k x/L)$ to obtain

$$f(x)\sin(w_k x/L) = \sum_{n=1}^{\infty} c_n \sin(w_n x/L)\sin(w_k x/L).$$

Integrate both sides of this equation:

$$\int_0^L f(x)\sin(w_k x/L)\,dx = \sum_{n=1}^{\infty} c_n \int_0^L \sin(w_n x/L)\sin(w_k x/L)\,dx.$$

Now look at the terms

$$\int_0^L \sin(w_n x/L)\sin(w_k x/L)\,dx \text{ when } n \neq k.$$

A straightforward integration yields

$$\int_0^L \sin(w_n x/L)\sin(w_k x/L)\,dx =$$
$$\frac{L}{w_k^2 - w_n^2}(w_n \cos(w_n)\sin(w_n) - w_k \sin(w_n)\cos(w_k)).$$

The numerator in this expression can be written

$$\cos(w_n)\cos(w_k)\left[w_n \tan(w_k) - w_k \tan(w_n)\right].$$

Now recall that

$$\tan(w_n) = -\frac{w_n}{AL} \text{ and } \tan(w_k) = -\frac{w_k}{AL}.$$

Then

$$\int_0^L \sin(w_n x/L)\sin(w_k x/L)\,dx =$$
$$= \cos(w_n)\cos(w_k)\left[-w_n\frac{w_k}{AL} + w_k\frac{w_n}{AL}\right] = 0$$

if $n \neq k$.

This is just like the orthogonality relationship that Euler used in obtaining the Fourier coefficients. In view of this, every term in the summation of the integrals is zero except the $n = k$ term, and we obtain

$$\int_0^L f(x)\sin(w_k x/L)\,dx = c_k \int_0^L \sin^2(w_k x/L)\,dx.$$

Then

$$c_k = \frac{\int_0^L f(x)\sin(w_k x/L)\,dx}{\int_0^L \sin^2(w_k x/L)\,dx}.$$

Upon carrying out the integration in the denominator, we have

$$c_k = \left(\frac{2w_k}{L}\right) \frac{\int_0^L f(x)\sin(w_k x/L)\,dx}{w_k - \cos(w_k)\sin(w_k)}.$$

This suggests that the solution of the problem is

$$u(x,t) = \sum_{n=1}^{\infty} c_n \sin\left(\frac{w_n x}{L}\right) e^{-kw_n^2 t/L^2},$$

in which we use the formula derived for the c_n's.

Given A and L, we can approximate the first few terms of the solution as closely as we like. As an example, first solve the eigenvalue equation numerically to approximate $w_1, w_2, \cdots w_5$. With $A = L = 1$, we get

$$w_1 \approx 2.028757838, w_2 \approx 4.913180439, w_3 \approx 7.978865712,$$
$$w_4 \approx 11.08553841, w_5 \approx 14.2073673.$$

Take $f(x) = x$. The coefficients in the series solution are

$$c_k = 2w_k \frac{\int_0^1 x\sin(w_k x)\,dx}{w_k - \cos(w_k)\sin(w_k)}.$$

Insert the approximate w_k's and perform numerical integrations to obtain

$$c_1 \approx 0.7291747434, c_2 \approx -0.1561635161, c_3 \approx 0.06144180676,$$
$$c_4 \approx -0.03215845050, c_5 \approx 0.01967069622.$$

The approximate solution is

$$u(x,t) \approx (0.7291747434)\sin(2.028757838x)e^{-0.7291747434kt}$$
$$-\ (0.1561635161)\sin(4.913180439x)e^{-4(4.913180439)kt}$$
$$(0.06144180676)\sin(7.978865712x)e^{-9(7.978865712)kt}$$
$$-\ (0.03215845050)\sin(11.08553841x)e^{-(16)11.08553841kt}$$
$$+\ 0.01967069622\sin(14.2074343673x)e^{-(25)14.2074343673kt}.$$

Problem 17 for this section provides another example in which the eigenfunctions are not the usual sines and cosines that form the basis for Fourier expansions.

Problems for Section 2.1

In each of problems 1–7, solve $u_t = ku_{xx}$ with the given boundary and initial conditions.

1. $u(0,t) = u(1,t) = 0$ for $t > 0$; $u(x,0) = \sin(\pi x)$ for $0 < x < 1$.

2. $u(0,t) = u(3,t) = T$ for $t > 0; u(x,0) = T$ for $0 < x < 3$.

3. $u_x(0,t) = u_x(4,t) = 0$ for $t > 0; u(x,0) = x^2$ for $0 < x < 4$.

4. $u(0,t) = u(2,t) = 0$ for $t > 0; u(x,0) = \sin(\pi x)$ for $0 < x < 2$.

5. $u_x(0,t) = u_x(6,t) = 0$ for $t > 0; u(x,0) = e^{-x}$ for $0 < x < 6$.

6. $u(0,t) = 3, u(5,t) = \sqrt{7}$ for $t > 0; u(x,0) = x^2$ for $0 < x < 5$.

7. $u_x(0,t) = u_x(L,t) = 0$ for $t > 0; u(x,0) = L - x^2$ for $0 < x < L$.

8. Solve

$$\begin{aligned} u_t &= 7u_{xx} \text{ for } 0 < x < 5, t > 0 \\ u(0,t) &= 1, u(5,t) = 4 \text{ for } t > 0 \\ u(x,0) &= e^{-x} \text{ for } 0 < x < 5. \end{aligned}$$

Graph the solution for selected values of t.

9. Solve

$$\begin{aligned} u_{tt} &= 5u_{xx} \text{ for } 0 < x < 4, t > 0, \\ u(0,t) &= 0, u(4,t) = 12 \text{ for } t > 0, \\ u(x,0) &= x^2(4-x) \text{ for } 0 < x < 4. \end{aligned}$$

10. Solve

$$\begin{aligned} u_t &= ku_{xx} - hu_x \text{ for } 0 < x < L, t > 0 \\ u(0,t) &= u(L,t) = 0 \text{ for } t > 0 \\ u(x,0) &= f(x) \text{ for } 0 < x < L, \end{aligned}$$

in which h is a positive number.

11. Solve

$$\begin{aligned} u_t &= 4u_{xx} - 2u_x \text{ for } 0 < x < L, t > 0 \\ u(0,t) &= u(L,t) = 0 \text{ for } t > 0 \\ u(x,0) &= 1 \text{ for } 0 < x < L. \end{aligned}$$

Graph the solution for some values of the time, with $L = \pi$.

12. Solve

$$\begin{aligned} u_t &= u_{xx} - 6u_x \text{ for } 0 < x < \pi, t > 0 \\ u(0,t) &= u(\pi,t) = 0 \text{ for } t > 0 \\ u(x,0) &= \sin(x) \text{ for } 0 < x < \pi. \end{aligned}$$

Graph the solution for selected values of t.

13. Solve

$$\begin{aligned} u_t &= ku_{xx} - hu \text{ for } 0 < x < L, t > 0 \\ u_x(0,t) &= u_x(L,t) = 0 \text{ for } t > 0 \\ u(x,0) &= x - x^2/L \text{ for } 0 < x < L, \end{aligned}$$

in which h is a positive constant.

14. Solve

$$\begin{aligned} u_t &= 4u_{xx} - 8u \text{ for } 0 < x < 2\pi, t > 0 \\ u_x(0,t) &= u_x(2\pi,t) = 0 \text{ for } t > 0 \\ u(x,0) &= x(2\pi - x) \text{ for } 0 < x < 2\pi. \end{aligned}$$

Generate some graphs of the solution for selected values of t.

15. Solve

$$\begin{aligned} u_t &= k(u_{xx} - 3u_x + 2u) \text{ for } 0 < x < L, t > 0 \\ u(0,t) &= u(L,t) = 0 \text{ for } t > 0 \\ u(x,0) &= f(x) \text{ for } 0 < x < L. \end{aligned}$$

16. A thin, homogeneous bar of length L has insulated ends and initial temperature B, a constant. Find the temperature distribution in the bar.

17. The problem

$$\begin{aligned} u_t &= ku_{xx} \text{ for } 0 < x < L, t > 0 \\ u(0,t) &= u_x(L,t) = 0 \text{ for } t > 0 \\ u(x,0) &= f(x) \text{ for } 0 < x < L. \end{aligned}$$

models heat conduction in a bar of length L with the left end kept at temperature zero but with an insulation condition on the right end. Using separation of variables $u(x,t) = X(x)T(t)$, show that the problem to solve for X is

$$X'' + \lambda X = 0; X(0) = X'(L) = 0.$$

By considering cases on λ, show that this problem has eigenvalues

$$\lambda_n = \frac{(2n-1)^2\pi^2}{4L^2}$$

for $n = 1, 2, \cdots$, and corresponding eigenfunctions

$$X_n(x) = \sin\left(\frac{(2n-1)\pi x}{2L}\right).$$

For each λ_n, show that corresponding solutions for T are constant multiples of

$$T_n(t) = e^{-(2n-1)^2\pi^2 kt/4L^2}.$$

Show that for $n = 1, 2, \cdots$, the functions

$$u_n(x,t) = b_n \sin\left(\frac{(2n-1)\pi x}{2L}\right) e^{-(2n-1)^2\pi^2 kt/4L^2}$$

are solutions of the heat equation satisfying both boundary conditions. To satisfy the initial condition, attempt a superposition

$$u(x,t) = \sum_{n=1}^{\infty} b_n \sin\left(\frac{(2n-1)\pi x}{2L}\right) e^{-(2n-1)^2\pi^2 kt/4L^2}.$$

We require that

$$u(x,0) = f(x) = \sum_{n=1}^{\infty} b_n \sin\left(\frac{(2n-1)\pi x}{2L}\right).$$

Explain why this is not a Fourier sine expansion of f on $[0, L]$. However, it is an expansion of f in a series of eigenfunctions of a boundary value problem for X. Show that

$$\int_0^L \sin\left(\frac{(2n-1)\pi x}{2L}\right) \sin\left(\frac{(2m-1)\pi x}{2L}\right) dx = 0$$

if n and m are distinct positive integers, and derive a formula for the b_n's by reasoning informally as in the problem with one radiating end.

18. A thin, homogeneous bar of length L has initial temperature equal to a constant, B, and the right end ($x = L$) is insulated while the left end is kept at temperature zero. Find the temperature distribution in the bar.

19. A thin, homogeneous bar of thermal diffusivity 9, length 2 cm, and insulated sides has its left end maintained at temperature zero while the right end is insulated. The bar has an initial temperature given by $f(x) = x^2$. Determine the temperature distribution in the bar. Calculate $\lim_{t\to\infty} u(x,t)$.

20. Suppose that we have a long, thin, homogeneous bar of length L, with sides poorly insulated. Heat radiates freely from the bar along its length. Assuming a positive transfer coefficient A and a constant temperature, T, in the surrounding medium, the equation for the temperature distribution is

$$u_t = ku_{xx} - A(u - T).$$

Assume insulated ends and an initial temperature of $f(x)$. Solve for $u(x,t)$. *Hint*: Let $w = u - T$.

21. Solve

$$\begin{aligned} u_t &= ku_{xx} \text{ for } -L < x < L, t > 0 \\ u(-L,t) &= u(L,t), u_x(-L,t) = u_x(L,t) \text{ for } t > 0 \\ u(x,0) &= f(x) \text{ for } 0 < x < L. \end{aligned}$$

This initial-boundary value problem models heat conduction in an insulated wire of length $2L$, which is thought of as bent into the form of a ring, with the ends joined. The boundary conditions identify the points $(-L,t)$ and (L, t) in the sense that the temperature and flux at these points are the same. This problem is known as *Fourier's ring*.

22. Let u be a solution of the heat equation

$$u_t = ku_{xx} \text{ for } 0 < x < L, t > 0.$$

Let $0 < a < b < L$. Prove that for any positive t,

$$\frac{d}{dt}\int_a^b \frac{1}{2}u(x,t)^2\,dx = kuu_x]_a^b - \int_a^b ku_x(x,t)^2\,dx.$$

This is the analogue for the heat equation of the energy integral defined previously for the wave equation. *Hint*: Multiply the heat equation by u and integrate both sides from a to b, applying integration by parts where appropriate.

2.2 A Nonhomogeneous Problem

Consider the nonhomogeneous initial-boundary value problem

$$\begin{aligned} u_t &= ku_{xx} + F(x,t) \text{ for } 0 < x < L, t > 0, \\ u(0,t) &= u(L,t) = 0 \text{ for } t > 0, \\ u(x,0) &= f(x) \text{ for } 0 < x < L. \end{aligned} \tag{2.6}$$

This is nonhomogeneous because of $F(x,t)$, which can account for a source of loss of energy within the medium.

In the case that $F(x,t) = 0$, this is problem 2.1, with a solution

$$\sum_{n=1}^{\infty} b_n \sin\left(\frac{n\pi x}{L}\right) e^{-n^2\pi^2 kt/L^2},$$

in which the b_n's are the Fourier sine coefficients of the initial temperature function on the interval.

This suggests that we attempt a solution of the present problem of the form

$$u(x,t) = \sum_{n=1}^{\infty} T_n(t)\sin\left(\frac{n\pi x}{L}\right). \tag{2.7}$$

Now focus on determining the functions $T_n(t)$ so that 2.7 is a solution of the problem. The strategy is to determine an ordinary differential equation for $T_n(t)$.

If t is fixed, then equation 2.7 can be thought of as a Fourier sine expansion of $u(x,t)$, as a function of x, on $[0, L]$. From this perspective, $T_n(t)$ is the nth Fourier sine coefficient of $u(x,t)$ on $[0, L]$. Of course, we do not know $u(x,t)$ yet, but at least formally, this Fourier sine coefficient would be

$$T_n(t) = \frac{2}{L}\int_0^L u(\xi,t)\sin\left(\frac{n\pi\xi}{L}\right)d\xi. \tag{2.8}$$

Assume that, for any $t \geq 0$, $F(x,t)$, as a function of x, can also be expanded in a Fourier sine series on $[0, L]$:

$$F(x,t) = \sum_{n=1}^{\infty} B_n(t)\sin\left(\frac{n\pi x}{L}\right), \tag{2.9}$$

where this coefficient (which depends on the choice of t), is

$$B_n(t) = \frac{2}{L}\int_0^L F(\xi,t)\sin\left(\frac{n\pi\xi}{L}\right)d\xi. \tag{2.10}$$

Assuming that we can differentiate equation 2.8 under the integral, we have

$$T_n'(t) = \frac{2}{L}\int_0^L u_t(\xi,t)\sin\left(\frac{n\pi\xi}{L}\right)d\xi. \tag{2.11}$$

Substitute $u_t = ku_{xx} + F(x,t)$ into equation 2.11 to obtain

$$T_n'(t) = \frac{2k}{L}\int_0^L u_{xx}(\xi,t)\sin\left(\frac{n\pi\xi}{L}\right)d\xi + \frac{2}{L}\int_0^L F(\xi,t)\sin\left(\frac{n\pi\xi}{L}\right)d\xi.$$

In view of equation 2.10, this is

$$T_n'(t) = \frac{2k}{L}\int_0^L u_{xx}(\xi,t)\sin\left(\frac{n\pi\xi}{L}\right)d\xi + B_n(t). \tag{2.12}$$

Apply integration by parts to the integral term of equation 2.12, using the boundary conditions of the problem at the last stage:

$$\begin{aligned}
&\int_0^L u_{xx}(\xi,t)\sin\left(\frac{n\pi\xi}{L}\right)\,d\xi \\
&= \sin\left(\frac{n\pi\xi}{L}\right)u_x(\xi,t)\Big]_0^L - \int_0^L u_x(\xi,t)\frac{n\pi}{L}\cos\left(\frac{n\pi\xi}{L}\right)\,d\xi \\
&= -\int_0^L \frac{n\pi}{L}u_x(\xi,t)\cos\left(\frac{n\pi\xi}{L}\right)\,d\xi \\
&= -\frac{n\pi}{L}u(\xi,t)\cos\left(\frac{n\pi\xi}{L}\right)\Big]_0^L + \frac{n\pi}{L}\int_0^\infty u(\xi,t)\left(-\frac{n\pi}{L}\right)\sin\left(\frac{n\pi\xi}{L}\right)\,d\xi \\
&= -\frac{n^2\pi^2}{L^2}\int_0^L u(\xi,t)\sin\left(\frac{n\pi\xi}{L}\right)\,d\xi \\
&= -\frac{n^2\pi^2}{L^2}\frac{L}{2}T_n(t) = -\frac{n^2\pi^2}{2L}T_n(t).
\end{aligned}$$

Substitute this result into equation 2.12 to obtain

$$T_n'(t) = -k\frac{n^2\pi^2}{L^2}T_n(t) + B_n(t).$$

For $n = 1, 2, \cdots$, this is a first-order ordinary differential equation for $T_n(t)$:

$$T_n'(t) + k\frac{n^2\pi^2}{L^2}T_n(t) = B_n(t). \tag{2.13}$$

We want to solve for $T_n(t)$, subject to the initial condition

$$\begin{aligned}
T_n(0) &= \frac{2}{L}\int_0^L u(\xi,0)\sin\left(\frac{n\pi\xi}{L}\right)\,d\xi \\
&= \frac{2}{L}\int_0^L f(\xi)\sin\left(\frac{n\pi\xi}{L}\right)\,d\xi = b_n.
\end{aligned}$$

Equation 2.13 is a special case of

$$y' + p(x)y = q(x),$$

which is a linear first-order differential equation. To solve this differential equation, multiply it by the integrating factor

$$e^{\int_0^x p(\xi)\,d\xi}$$

to obtain

$$\frac{d}{dx}\left(ye^{\int_0^x p(\xi)\,d\xi}\right) = B_n(t)e^{\int_0^x p(\xi)\,d\xi},$$

which can be integrated and solved for y. When this is done with equation 2.13, and the initial condition $T_n(0) = b_n$ is used, we obtain:

$$T_n(t) = \int_0^t e^{-kn^2\pi^2(t-\tau)/L^2}B_n(\tau)\,d\tau + b_n e^{-n^2\pi^2kt/L^2}. \tag{2.14}$$

Finally, substitute this expression into equation 2.7 to obtain the solution:

$$u(x,t) = \sum_{n=1}^{\infty} \left(\int_0^t e^{-kn^2\pi^2(t-\tau)/L^2} B_n(\tau)\, d\tau \right) \sin\left(\frac{n\pi x}{L}\right)$$
$$+ \sum_{n=1}^{\infty} b_n \sin\left(\frac{n\pi x}{L}\right) e^{-n^2\pi^2 kt/L^2}. \qquad (2.15)$$

In calculating this solution for specific $F(x,t)$ and $f(x)$, compute $B_n(t)$ from equation 2.10, which can be done because $F(\xi,t)$ is given, and then compute

$$b_n = \frac{2}{L}\int_0^L f(\xi)\sin\left(\frac{n\pi\xi}{L}\right) d\xi.$$

Example 2.5 We will solve the problem

$$u_t = 4u_{xx} + t^2\cos(x/2) \text{ for } 0 < x < \pi, t > 0,$$
$$u(0,t) = u(\pi,t) = 0 \text{ for } t > 0,$$
$$u(x,0) = \begin{cases} 0 & \text{for } 0 < x \le \pi/2, \\ 25 & \text{for } \pi/2 < x < \pi. \end{cases}$$

In the notation of the discussion, $F(x,t) = t^2\cos(x/2)$, $k = 4$ and $L = \pi$. Compute the terms appearing in the solution 2.15.

First,

$$B_n(t) = \frac{2}{\pi}\int_0^{\pi} t^2\cos(\xi/2)\sin(n\xi)\, d\xi = \frac{8}{\pi}\frac{2n}{4n^2-1}t^2.$$

Next,

$$\int_0^t e^{-kn^2\pi^2(t-\tau)/L^2} B_n(\tau)\, d\tau$$
$$= \int_0^t \frac{8}{\pi}\frac{2n}{4n^2-1}\tau^2 e^{-4n^2(t-\tau)}\, d\tau$$
$$= \frac{1}{2}\frac{-4n^2t + 8n^4t^2 + 1 - e^{-4n^2t}}{n^5\pi(4n^2-1)}.$$

Finally,

$$b_n = \frac{2}{\pi}\int_0^{\pi} f(\xi)\sin(n\xi)\, d\xi$$
$$= \frac{2}{\pi}\int_{\pi/2}^{\pi} 25\sin(n\xi)\, d\xi$$
$$= \frac{50}{n\pi}(\cos(n\pi/2) - (-1)^n).$$

Insert these pieces into the solution (2.15) to obtain

$$u(x,t) = \sum_{n=1}^{\infty} \left(\frac{1}{2} \frac{-4n^2t + 8n^4t^2 + 1 - e^{-4n^2t}}{n^5\pi(4n^2-1)} \right) \sin(nx)$$
$$+ \sum_{n=1}^{\infty} \frac{50}{n\pi}(\cos(n\pi/2) - (-1)^n) \sin(nx) e^{-4n^2t}.$$

The second term in $u(x,t)$ is the solution of the problem if the source term $F(x,t)$ is omitted. Denote this solution $u_h(x,t)$ (h for homogeneous):

$$u_h(x,t) = \sum_{n=1}^{\infty} \frac{50}{n\pi}(\cos(n\pi/2) - (-1)^n) \sin(nx) e^{-4n^2t}.$$

Then

$$u(x,t) = u_h(x,t)$$
$$+ \sum_{n=1}^{\infty} \left(\frac{1}{2} \frac{-4n^2t + 8n^4t^2 + 1 - e^{-4n^2t}}{n^5\pi(4n^2-1)} \right) \sin(nx).$$

This way of writing the solution clarifies the effect of $F(x,t)$ on the diffusion process. Compare graphs of the terms of this solution as surfaces in x, t, u–space. Figure 2.13 is $u_h(x,t)$, Figure 2.14 is $u(x,t) - u_h(x,t)$ (just the contribution from $F(x,t)$), and Figure 2.15 is $u(x,t)$, for $0 \le x \le \pi$ and $0 \le t \le 3$. The form of $F(x,t)$ suggests that its effect on the solution becomes greater as t increases, and this is borne out by comparing the surfaces for the solution with and without this term.

Problems for Section 2.2

In each of problems 1–5, solve the problem 2.6 with the given information, for arbitrary k and L.

1. $F(x,t) = t, f(x) = x(L-x)$.
2. $F(x,t) = x\sin(t), f(x) = 1$.
3. $f(x,t) = \cos(x), f(x) = x^2(L-x)$.
4. $F(x) = \begin{cases} K & \text{for } 0 \le x \le L/2, \\ 0 & \text{for } L/2 < x \le L, \end{cases}$ and $f(x) = \sin(\pi x/L)$.
5. $F(x,t) = xt, f(x) = K$.
6. Adapt the method of solution of problem 2.6 to derive a solution of:

$$u_t = ku_{xx} + F(x,t) \text{ for } 0 < x < L, t > 0,$$
$$u_x(0,t) = u_x(L,t) = 0 \text{ for } t > 0,$$
$$u(x,0) = f(x) \text{ for } 0 < x < L.$$

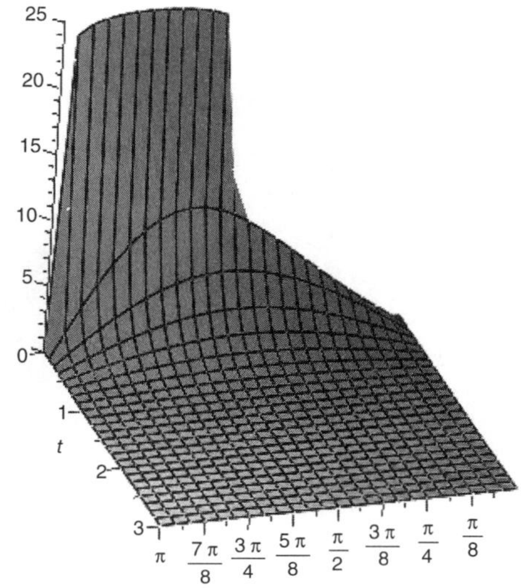

Figure 2.13: $u_h(x,t)$ in example 2.5.

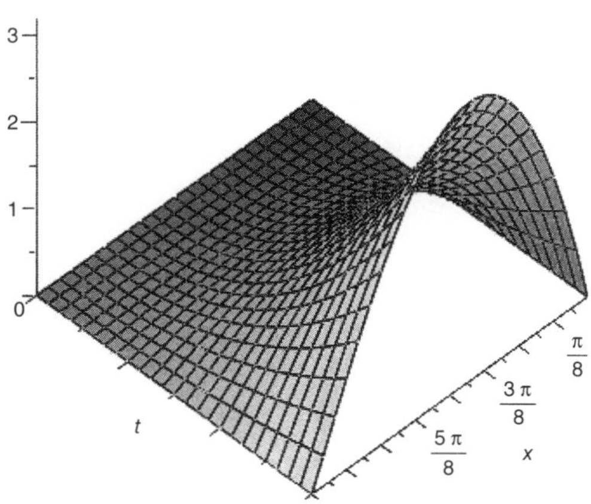

Figure 2.14: $u(x,t) - u_h(x,t)$.

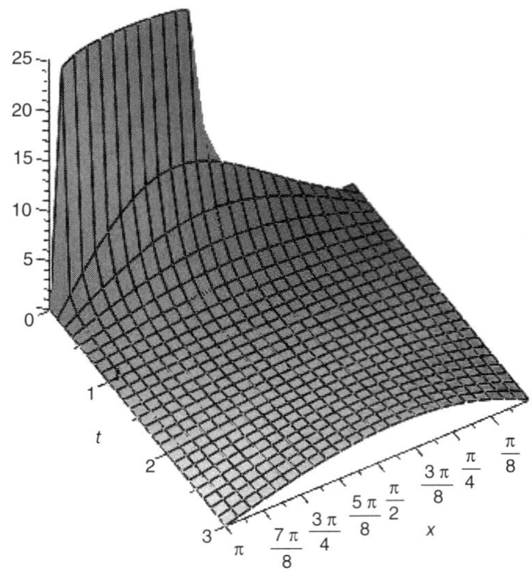

Figure 2.15: $u(x,t)$.

Hint: Begin with

$$u(x,t) = \frac{1}{2}T_0(t) + \sum_{n=1}^{\infty} T_n(t)\cos\left(\frac{n\pi x}{L}\right).$$

Then

$$T_n(t) = \frac{2}{L}\int_0^L u_(\xi,t)\cos\left(\frac{n\pi x}{L}\right)\,d\xi.$$

Now write

$$F(x,t) = \frac{1}{2}A_0(t) + \sum_{n=1}^{\infty} A_n(t)\cos\left(\frac{n\pi x}{L}\right),$$

where

$$A_n(t) = \frac{2}{L}\int_0^L F(\xi,t)\cos\left(\frac{n\pi\xi}{L}\right)\,d\xi.$$

7. Solve the initial-boundary value problem of problem 6 with $F(x,t) = xt$ and $f(x) = 1$.

8. Solve

$$u_t = ku_{xx} \text{ for } 0 < x < L, t > 0,$$
$$u(0,t) = \alpha(t), u_x(L,t) = \beta(t) \text{ for } t > 0,$$
$$u(x,0) = 0 \text{ for } 0 < x < L.$$

Hint: Attempt a solution

$$u(x,t) = \sum_{n=1}^{\infty} T_n(t) \sin\left(\frac{(2n-1)\pi x}{2L}\right)$$

and note the hint to problem 17 of Section 2.1. Let

$$\lambda_n = \left(\frac{(2n-1)\pi}{2L}\right)^2.$$

Show that

$$T_n'(t) = k\lambda_n T_n(t) = b_n(t),$$

where

$$b_n(t) = \frac{2k}{L}\sqrt{\lambda_n}\alpha(t) + (-1)^{n+1}\beta(t).$$

Solve for $T_n(t)$, using the initial condition $T_n(0) = 0$.

9. Solve the initial-boundary value problem of problem 8 with $\alpha(t) = 1$ and $\beta(t) = t$.

10. Solve:

$$\begin{aligned} u_t &= ku_{xx} + F(x,t) \text{ for } 0 < x < L, t < 0, \\ u_x(0,t) &= \alpha(t), u_x(L,t) = \beta(t) \text{ for } t > 0, \\ u(x,0) &= 0 \text{ for } 0 < x < L. \end{aligned}$$

2.3 The Heat Equation in Two Space Variables

In two space variables, the heat equation is

$$u_t = k(u_{xx} + u_{yy}).$$

This equation models diffusion in a medium described by two independent variables, x and y. Sometimes problems involving the two-dimensional heat equation can be solved by separation of variables, followed by a Fourier expansion in each variable.

Example 2.6 We will model the temperature function $u(x,y,t)$ for a thin, flat, homogeneous plate occupying the region $0 \leq x \leq a$, $0 \leq y \leq b$, if the sides are kept at temperature zero, and the initial temperature is given. The initial-boundary value problem for u is:

$$\begin{aligned} u_t &= k(u_{xx} + u_{yy}) \text{ for } 0 < x < a, 0 < y < b, t > 0, \\ u(x,0,t) &= u(x,b,t) = 0 \text{ for } 0 < x < a, t > 0, \\ u(0,y,t) &= u(a,y,t) = 0 \text{ for } 0 < y < b, t > 0, \\ u(x,y,0) &= f(x,y). \end{aligned}$$

Try to separate the variables, with $u(x, y, t) = X(x)Y(y)T(t)$:

$$XYT' = k(X''YT + XY''T).$$

Divide both sides by $kXYT$ to obtain

$$\frac{T'}{kT} = \frac{X''}{X} + \frac{Y''}{Y}.$$

which we write as

$$\frac{X''}{X} = \frac{T'}{kT} - \frac{Y''}{Y}.$$

The left side depends only on x, the right side on y and t. Because these variables are independent, both sides must equal some constant:

$$\frac{X''}{X} = \frac{T'}{kT} - \frac{Y''}{Y} = -\lambda.$$

Then

$$\frac{X''}{X} = -\lambda \text{ and } \frac{T'}{kT} - \frac{Y''}{Y} = -\lambda.$$

Write these equations as

$$X' + \lambda X = 0 \text{ and } \frac{T'}{kT} + \lambda = \frac{Y''}{Y}.$$

Because y and t are independent, both sides of the last equation must equal some constant:

$$\frac{T'}{kT} + \lambda = \frac{Y''}{Y} = -\mu.$$

Then

$$Y'' + \mu Y = 0 \text{ and } T' + (\lambda + \mu)T = 0.$$

As we might expect, it takes two separation constants to separate three variables.

Arguing as in the one-dimensional case, the boundary conditions imply that

$$X(0) = X(a) = 0 \text{ and } Y(0) = Y(b) = 0.$$

The problems for X and Y are

$$X'' + \lambda X = 0; X(0) = X(a) = 0,$$

and

$$Y'' + \mu Y = 0; Y(0) = Y(b) = 0.$$

These have eigenvalues and eigenfunctions

$$\lambda_n = \frac{n^2\pi^2}{a^2}, X_n(x) = \sin\left(\frac{n\pi x}{a}\right)$$

and

$$\mu_m = \frac{m^2\pi^2}{L^2}, Y_m(y) = \sin\left(\frac{m\pi y}{b}\right),$$

in which n and m independently take on all positive integer values.

Now the problem for $T(t)$ is

$$T'(t) + k\left(\frac{n^2\pi^2}{a^2} + \frac{m^2\pi^2}{b^2}\right)T(t) = 0.$$

With

$$\alpha_{nm} = \frac{n^2\pi^2}{a^2} + \frac{m^2\pi^2}{b^2},$$

the solution is

$$T_{nm}(t) = e^{-\alpha_{nm}kt}.$$

For each positive integer n and m, the function

$$u_{nm}(x, y, t) = \sin\left(\frac{n\pi x}{a}\right)\sin\left(\frac{m\pi y}{b}\right)e^{-\alpha_{nm}kt}$$

satisfies the two-dimensional heat equation and the boundary conditions. To satisfy the initial condition, we must generally attempt a double superposition

$$u(x, y, t) = \sum_{n=1}^{\infty}\sum_{m=1}^{\infty} c_{nm}u_{nm}(x, y, t).$$

The initial condition requires that

$$u(x, y, 0) = \sum_{n=1}^{\infty}\sum_{m=1}^{\infty} c_{nm}\sin\left(\frac{n\pi x}{a}\right)\sin\left(\frac{m\pi y}{b}\right) = f(x, y).$$

This is a double Fourier series of $f(x, y)$ on the rectangle $0 \le x \le a, 0 \le y \le b$. We can determine the coefficients c_{nm} by an observation. If y is fixed as any number between 0 and b, we may think of $u(x, y)$ as a function of just x, and the last equation is a Fourier sine expansion of this function on $[0, a]$. To emphasize how we want to think about this expansion, write it as

$$u(x, y, 0) = \sum_{n=1}^{\infty}\left[\sum_{m=1}^{\infty} c_{nm}\sin\left(\frac{m\pi y}{b}\right)\right]\sin\left(\frac{n\pi x}{a}\right).$$

The entire sum in square brackets is the nth Fourier coefficient of the expansion of $u(x, y, 0)$, as a function of x, on $[0, a]$. Therefore,

$$\sum_{m=1}^{\infty} c_{nm}\sin\left(\frac{m\pi y}{b}\right) = \frac{2}{a}\int_0^a f(\xi, y)\sin\left(\frac{n\pi\xi}{a}\right)d\xi.$$

In this expression, the integral on the right is a function of y, and on the left is its Fourier sine expansion on $[0, b]$. For any given n, the mth coefficient c_{nm} is

$$c_{nm} = \frac{2}{b}\int_0^b\left[\frac{2}{a}\int_0^a f(\xi, \eta)\sin\left(\frac{n\pi\xi}{a}\right)d\xi\right]\sin\left(\frac{m\pi\eta}{b}\right)d\eta.$$

This is a double integral over the rectangle $0 \leq x \leq a$, $0 \leq y \leq b$.

To illustrate, suppose $f(x,y) = x(a-x)y\cos(\pi y/2b)$, a product of a function of x with a function of y. In such a case, the double integral of $f(x,y)$ is a product of integrals with respect to x and y. The individual integrals are

$$\int_0^a \xi(a-\xi)\sin(n\pi\xi/a)\,d\xi = \frac{2a^3(1-(-1)^n)}{n^3\pi^3}$$

and

$$\int_0^b \eta\cos(\pi\eta/2b)\sin(n\pi\eta/b)\,d\eta = \frac{16b^2(-1)^{m+1}}{(4m^2-1)^2}.$$

Then c_{nm} is $4/ab$ times the product of these integrals, and we have the solution

$$u(x,y) = \sum_{n=1}^{\infty}\sum_{m=1}^{\infty} \frac{128(-1)^{m+1}(1-(-1)^n)a^2b}{n^3\pi^3(4m^2-1)^2}\sin(n\pi x/a)\sin(m\pi y/b)e^{-\alpha_{nm}kt}.$$

Problems for Section 2.3

1. Solve

$$\begin{aligned}
u_t &= u_{xx} + u_{yy} \text{ for } 0 < x < \pi, 0 < y < \pi, t > 0,\\
u(x,0,t) &= u(x,\pi t) = 0 \text{ for } 0 < x < \pi, t > 0,\\
u(0,y,t) &= u(\pi,y,t) = 0 \text{ for } 0 < y < \pi, t > 0,\\
u(x,y,0) &= \sin(x)y(\pi - y) \text{ for } 0 < x < \pi, 0 < y < \pi, t > 0.
\end{aligned}$$

2. Solve

$$\begin{aligned}
u_t &= u_{xx} + u_{yy} \text{ for } 0 < x < 1, 0 < y < 1, t > 0,\\
u_x(0,y,t) &= u_x(1,y,t) = 0 \text{ for } 0 < y < 1, t > 0,\\
u_y(x,0,t) &= u_y(x,1,t) = 0 \text{ for } 0 < x < 1, t > 0,\\
u(x,y,0) &= K \text{ for } 0 < x < 1, 0 < y < 1.
\end{aligned}$$

3. Solve

$$\begin{aligned}
u_t &= k(u_{xx} + u_{yy}) \text{ for } 0 < x < \pi, 0 < y < \pi, t > 0,\\
u(x,0,t) &= u(x,\pi,t) = 0 \text{ for } 0 < x < \pi, t > 0,\\
u_x(0,y,t) &= u_x(\pi,y,t) = 0 \text{ for } 0 < y < \pi, t > 0\\
u(x,y,0) &= x\cos(x/2)y(\pi - y) \text{ for } 0 < x < \pi, 0 < y < \pi.
\end{aligned}$$

2.4 The Weak Maximum Principle

We will establish an important property shared by every continuous solution of the heat equation in one space dimension, regardless of initial and boundary conditions.

From calculus, we know that any real-valued function $u(x,t)$ that is continuous on the closed rectangle $0 \leq x \leq L, 0 \leq t \leq T$ must achieve a maximum value on this rectangle. This maximum value may occur at any point of the rectangle, and may occur at many points (consider, for example, a constant function).

However, we claim that, if u is also a solution of a heat equation $u_t = ku_{xx}$, then this maximum value is achieved at a point on the lower horizontal side or on a vertical side of the rectangle. This is the weak maximum principle for the heat equation.

While the principle says that $u(x,t)$ must achieve its maximum at a point of the lower or vertical sides, it does not exclude the possibility that this maximum is also achieved at an interior point of the strip. This is why the adjective "weak" is used in naming this maximum principle.

Theorem 2.1 (Weak Maximum Principle for the Heat Equation) Let $u(x,t)$ be continuous on the closed rectangle R in the $x,t-$plane, consisting of points (x,t) with

$$0 \leq x \leq L, 0 \leq t \leq T.$$

Suppose that $u_t = ku_{xx}$ on this rectangle. Then $u(x,t)$ assumes its maximum value on R at a point on the base $t = 0$, or on a vertical side $x = 0$ or $x = L$.

There is a physically based rationale for this conclusion. A solution of the heat equation on $[0, L]$ may be interpreted as a temperature distribution in a bar of length L. In the absence of a source of energy, we would not expect the temperature in the bar at any later time to exceed either the initial temperature $(t = 0)$ throughout the bar, or the temperature at the ends of the bar ($x = 0$ or $x = L$).

Here is a proof of the theorem.

Proof Let C consist of points $(x, 0)$ with $0 \leq x \leq L$, and points $(0,t)$ and (L,t) with $0 \leq t \leq T$. These are the lower and vertical sides of R (Figure 2.16).

We know that, by continuity, $u(x,t)$ must achieve a maximum value M on R, and also a maximum M_C on the closed, bounded set C. Because C is part of R, then certainly $M_C \leq M$. We will show that $M \leq M_C$, which will imply that $M = M_C$ and prove the theorem.

Let $\epsilon = M - M_c$, and suppose that $\epsilon > 0$. Choose a point (x_0, t_0) in R such that $u(x_0, t_0) = M$. Because $M - M_C = \epsilon > 0$, then (x_0, t_0) is not in C. Therefore

$$0 < x < L \text{ and } 0 < t \leq T.$$

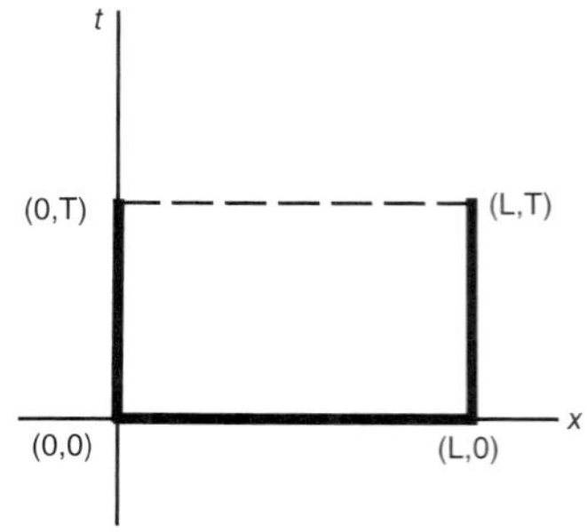

Figure 2.16: $u(x,t)$ has a maximum on a lower or vertical side.

Define a new function

$$w(x,t) = u(x,t) + \frac{\epsilon}{4L^2}(x - x_0)^2.$$

Consider $w(x,t)$ at points of C. For $0 \leq x \leq L$,

$$\begin{aligned} w(x,0) &= u(x,0) + \frac{\epsilon}{4L^2}(x - x_0)^2 \leq M - \epsilon + \frac{\epsilon}{4L^2}L^2 \\ &= M - \frac{3}{4}\epsilon < M. \end{aligned}$$

Similarly,

$$w(0,t) < M \text{ and } w(L,t) < M \text{ for } 0 \leq t \leq T.$$

But $w(x_0, t_0) = u(x_0, t_0) = M$. Therefore the maximum of $w(x,t)$ on R is at least M and is achieved at some point (x_1, t_1) of R not on C. Because $0 < x_1 < L$ and $0 \leq t_1 \leq T$, we must have

$$w_t(x_1, t_1) \geq 0 \text{ and } w_{xx}(x_1, t_1) \leq 0.$$

Then

$$w_t(x_1, t_1) - kw_{xx}(x_1, t_1) \geq 0.$$

But

$$\begin{aligned} w(x_1, t_1) - kw_{xx}(x_1, t_1) &= u_t(x_1, t_1) - ku_{xx}(x_1, t_1) - k\frac{\epsilon}{2L^2} \\ &= -\frac{k\epsilon}{2L^2} < 0. \end{aligned}$$

This is a contradiction, and we conclude that $M = M_C$, completing the proof.

By considering $-u(x,t)$, we can show that a continuous solution of the heat equation also achieves its minimum value at a point on the lower or vertical sides of R.

We will use the weak maximum principle to derive two important properties of continuous solutions of the general initial-boundary value problem:

$$\begin{aligned} u_t &= ku_{xx} + F(x,t) \text{ for } 0 < x < L, t > 0, \\ u(0,t) &= g(t), u(L,t) = h(t) \text{ for } t > 0, \\ u(x,0) &= f(x) \text{ for } 0 \le x \le L, \end{aligned} \tag{2.16}$$

It is assumed here that f, g and h and F are continuous.

First we prove a uniqueness result for solutions of this problem.

Theorem 2.2 (Uniqueness of Continuous Solutions) The initial-boundary value problem 2.16 can have only one continuous solution.

Proof Suppose u_1 and u_2 are continuous solutions. Let $w = u_1 - u_2$. It is routine to check that

$$\begin{aligned} w_t &= kw_{xx} \text{ for } 0 < x < L, t > 0, \\ w(0,t) &= w(L,t) = 0 \text{ for } t > 0, \\ w(x,0) &= 0 \text{ for } 0 < x < L. \end{aligned}$$

Let $T > 0$ and let R be the rectangle consisting of points (x,t) with $0 \le x \le L, 0 \le t \le T$. The initial and boundary conditions imply that $w(x,t) = 0$ on the lower and vertical sides of R. But $w(x,t)$ achieves its maximum at a point of these sides, so $w(x,t) \le 0$ on R.

But $w(x,t)$ also achieves its minimum on R at a point of these sides, so $w(x,t) \ge 0$ on R. We conclude that $w(x,t) = 0$ at all points of R. Because T is any positive number, this means that $w(x,t)$ is identically zero, so $u_1(x,t) = u_2(x,t)$ for $0 \le x \le L, t \ge 0$, as was to be shown.

Finally, we will show that continuous solutions of problem 2.16 depend continuously on the initial and boundary data. This means that small changes in $g(t), h(t)$ and $f(x)$ result in correspondingly small changes in the solutions.

Theorem 2.3 (Continuous Dependence on Initial-Boundary Data) Let $u_1(x,t)$ and $u_2(x,t)$ be continuous solutions of the problem 2.16 with, respectively, initial temperature and boundary conditions $f_1(x), g_1(t), h_1(t)$ and $f_2(x), g_2(t), h_2(t)$. Let ϵ be a positive number, and suppose that

$$\begin{aligned} &|f_1(x) - f_2(x)| \le \epsilon \text{ for } 0 \le x \le L, \\ &|g_1(x) - g_2(x)| \le \epsilon \text{ for } t \ge 0, \\ &|h_1(t) - h_2(t)| \le \epsilon \text{ for } t \ge 0. \end{aligned}$$

Then,

$$|u_1(x,t) - u_2(x,t)| \le \epsilon \text{ for } 0 \le x \le L, t \ge 0.$$

Proof Let $w = u_1 - u_2$. Then

$$w_t = kw_{xx} \text{ for } 0 < x < L, t > 0,$$
$$w(x,0) = f_1(x) - f_2(x) \text{ for } 0 < x < L,$$
$$w(0,t) = g_1(t) - g_2(t), w(L,t) = h_1(t) - h_2(t) \text{ for } t > 0.$$

Let T be any positive number and consider $w(x,t)$ on the lower and vertical sides of $R : 0 \leq x \leq L, 0 \leq t \leq T$. On the lower side of R,

$$|w(x,0)| = |f_1(x) - f_2(x)| \leq \epsilon.$$

On the left vertical side,

$$|w(0,t)| = |g_1(t) - g_2(t)| \leq \epsilon.$$

And on the right vertical side,

$$|w(L,t)| = |h_1(t) - h_2(t)| \leq \epsilon.$$

By the weak maximum principle, $|w(x,t)| \leq \epsilon$ on R. Because T can be any positive number, then $|w(x,t)| \leq \epsilon$ for $0 \leq x \leq L, t \geq 0$.

In particular, solutions for different boundary and initial data will be at least as close as these initial data.

An initial-boundary value problems (such as 2.16) is said to be *well posed* if it has a unique solution and solutions depend continuously on the initial and boundary information. Although we have not proved an existence theorem for problem 2.16, we have shown that continuous solutions must be unique and depend continuously on the initial and boundary conditions.

Problems for Section 2.4

1. Fill in the details of the following alternate proof of the weak maximum principle. Let
$$M = \max_{(x,t)} \text{ in } _R u(x,t).$$
There is at least one point (x_0, t_0) in R at which $u(x_0, t_0) = M$. If $M_C < M$, this point is not in C. Denote
$$w(t) = u(x,t) - \epsilon(t - t_0),$$
with ϵ any positive number. Then w is continuous on R and $w(x_0, t_0) = u(x_0, t_0) = M > M_C$. Choose ϵ sufficiently small that $w(x_0, t_0)$ is greater tan the maximum achieved by $w(x,t)$ on C. Then the maximum of $w(x,t)$ on R is achieved at a point (x_1, t_1) not in C. Now reason as in the proof of the theorem.

2. Prove the following theorem. Suppose

$$u_t = ku_{xx} \text{ for } -\infty < x < \infty, t > 0,$$
$$u(x,0) = f(x) \text{ for } -\infty < x < \infty.$$

Suppose $f(x)$ is continuous on the whole real line. Let $u(x,t) \to 0$ uniformly in t as $x \to \infty$ and as $x \to -\infty$. Then

$$|u(x,t)| \leq \max|f(x)|.$$

Hint: Apply the weak maximum principle on $-a \leq x \leq a, t \geq 0$, and then let $a \to \infty$.

3. Let C be a simple closed curve in the $x, y-$ plane, bounding a region D. Formulate and prove a weak maximum principle for the two-dimensional heat equation

$$u_t = k(u_{xx} + u_{yy})$$

on the three-dimensional region consisting of all (x, y, t) with $(x, y) \in D$ and $0 \leq t \leq T$, where $T > 0$.

4. Let D be a region of the $x, y - -$plane bounded by a simple closed curve C. Consider the problem

$$u_t = k(u_{xx} + u_{yy}) + F(x,t) \text{ for } (x,y) \text{ in } D, t > 0,$$
$$u(x,y,0) = \varphi(x,y) \text{ for } (x,y) \text{ in } D,$$
$$u(x,y,t) = \psi(x,y,t) \text{ for } (x,y) \text{ on } C, t > 0.$$

Let φ, ψ and F be continuous. Assuming that this problem has a solution, prove that this solution is unique and depends continuously on the initial and boundary data.

Chapter 3

Solutions of the Wave Equation

This chapter deals with models of wave motion under a variety of conditions, including fixed ends, moving ends, forcing terms, and damping effects. We also delve into the structure of solutions of the wave equation, continuous dependence of solutions on initial and boundary data, and solutions in unbounded media and in higher dimensions.

3.1 Solutions on Bounded Intervals

3.1.1 Fixed Ends

Consider a homogeneous string or wire of length L, fixed at its ends. At time zero it is displaced and released with a given initial velocity. The initial-boundary-value problem for the function describing the wave motion is:

$$\begin{aligned} u_{tt} &= c^2 u_{xx} \text{ for } 0 < x < L, t > 0, \\ u(0,t) &= u(L,t) = 0 \text{ for } t > 0, \\ u(x,0) &= \varphi(x), u_t(x,0) = \psi(x) \text{ for } 0 < x < L. \end{aligned} \tag{3.1}$$

Separate the variables in problem 3.1. Put $u(x,t) = X(x)T(t)$ into the wave equation to get

$$XT'' = c^2 X''T,$$

so

$$\frac{X''}{X} = \frac{T''}{c^2 T}.$$

Both sides must be constant, because x and t are independent and either could be fixed while the other varies in this equation. Therefore, for some λ to be determined,

$$\frac{X''}{X} = \frac{T''}{c^2T} = -\lambda.$$

Then

$$X'' + \lambda X = 0, T'' + \lambda c^2 T = 0.$$

Reasoning as we did with the heat equation, the boundary condition

$$u(0,t) = X(0)T(t) = 0$$

implies that $X(0) = 0$. Similarly, $X(L) = 0$ and the problem for X is a familiar one:

$$X'' + \lambda X = 0; X(0) = X(L) = 0,$$

with eigenvalues and eigenfunctions

$$\lambda_n = \frac{n^2\pi^2}{L^2}, X_n(x) = \sin\left(\frac{n\pi x}{L}\right)$$

for $n = 1, 2, 3, \cdots$. For each positive integer n, the problem for T is

$$T'' + \frac{n^2\pi^2c^2}{L^2}T = 0,$$

with solutions

$$T_n(t) = c_n \cos\left(\frac{n\pi ct}{L}\right) + d_n \sin\left(\frac{n\pi ct}{L}\right).$$

For each positive integer n, we now have a function

$$u_n(x,t) = \left[c_n \cos\left(\frac{n\pi ct}{L}\right) + d_n \sin\left(\frac{n\pi ct}{L}\right)\right] \sin\left(\frac{n\pi x}{L}\right).$$

To satisfy the initial position and velocity conditions, we must usually write a superposition

$$\begin{aligned} u(x,t) &= \sum_{n=1}^{\infty} u_n(x,t) \\ &= \sum_{n=1}^{\infty} \left[c_n \cos\left(\frac{n\pi ct}{L}\right) + d_n \sin\left(\frac{n\pi ct}{L}\right)\right] \sin\left(\frac{n\pi x}{L}\right) \end{aligned} \tag{3.2}$$

and choose the coefficients to fit these conditions.

From the initial position function, we require that

$$u(x,0) = \sum_{n=1}^{\infty} c_n \sin\left(\frac{n\pi x}{L}\right) = \varphi(x).$$

This is a Fourier sine expansion of $\varphi(x)$ on $[0, L]$, so

$$c_n = \frac{2}{L}\int_0^L \varphi(\xi) \sin\left(\frac{n\pi\xi}{L}\right) d\xi. \tag{3.3}$$

Now turn to the initial velocity condition. Assuming that we can differentiate the series term by term, compute

$$u_t(x,t) = \sum_{n=1}^{\infty} \frac{n\pi c}{L}\left[-c_n \sin\left(\frac{n\pi ct}{L}\right) + d_n \cos\left(\frac{n\pi ct}{L}\right)\right] \sin\left(\frac{n\pi x}{L}\right).$$

We require that

$$u_t(x,0) = \sum_{n=1}^{\infty} \frac{n\pi c}{L} d_n \sin\left(\frac{n\pi x}{L}\right) = \psi(x).$$

This is a Fourier sine expansion of $\psi(x)$ on $[0, L]$. However, because of the differentiation, the constant d_n is multiplied by a factor of $n\pi c/L$ in this expansion. This product is the entire coefficient in this expansion, so let

$$\frac{n\pi c}{L} d_n = \frac{2}{L}\int_0^L \psi(\xi)\sin\left(\frac{n\pi \xi}{L}\right) d\xi.$$

This means that

$$d_n = \frac{2}{n\pi c}\int_0^L \psi(\xi)\sin\left(\frac{n\pi \xi}{L}\right) d\xi. \tag{3.4}$$

With the coefficients given by equations 3.3 and 3.4, equation 3.2 is the solution of the problem.

Example 3.1 Suppose a string has $c = 2$ and length $L = 1$. The ends are fixed. The segment of string between 1/3 and 2/3 is lifted horizontally, and the string is released from rest from that position. We want to describe the ensuing motion of the string.

This is the initial-boundary value problem

$$\begin{aligned} u_{tt} &= 4u_{xx} \text{ for } 0 < x < 1, t > 0, \\ u(0,t) &= u(1,t) = 0 \text{ for } t > 0, \\ u(x,0) &= \varphi(x), u_t(x,0) = 0 \text{ for } 0 < x < 1, \end{aligned}$$

where

$$\varphi(x) = \begin{cases} x & \text{for } 0 \le x \le 1/3, \\ 1/3 & \text{for } 1/3 \le x \le 2/3, \\ 1-x & \text{for } 2/3 \le x \le 1. \end{cases}$$

Figure 3.1 is a graph of the string at the time of release, $t = 0$.

The solution is given by equation 3.2, with $d_n = 0$ because the string is

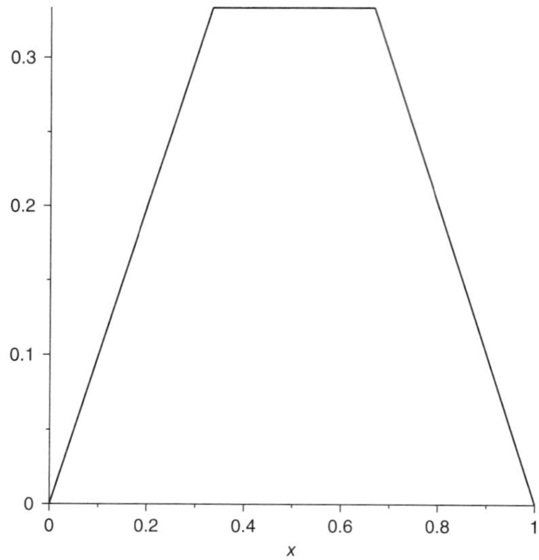

Figure 3.1: Initial position of the string in example 3.1.

released from rest. From equation 3.3,

$$
\begin{aligned}
c_n &= 2\int_0^1 \varphi(\xi)\sin(n\pi\xi)\,d\xi \\
&= 2\int_0^{1/3} \xi\sin(n\pi\xi)\,d\xi + 2\int_{1/3}^{2/3}\frac{1}{3}\sin(n\pi\xi)\,d\xi \\
&+ 2\int_{2/3}^{1}(1-\xi)\sin(n\pi\xi)\,d\xi \\
&= \frac{2}{n^2\pi^2}(\sin(n\pi/3)+\sin(2n\pi/3)).
\end{aligned}
$$

The solution is

$$
u(x,t) = \frac{2}{\pi^2}\sum_{n=1}^{\infty}\frac{1}{n^2}(\sin(n\pi/3)+\sin(2n\pi/3))\sin(n\pi x)\cos(2n\pi t).
$$

Figure 3.2 shows the position of the string at times $t = 0.05, 0.1, 1.4$ and 1.68. In this time frame the wave moves downward from the initial position as t increases. The wave at time t_0 is the graph of $u(x, t_0)$ as a curve in the $x, u-$plane. The influence of the initial position is revealed in these string profiles at times close to zero. The wave motion can also be displayed as a surface in $x, t, u-$space, as in Figure 3.3. The intersection of this surface with the plane $t = t_0$ is the shape of the string at this time.

Example 3.2 Suppose a string of length 1 is not displaced initially, but is set in motion at time $t = 0$ with an initial velocity $\psi(x) = 3(1-x)$. With $c = 6$,

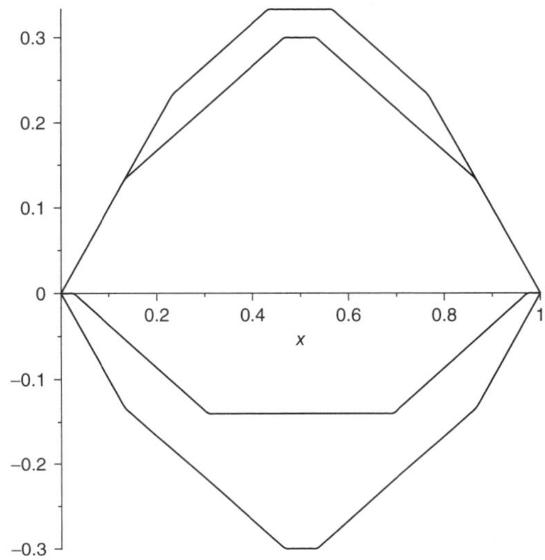

Figure 3.2: Wave profiles at $t = 0.05, 0.1, 1.4$, and 1.68 in example 3.1.

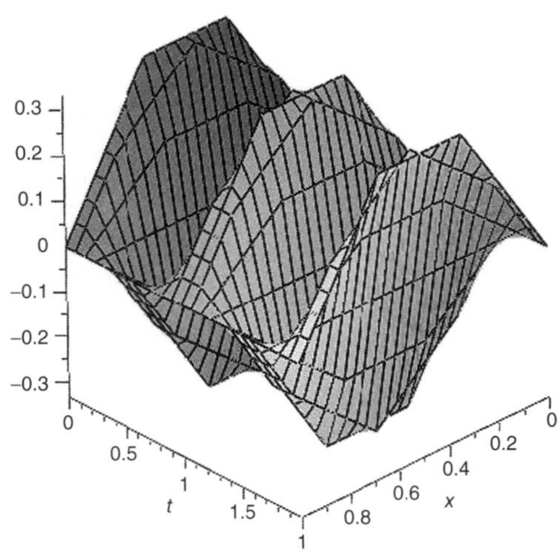

Figure 3.3: Wave surface of example 3.1 in $x, t, u-$space.

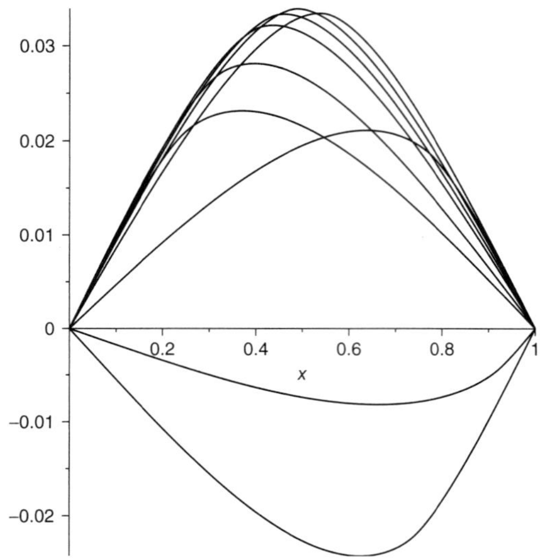

Figure 3.4: Wave profiles at $t = 0.041, 0.053, 0.067, 0.074, 0.081, 0.092, 1.13, 1.18,$ and 1.21 in example 3.2.

the motion is modeled by the initial-boundary value problem

$$u_{tt} = 36u_{xx} \text{ for } 0 < x < 1, t > 0,$$
$$u(0,t) = u(1,t) = 0 \text{ for } t > 0,$$
$$u(x,0) = 0, u_t(x,0) = \psi(x) = 3(1-x) \text{ for } 0 < x < 1.$$

Equation 3.2 is the solution, with each $c_n = 0$ because there is no initial displacement. From equation 3.4, compute

$$d_n = \frac{1}{n\pi}\int_0^1 (1-\xi)\sin(n\pi\xi)\,d\xi = \frac{1}{n^3\pi^3}.$$

The solution is

$$u(x,t) = \frac{1}{\pi^3}\sum_{n=1}^{\infty}\frac{1}{n^3}\sin(n\pi x)\sin(6n\pi t).$$

Figure 3.4 shows the position of the string at times

$$t = 0.041, 0.053, 0.067, 0.074, 0.081, 0.092, 1.13, 1.18, \text{ and } 1.21.$$

For $t = 0.041$ to 0.081, the string is moving to the right and increasing in amplitude. For the last three times, the waves are moving downward. Figure 3.5 displays the wave function as a surface in 3− space.

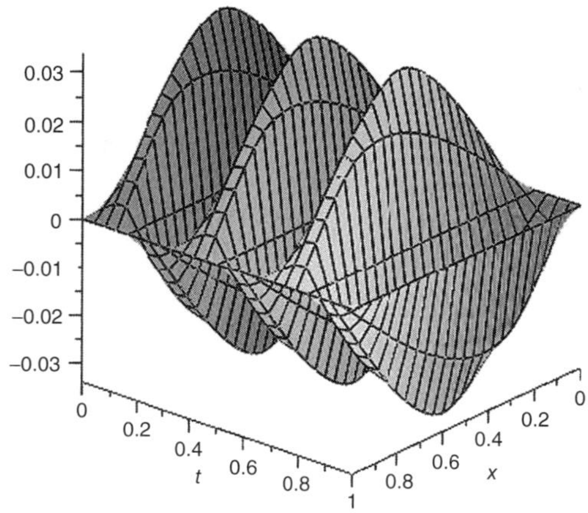

Figure 3.5: Wave function as a surface in 3−space in example 3.2.

Example 3.3 Consider the problem

$$u_{tt} = 9u_{xx} \text{ for } 0 < x < \pi, t > 0,$$
$$u(0,t) = u(\pi,t) = 0 \text{ for } t \geq 0,$$
$$u(x,0) = x^2(\pi - x), u_t(x,0) = \sin(x) \text{ for } 0 < x < \pi.$$

This string has length π and fixed ends, and is released from the given position with the given velocity. Compute the coefficients in the solution:

$$c_n = \frac{2}{\pi}\int_0^\pi \xi^2(\pi - \xi)\sin(n\xi)\,d\xi = -\frac{4}{n^3}(1 + 2(-1)^n)$$

and

$$d_n = \frac{2}{3n\pi}\int_0^\pi \sin(\xi)\sin(n\xi)\,d\xi = \begin{cases} 0 & \text{if } n = 2,3,4,\cdots; \\ 1/2 & \text{if } n = 1. \end{cases}$$

the solution is

$$u(x,t) = \sum_{n=1}^{\infty} -\frac{4}{n^3}(1 + 2(-1)^n)\sin(nx)\cos(3nt) + \frac{1}{3}\sin(x)\sin(3t).$$

The summation term is the solution with the given initial displacement, but zero initial velocity. The other term is the solution with no initial displacement, and the given initial velocity. Because this initial velocity is just $\sin(x)$, this part of $u(x,t)$ is a single term.

Figure 3.6 shows the shape of the string at times $t = 0.013, 0.27$, 0.41, 0.57, 0.74, and 0.91.

Over these times, the wave peaks move downward. For the last two times the graph is entirely below the horizontal. Figure 3.7 displays the solution as a surface in 3−space, for $0 \leq x \leq \pi$, $0 \leq t \leq 3$.

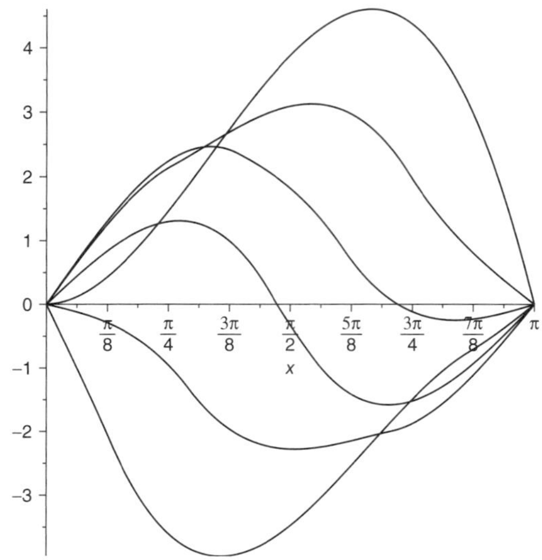

Figure 3.6: Wave profiles in example 3.3 at times $t = 0.013, 0.27, 0.41, 0.57, 0.74,$ and 0.91.

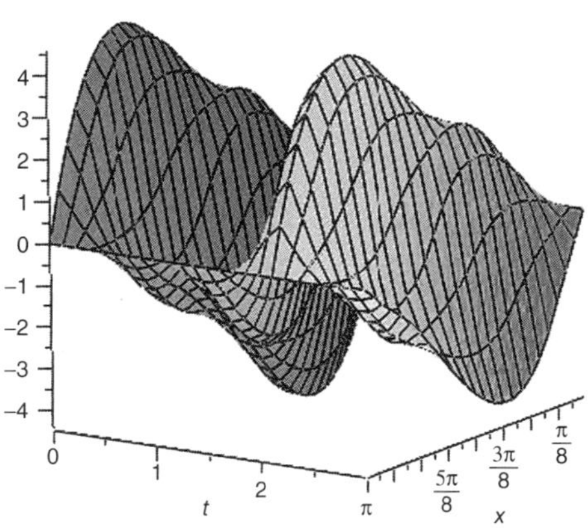

Figure 3.7: Wave function as a surface in 3−space in example 3.3.

3.1.2 Fixed Ends with a Forcing Term

If the ends are fixed, but there is an external force acting on the string, then the initial-boundary value problem describing the motion can sometimes (depending on the force) be solved by a transformation of the problem to one we have already solved. Here are two illustrations of this idea.

Example 3.4 We will solve the problem

$$\begin{aligned} u_{tt} &= c^2 u_{xx} + Ae^x \text{ for } 0 < x < L, t > 0, \\ u(0,t) &= u(L,t) = 0 \text{ for } 0 < x < L, t > 0, \\ u(x,0) &= \varphi(x), u_t(x,0) = 0 \text{ for } 0 < x < L. \end{aligned}$$

The initial velocity is zero and A is a positive constant.

If $u(x,t) = X(x)T(t)$ is substituted into this wave equation, we obtain

$$XT'' = c^2 X''T + Ae^x,$$

and it is impossible to separate the variables.

Try putting

$$u(x,t) = v(x,t) + f(x),$$

as we did with the heat equation when the ends of the bar were kept at different temperatures. Now

$$u_{tt} = v_{tt} = c^2 u_{xx} = c^2(v_{xx} + f''(x)) + Ae^x.$$

This will be a standard wave equation $v_{tt} = c^2 v_{xx}$ for $v(x,t)$ if we choose $f(x)$ so that

$$c^2 f''(x) + Ae^x = 0.$$

We need

$$f''(x) = -\frac{A}{c^2}e^x.$$

Integrate twice to get

$$f(x) = -\frac{A}{c^2}e^x + K_1 x + K_2,$$

in which K_1 and K_2 have to be chosen to make this method work.

Now look at the boundary conditions. First,

$$u(0,t) = 0 = v(0,t) + f(0) = v(0,t) - \frac{A}{c^2} + K_2.$$

This will allow us to set $v(0,t) = 0$ if we choose

$$K_2 = \frac{A}{c^2}.$$

So far

$$f(x) = -\frac{A}{c^2}e^x + K_1 x + \frac{A}{c^2}.$$

Next

$$\begin{aligned} u(L,t) &= v(L,t) + f(L) \\ &= v(L,t) - \frac{A}{c^2}e^L + K_1 L + \frac{A}{c^2} = 0 \end{aligned}$$

and we can set $v(L,t) = 0$ if

$$-\frac{A}{c^2}e^L + K_1 L + \frac{A}{c^2} = 0.$$

This leads us to set

$$K_1 = \frac{A}{Lc^2}e^L - \frac{A}{Lc^2} = \frac{A}{Lc^2}\left(e^L - 1\right).$$

We now have

$$f(x) = -\frac{A}{c^2}e^x + \frac{A}{Lc^2}\left(e^L - 1\right)x + \frac{A}{c^2}.$$

Finally, look at the initial conditions. First,

$$u(x,0) = v(x,0) + f(x) = \varphi(x),$$

so

$$v(x,0) = \varphi(x) - f(x).$$

And

$$u_t(x,0) = 0 = v_t(x,0).$$

The problem for $v(x,t)$ is, therefore,

$$\begin{aligned} v_{tt} &= c^2 v_{xx} \text{ for } 0 < x < L, t > 0, \\ v(0,t) &= v(L,t) = 0 \text{ for } t > 0, \\ v(x,0) &= \varphi(x) - f(x), v_t(x,0) = 0. \end{aligned}$$

This problem has the solution

$$v(x,t) = \sum_{n=1}^{\infty} c_n \sin\left(\frac{n\pi x}{L}\right)\cos\left(\frac{n\pi ct}{L}\right),$$

with

$$c_n = \frac{2}{L}\int_0^L (\varphi(\xi) - f(\xi))\sin\left(\frac{n\pi x}{L}\right)d\xi.$$

The solution of the forced problem for $u(x,t)$ is

$$u(x,t) = v(x,t) - \frac{A}{c^2}e^x + \frac{A}{Lc^2}\left(e^L - 1\right)x + \frac{A}{c^2}.$$

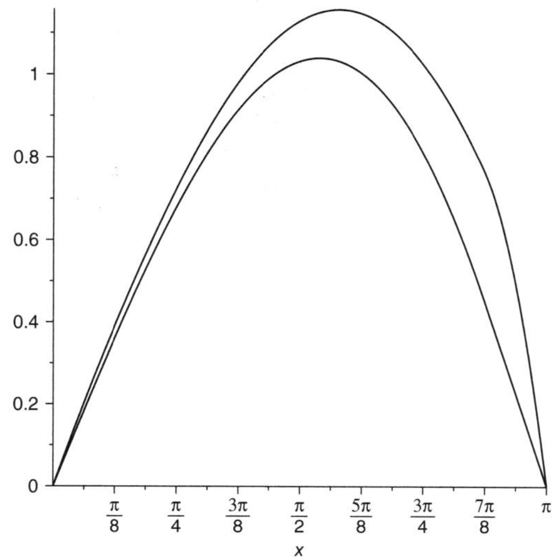

Figure 3.8: Comparison of unforced and unforced motion at $t = 0.2$ in example 3.4.

For comparison, the solution of the problem with the forcing term Ae^x removed is

$$u_0(x,t) = \sum_{n=1}^{\infty} k_n \sin\left(\frac{n\pi x}{L}\right) \cos\left(\frac{n\pi ct}{L}\right),$$

where

$$k_n = -\frac{2}{L}\int_0^L \varphi(\xi) \sin\left(\frac{n\pi\xi}{L}\right) d\xi.$$

As an example, let $c = 2, A = 1, L = \pi$, and $\varphi(x) = x\cos(x/2)$. Now

$$f(x) = -\frac{1}{4}e^x + \frac{1}{4\pi}\left(e^\pi - 1\right)x + \frac{1}{4}.$$

Using the notation of the discussion, the coefficients for $v(x,t)$ are

$$c_n =$$

$$-\frac{\pi}{n}\frac{1}{(2n-1)^2(2n+1)^2(n^2+1)}\left[(4n^2-1)^2(1-e^\pi(-1)^n) + 64n^2(-1)^n(n^2+1)\right]$$

and, for $u_0(x,t)$,

$$k_n = \frac{32n(-1)^n}{(2n-1)^2(2n+1)^2\pi}.$$

Figure 3.8 compares graphs of $u(x, 0.2)$ (upper curve) and $u_0(x, 0.2)$ (lower curve). Figure 3.9 and 3.10 makes these comparisons for $t = 0.6$ and $t = 0.8$, in turn. Figure 3.11 is a graph of $u_0(x,t)$ as a surface, for $0 \leq x \leq \pi$ and

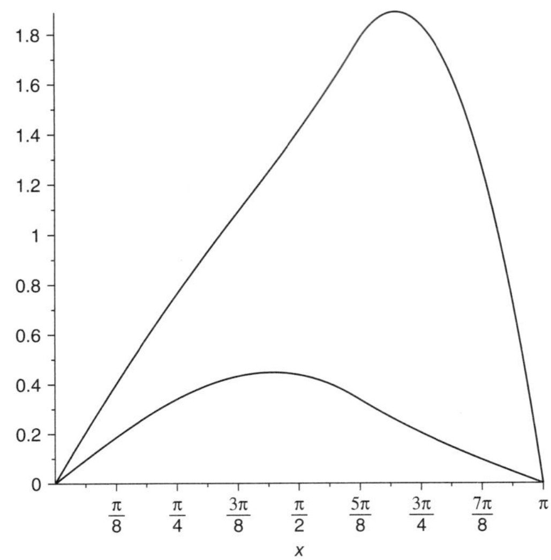

Figure 3.9: Forced and unforced wave profiles at $t = 0.6$ in example 3.4.

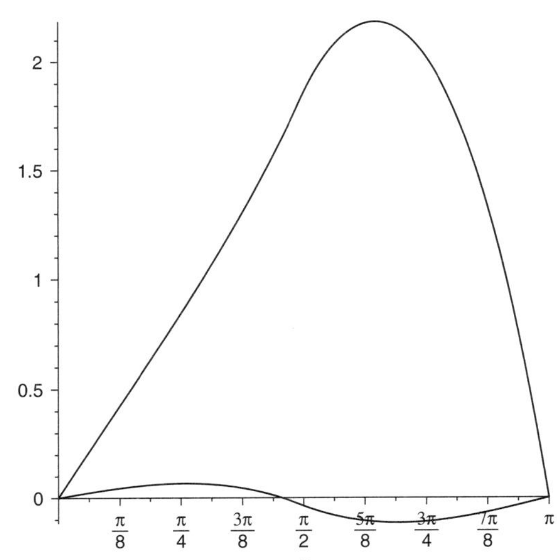

Figure 3.10: Forced and unforced wave profiles at $t = 0.8$ in example 3.4.

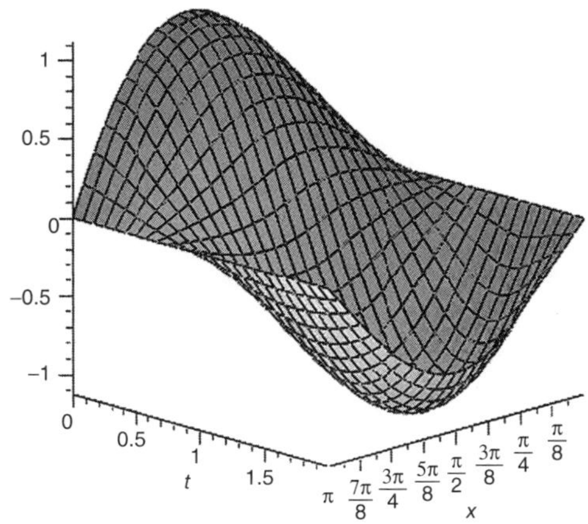

Figure 3.11: Unforced wave surface $u_0(x.t)$ in example 3.4.

$0 \leq t \leq 2$, while Figures 3.12 and 3.13 show $v(x,t)$ and $u(x,t)$, respectively. We can compare Figures 3.11 and 3.13 to see the influence of the forcing term on the motion, while Figure 3.12 is the solution $v(x,t)$ of the transformed problem used to obtain $u(x,t)$.

The method used with the forcing term Ae^x will work for many forcing terms of the form $g(x)$. However, if the forcing term depends on t, a more sophisticated approach is needed. We will solve problem with forcing terms $P(x,t)$ when we develop characteristics and d'Alembert's solution.

Example 3.5 We will solve the problem

$$\begin{aligned} u_{tt} &= c^2 u_{xx} + K\cos(4x) \text{ for } 0 < x < \pi, t > 0, \\ u(0,t) &= u(\pi,t) = 0 \text{ for } t > 0, \\ u(x,0) &= \varphi(x) = x(\pi - x), u_t(x,0) = 0 \text{ for } 0 < x < \pi. \end{aligned}$$

K is a positive constant.

Let $u(x,t) = v(x,t) + f(x)$. Then

$$u_{tt} = v_{tt} = c^2 u_{xx} = c^2(v_{xx} + f''(x)) + K\cos(4x),$$

so

$$v_{tt} = c^2 v_{xx} + c^2 f''(x) + K\cos(4x).$$

This will be a standard wave equation for $v(x,t)$ if $c^2 f''(x) + K\cos(4x) = 0$, or

$$f''(x) = -\frac{K}{c^2}\cos(4x).$$

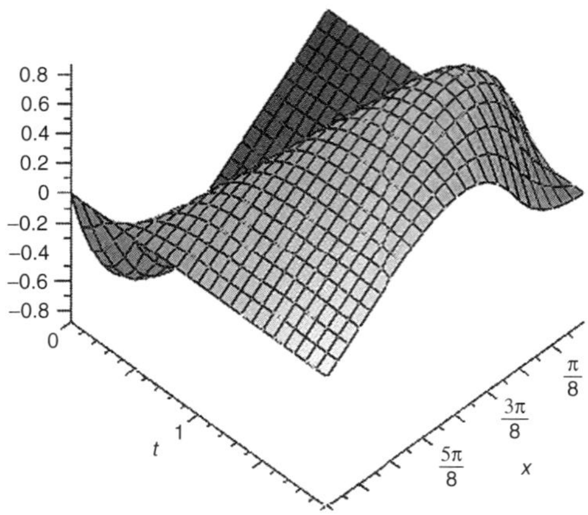

Figure 3.12: Solution $v(x,t)$ of the transformed problem in example 3.4.

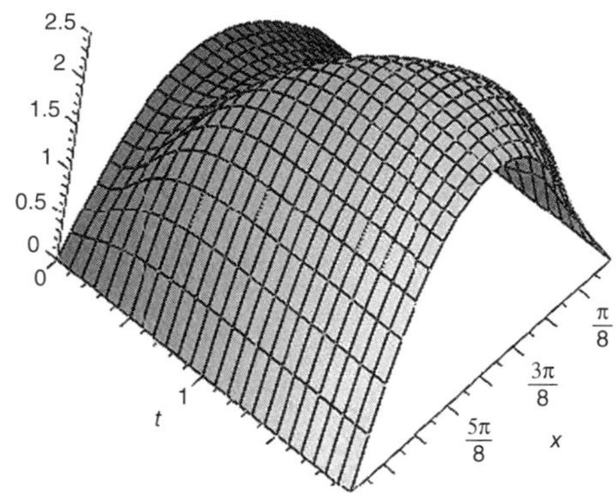

Figure 3.13: Forced wave surface $u(x,t)$ in example 3.4.

Integrate twice to get

$$f(x) = \frac{K}{16c^2}\cos(4x) + ax + b,$$

with a and b to be determined to form a tractable problem for $v(x,t)$. First,

$$u(0,t) = 0 = v(0,t) + f(0) = v(0,t) + \frac{K}{16c^2} + b.$$

This will be $v(0,t) = 0$ if

$$b = -\frac{K}{16c^2}.$$

So far,

$$f(x) = \frac{K}{16c^2}\cos(4x) + ax - \frac{K}{16c^2}.$$

Next,

$$\begin{aligned} u(\pi,t) = 0 &= v(\pi,t) + f(\pi) \\ &= v(\pi,t) + \frac{K}{16c^2} + a\pi - \frac{K}{16c^2} = v(\pi,t) + a\pi. \end{aligned}$$

This will give us $v(\pi,t) = 0$ if $a = 0$, so

$$f(x) = \frac{K}{16c^2}\cos(4x) - \frac{K}{16c^2} = \frac{K}{16c^2}(\cos(4x) - 1).$$

From the initial conditions,

$$u_t(x,0) = v_t(x,0) = 0$$

and

$$u(x,0) = v(x,0) + f(x) = \varphi(x)$$

imply that

$$v(x,0) = \varphi(x) - f(x).$$

The problem for $v(x,t)$ is

$$\begin{aligned} v_{tt} &= c^2 v_{xx} \text{ for } 0 < x < \pi, t > 0, \\ v(0,t) &= v(\pi,t) = 0 \text{ for } t > 0, \\ v(x,0) &= \varphi(x) - f(x), v_t(x,0) = 0 \text{ for } 0 < x < \pi. \end{aligned}$$

This has the solution

$$v(x,t) = \sum_{n=1}^{\infty} c_n \sin(nx)\cos(nct),$$

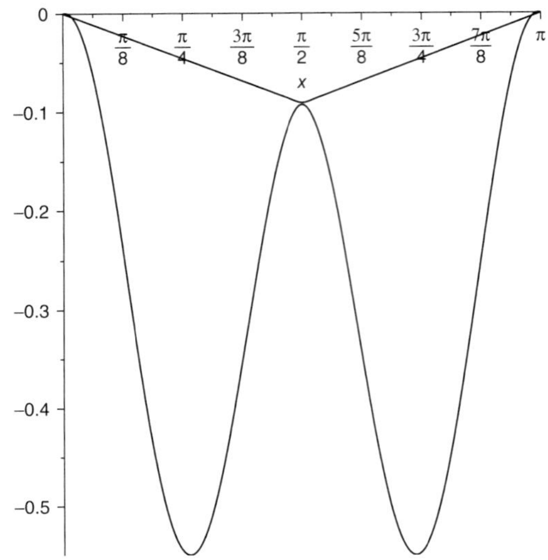

Figure 3.14: $u(x, 0.8)$ and $u_0(x, 0.8)$ in example 3.5.

where

$$\begin{aligned} c_n &= \frac{2}{\pi}\int_0^{\pi} (\varphi(\xi) - f(\xi))\sin(n\xi)\, d\xi \\ &= \frac{2}{\pi}\int_0^{\pi} \left[\xi(\pi-\xi) - \frac{K}{16c^2}(\cos(4\xi)-1)\right]\sin(n\xi)\, d\xi \\ &= \begin{cases} 0 & \text{if } n = 4, \\ \frac{2}{c^2 n^3 (n^2-16)\pi}\left((-1)^n - 1\right)\left(32c^2 + Kn^2 - 2(nc)^2\right) & \text{if } n \neq 4. \end{cases} \end{aligned}$$

The solution of the forced problem for $u(x, t)$ is

$$u(x, t) = v(x, t) + f(x).$$

The solution of the problem with the forcing term removed is

$$u_0(x, t) = \sum_{n=1}^{\infty} c_n^* \sin(nx)\cos(nct),$$

where

$$\begin{aligned} c_n^* &= \frac{2}{\pi}\int_0^{\pi} \xi(\pi-\xi)\sin(n\xi)\, d\xi \\ &= \frac{4}{\pi n^3}\left(1 - (-1)^n\right). \end{aligned}$$

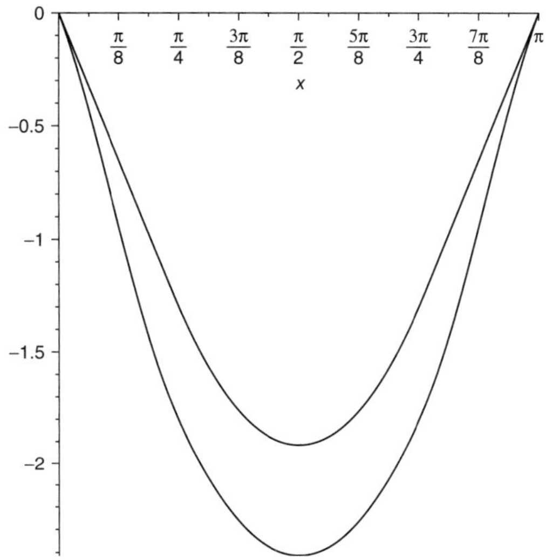

Figure 3.15: $u(x, 1.2)$ and $u_0(x, 1.2)$.

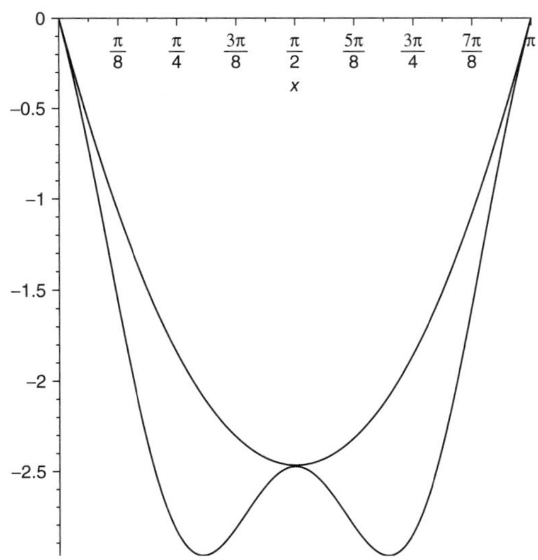

Figure 3.16: $u(x, 1.6)$ and $u_0(x, 1.6)$.

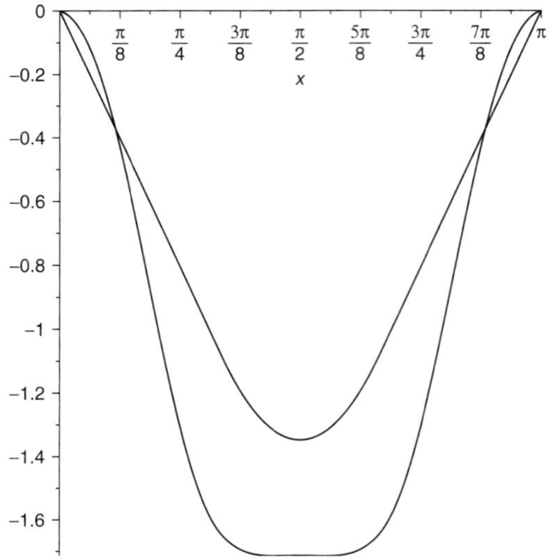

Figure 3.17: $u(x, 2.1)$ and $u_0(x, 2.1)$.

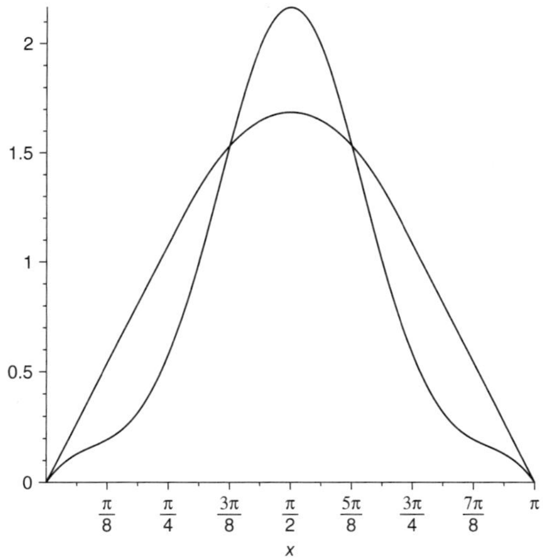

Figure 3.18: $u(x, 2.7)$ and $u_0(x, 2.7)$.

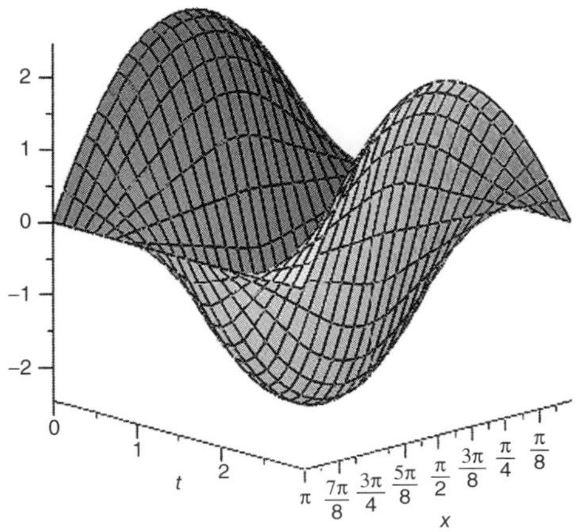

Figure 3.19: Unforced wave surface $u_0(x,t)$ in example 3.5.

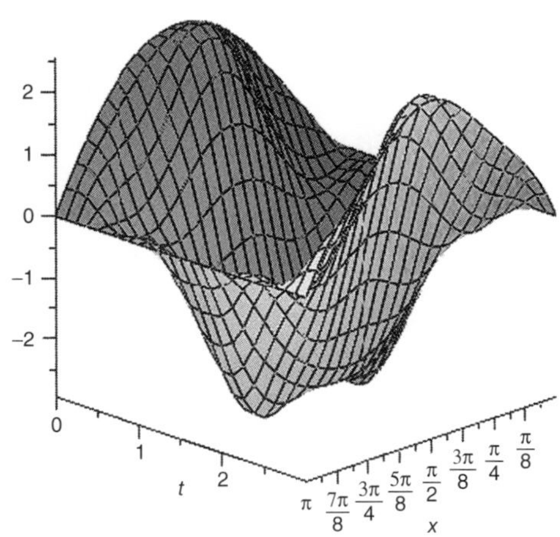

Figure 3.20: Surface $u(x,t)$ with the forcing term in example 3.5.

For example, let $K = 16$ and $c = 2$. Figures 3.14–3.18 compare the wave profiles $u(x,t)$ and $u_0(x,t)$ for $t = 0.8, 1.2, 1.6, 2.1$, and 2.7, respectively. In each, the forced wave has the larger amplitude. Figure 3.19 is part of the surface $u_0(x,t)$, and Figure 3.20 is part of the surface $u(x,t)$ for $0 \leq x \leq \pi, 0 \leq t \leq 3$.

3.1.3 Damped Wave Motion

We will consider the motion of a fixed end string, released from rest after an initial displacement, but also acted on by a damping force.. Damping can take many forms. We will treat the case that the damping force is proportional to the velocity of the string.

Let the string have length π. The initial-boundary value problem is

$$\begin{aligned} u_{tt} + ku_t &= c^2 u_{xx} \text{ for } 0 < x < \pi, t > 0, \\ u(0,t) &= u(\pi,t) = 0, \\ u(x,0) &= \varphi(x), u_t(x,0) = 0 \text{ for } 0 < x < \pi. \end{aligned}$$

Separate the variables by setting $u(x,t) = X(x)T(t)$ to obtain

$$XT'' + kXT' = c^2 X''T.$$

Upon dividing by c^2XT, we have

$$\frac{T''}{c^2T} + \frac{kT'}{c^2T} = \frac{X''}{X} = -\lambda.$$

Then

$$X'' + \lambda X = 0 \text{ and } T'' + kT' + \lambda c^2 T = 0.$$

From the boundary conditions, $X(0) = X(\pi) = 0$, so the problem for X has eigenvalues and eigenfunctions

$$\lambda_n = n^2, X_n(x) = \sin(nx).$$

The assumption of zero initial velocity means that $T'(0) = 0$, so the problem for T is

$$T'' + kT' + n^2c^2T = 0; T'(0) = 0.$$

To solve for λ and T, substitute $T(t) = e^{rt}$ into the differential equation to find the characteristic equation

$$r^2 + kr + n^2c^2 = 0.$$

The characteristic roots are

$$r = \frac{1}{2}\left(-k \pm \sqrt{k^2 - 4n^2c^2}\right).$$

A general treatment requires cases. We will restrict our attention to the case in which the damping eventually kills the motion, but is not so strong that it prevents some oscillations first. Suppose $k < 2c$. Then $k^2 - 4n^2c^2 < 0$ for every positive integer n, and the solution for $T(t)$ contains sine and cosine terms

$$T_n(t) = e^{-kt/2}\left(A_n \cos(\alpha_n ct) + B_n \sin(\alpha_n ct)\right),$$

for $n = 1, 2, \cdots$, where

$$\alpha_n = \sqrt{4n^2c^2 - k^2}.$$

The condition $T'(0) = 0$ requires that

$$-\frac{k}{2}A_n + B_n\alpha_n = 0,$$

so

$$B_n = \frac{k}{2\alpha_n}A_n.$$

The solutions for T are therefore constant multiples of

$$T_n(t) = e^{-kt/2}\left(\cos(\alpha_n ct) + \frac{k}{2\alpha_n}\sin(\alpha_n ct)\right).$$

Form a superposition of the functions $u_n(x,t) = X_n(x)T_n(t)$:

$$u(x,t) = \sum_{n=1}^{\infty} b_n e^{-kt/2}\left(\cos(\alpha_n ct) + \frac{k}{2\alpha_n}\sin(\alpha_n ct)\right)\sin(nx).$$

Finally, choose the coefficients to satisfy the initial position condition:

$$u(x,0) = \varphi(x) = \sum_{n=1}^{\infty} b_n \sin(nx).$$

This is the Fourier sine expansion of $\varphi(x)$ on $[0, \pi]$, so choose

$$b_n = \frac{2}{\pi}\int_0^{\pi} \varphi(\xi)\sin(n\xi)\,d\xi$$

to complete the solution.

Example 3.6 Suppose, in this model of damped motion, $k = c = 1$ and the initial position is given by

$$\varphi(x) = \begin{cases} 1 + \frac{3}{\pi}\left(x - \frac{\pi}{3}\right) & \text{for } 0 \le x \le \pi/3, \\ 1 & \text{for } \pi/3 \le x \le 2\pi/3, \\ 1 - \frac{3}{\pi}\left(x - \frac{2\pi}{3}\right) & \text{for } 2\pi/3 \le x \le \pi. \end{cases}$$

Now

$$\alpha_n = \sqrt{4n^2c^2 - k^2} = \sqrt{4n^2 - 1}$$

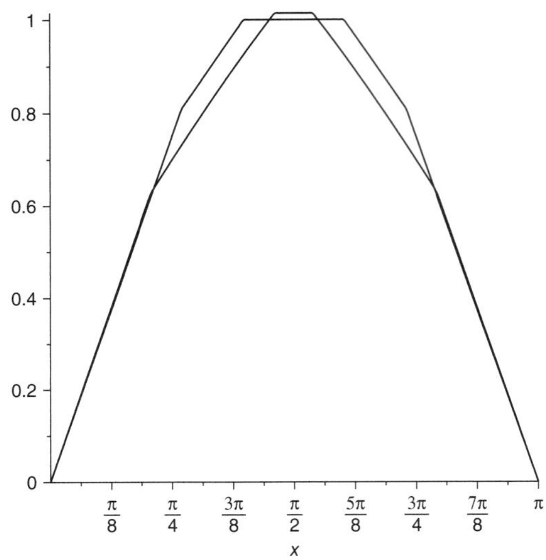

Figure 3.21: $u(x, 0.2)$ and $u_0(x, 0.2)$ in example 3.6.

and

$$\begin{aligned} b_n &= \frac{2}{\pi}\int_0^\pi \varphi(\xi)\sin(n\xi)\,d\xi \\ &= \frac{6}{n^2\pi^2}\left(\sin\left(\frac{n\pi}{3}\right) + \sin\left(\frac{2n\pi}{3}\right)\right). \end{aligned}$$

The solution is

$$u(x,t) =$$
$$\sum_{n=1}^{\infty} b_n e^{-t/2}\left(\cos(\sqrt{4n^2-1}t) + \frac{1}{2\sqrt{4n^2-1}}\sin(\sqrt{4n^2-1}t)\right)\sin(nx).$$

If the damping term, ku_t, is omitted, the problem is a standard one with solution

$$u_0(x,t) = \sum_{n=1}^{\infty} b_n \sin(nx)\cos(nt).$$

Figures 3.21–3.24 compare $u(x,t)$ and $u_0(x,t)$ at times $t = 0.2, 0.5, 2.5$ and 5.1, respectively. Figure 3.25 shows the (damped) solution surface $u(x,t)$, while for comparison, Figure 3.26 shows the solution surface $u_0(x,t)$ for the undamped problem.

Problems for Section 3.1

In each of problems 1–11, write the solution of problem 3.1 for the given information. Graph wave profiles at a selection of times.

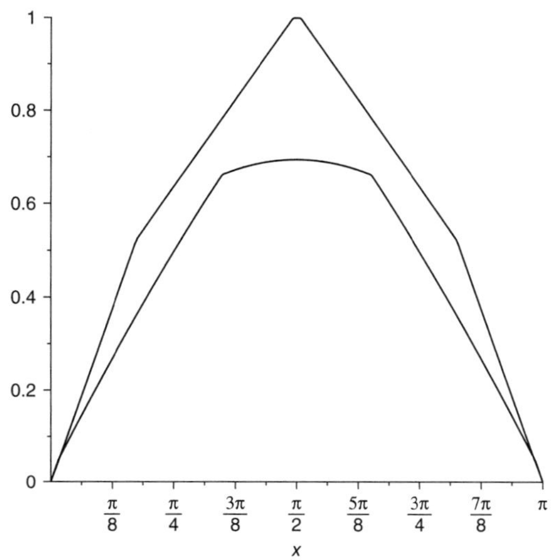

Figure 3.22: $u(x, 0.5)$ and $u(x, 0.5)$.

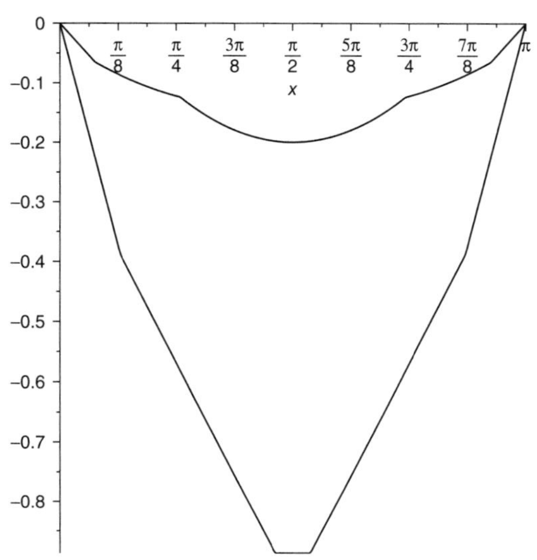

Figure 3.23: $u(x, 2.5)$ and $u_0(x, 2.5)$.

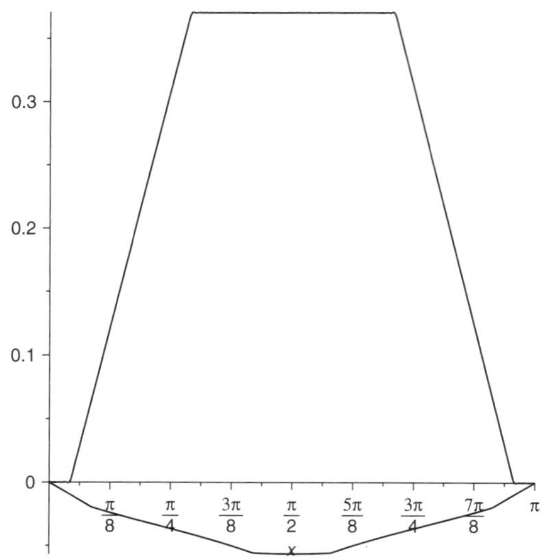

Figure 3.24: $u(x, 5.1)$ and $u_0(x, 5.1)$.

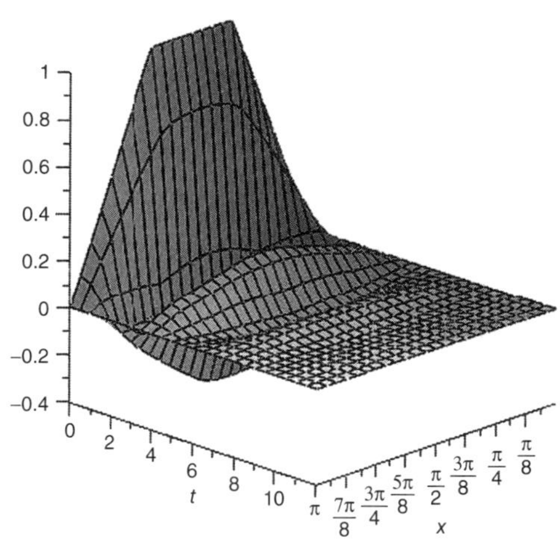

Figure 3.25: Surface graph of the damped solution $u(x,t)$ for $0 \leq t \leq 12$.

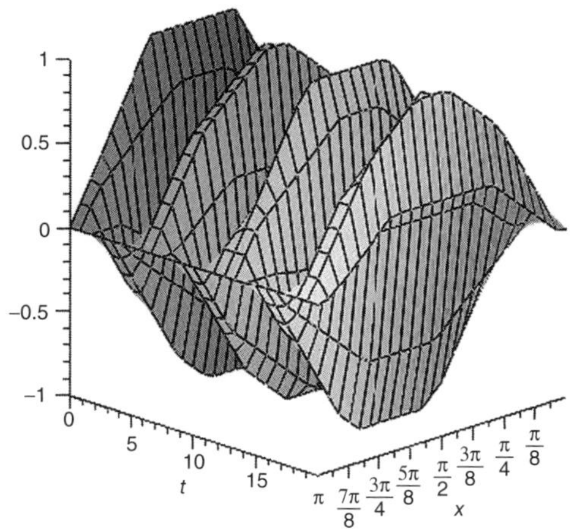

Figure 3.26: Surface graph of the undamped solution $u_0(x,t)$ for $0 \le t \le 20$.

1. $\varphi(x) = 1 - \cos(x), \psi(x) = 0, c = 3, L = 2\pi$
2. $\varphi(x) = 0, \psi(x) = x, c = 2, L = 1$.
3. $\varphi(x) = x(1-x), \psi(x) = 0, c = 6, L = 1$.
4. $\varphi(x) = \sin(x), \psi(x) = 0, c = 4, L = \pi$.
5. $\varphi(x) = 0, \psi(x) = e^{-x}, c = 3, L = 2$.
6. $\varphi(x) = 1 - \cos(x), \psi(x) = 0, c = 4, L = 2\pi$.
7. $\varphi(x) = \sin(x), \psi(x) = x, c = 3, L = \pi$.
8. $\varphi(x) = x^2(2-x), \psi(x) = x^2, c = 5, L = 2$.
9. $\varphi(x) = \sin^2(x), \psi(x) = 1, c = 1, L = \pi$.
10. $\varphi(x) = x(1-x^2), \psi(x) = \cos(x), c = 3, L = 1$.
11. It is interesting to observe the effect of c on the wave motion. Adjust the solution found in example 3.1 for the cases that $c = 0.2, 0.8, 1.5, 3$, and 6. Graph the solution for each of these values of c on the same set of axes, for the times used in the example. What is the effect of c on the motion of the string?
12. Carry out the program of problem 11 for example 3.2.
13. Carry out the program of problem 11 for example 3.3.

14. Use separation of variables to solve the *telegraph equation*

$$u_{tt} + Au_t + Bu = c^2 U_{xx} \text{ for } 0 < x < L, t > 0,$$

with A, B, and C positive constants. The boundary conditions are

$$u(0, t) = u(L, t) = 0 \text{ for } t > 0,$$

and the initial conditions are

$$u(x, 0) = \varphi(x), u_t(x, 0) = 0 \text{ for } 0 \leq x \leq L.$$

Assume that $A^2L^2 < 4(BL^2 + c^2\pi^2)$. This plays a role in solving the differential equation for $T(t)$, which results from separating the variables.

15. The current $i(x, t)$ and voltage $v(x, t)$ at time t and distance x from one end of a transmission line satisfy the system:

$$-v_x = Ri + Li_t,$$
$$-i_x = Sv + Kv_t,$$

where R is the resistance, L the inductance, S the leakage conductance, and K the capacitance to ground, all per unit length and all assumed constant. By differentiating appropriately and eliminating t, show that i satisfies a telegraph equation (problem 14). Show that v also satisfies a telegraph equation.

16. Let $\theta(x, t)$ be the angular displacement at time t of the cross-section at x of a homogeneous cylindrical shell about the $x-$axis. It can be shown that

$$\theta_{tt} = \alpha^2 \theta_{xx} \text{ for } 0 < x < L, t > 0.$$

Solve this equation subject to the conditions

$$\theta(x, 0) = \varphi(x), \theta_t(x, 0) = 0 \text{ for } 0 < x < L,$$

assuming that the ends of the shell are fixed elastically. This means that

$$\theta_x(0, t) - \alpha\theta(0, t) = 0, \theta_x(0, t) + \alpha\theta(L, t) = 0,$$

for some positive number α.

17. In example 3.5, write the solutions for $K = 0.4, 2.5, 5.6, 10.2$, and 15.1. Let $c = 2$. For each of the times given in the example, graph the solution for each of these values of K on the same set of axes. This will suggest the effect of this constant on the resulting motion.

18. Solve

$$u_{tt} = 9u_{xx} + Ax^2 \text{ for } 0 < x < 1, t > 0,$$
$$u(0, t) = u(1, t) = 0 \text{ for } t > 0,$$
$$u(x, 0) = u_t(x, 0) = 0 \text{ for } 0 < x < 1.$$

A is a positive constant. Generate graphs of the wave profile at different times, first with $A = 1.5$, then with $A = 5.7$ and $A = 15.3$.

19. Solve

$$\begin{aligned} u_{tt} &= 4u_{xx} + \cos(x) \text{ for } 0 < x < \pi, t > 0, \\ u(0,t) &= u(\pi,t) = 0 \text{ for } t > 0, \\ u(x,0) &= u_t(x,0) = 0 \text{ for } 0 < x < \pi. \end{aligned}$$

Generate graphs of the solution at different times.

20. Write the solution of problem 19 with $\cos(\alpha x)$ in place of $\cos(x)$ as a forcing term. Choose some values of t and, for each, graph the solutions corresponding to $\alpha = 0.5, 1.2, 1.6, 2.1$, and 2.8 on the same set of axes. This suggests the effect of α on the motion of the string.

21. Solve

$$\begin{aligned} u_{tt} &= u_{xx} - 3e^{-x} \text{ for } 0 < x < 2, t > 0, \\ u(0,t) &= u(2,t) = 0 \text{ for } t > 0, \\ u(x,0) &= u_t(x,0) = 0 \text{ for } 0 < x < 2. \end{aligned}$$

Graph the solution at various times.

22. Solve

$$\begin{aligned} u_{tt} &= 16u_{xx} - e^{-x} \text{ for } 0 < x < 3, t > 0, \\ u(0,t) &= u(3,t) = 0 \text{ for } t > 0, \\ u(x,0) &= u_t(x,0) = 0 \text{ for } 0 < x < 3. \end{aligned}$$

23. Solve

$$\begin{aligned} u_{tt} &= 9u_{xx} + 4x \text{ for } 0 < x < 1, t > 0, \\ u(0,t) &= u(1,t) = 0 \text{ for } t > 0, \\ u(x,0) &= 0, u_t(x,0) = 1 \text{ for } 0 < x < 1. \end{aligned}$$

24. Solve

$$\begin{aligned} u_{tt} &= 4u_{xx} \text{ for } 0 < x < 9, t > 0, \\ u(0,t) &= 0, u(9,t) = 1 \text{ for } t > 0, \\ u(x,0) &= 0, u_t(x,0) = x \text{ for } 0 < x < 9. \end{aligned}$$

Generate graphs showing the shape of the wave profile at different times.

25. Try to solve

$$\begin{aligned} u_{tt} &= 4u_{xx} + t^2 \text{ for } 0 < x < 1, t > 0, \\ u(0,t) &= u(1,t) = 0 \text{ for } t > 0, \\ u(x,0) &= u_t(x,0) = 0 \text{ for } 0 < x < 1. \end{aligned}$$

26. Consider the general wave equation

$$u_{tt} = c^2 u_{xx} + g(x,t) \text{ for } 0 < x < L, t > 0.$$

Suppose $u(x,t)$ is a continuous solution with continuous first and second partial derivatives. Prove that

$$\frac{d}{dt}\int_a^b \frac{1}{2}(u_t^2 + c^2 u_x^2)\,dx = c^2 u_t u_x\Big]_a^b + \int_a^b g u_t\,dt$$

if $0 < a < x < b < L$ and $t > 0$. *Hint*: Multiply the wave equation by u_t, integrate the resulting equation from a to b, and use integration by parts on one of the resulting integrals. It will also be useful to notice that

$$u_t u_{tt} = \frac{1}{2}\frac{\partial}{\partial t}u_t^2,$$

with a similar relationship for $u_x u_{xt}$.

27. Assume that the problem

$$u_{tt} = c^2 u_{xx} + f(x,t) \text{ for } 0 < x < L, t > 0,$$
$$u_x(0,t) = A(t), u_x(L,t) = B(t) \text{ for } t > 0,$$
$$u(x,0) = \varphi(x), u_t(x,0) = \psi(x) \text{ for } 0 < x < L$$

has a solution. It is assumed that f, A, B, φ and ψ are continuous. Show that this solution is unique. *Hint*: Suppose there are solutions u and v and let $w = u - v$. Determine an initial-boundary value problem satisfied by w. Let

$$E(t) = \frac{1}{2}\int_0^L (w_t^2 + c^2 w_x^2)\,dx$$

for $t \geq 0$. Calculate $E'(t)$, using an integration by parts on the term involving w_{xt}, and show that $E'(t) = 0$, hence that $E(t)$ is constant. Finally, show that this constant must be zero to conclude that $w_t^2 + c^2 w_x^2$ is identically zero. Use this to show that $w(x,t) = 0$.

In each of problems 28–32, write the solution for damped wave motion (fixed ends of the string, released from rest), for the given information.

28. $k = 1, c = 3, \varphi(x) = \sin(x)$.

29. $k = 1/2, c = 2, \varphi(x) = x\cos(x/2)$.

30. $k = 1, c = 4, \varphi(x) = x\sin(x)$.

31. $k = 2, c = 4, \varphi(x) = x^2(\pi - x)$.

32. $k = 1/4, c = 1, \varphi(x) = x(\pi - x)^2$.

33. Redo problem 28 with k left unspecified. For a particular time, graph the solution for $k = \frac{1}{8}, \frac{1}{4}, \frac{1}{2}, \frac{2}{3}, 1$ to get some sense of the impact of the damping term on the solution. Try this for a selection of other times.

3.2 The Cauchy Problem

Jean-Baptiste le Rond d'Alembert (1717–1786) was the illegitimate son of a French artillery officer and a well-known author. He was left on the steps of the Saint-Jean-le-Rond de Paris church, and his name is after the patron saint of that church, by custom of the time. He was later adopted, and his paternal father paid for his education.

d'Alembert left his mark on mathematical physics by obtaining a transformation that reveals the structure of solutions of the wave equation and provides a powerful method for solving problems in wave motion, including some beyond the reach of separation of variables.

Begin with the standard wave equation

$$u_{tt} = c^2 u_{xx}.$$

The lines $x - ct = k_1$ and $x + ct = k_2$, with k_1 and k_2 arbitrary constants, are called *characteristics* of the wave equation. Using these, define a transformation

$$\xi = x - ct, \eta = x + ct$$

from the $x, t-$plane to the $\xi, \eta-$plane. This transformation is one to one, which means that it defines a unique correspondence sending each point in the $x, t-$plane to exactly one point in the $\xi, \eta-$plane, and conversely. The inverse transformation is

$$x = \frac{1}{2}(\xi + \eta), t = \frac{1}{2c}(\eta - \xi).$$

If we substitute these into $u(x, t)$, we obtain a function of ξ and η:

$$U(\xi, \eta) = u(x(\xi, \eta), t(\xi, \eta)).$$

Using the chain rule, compute the partial derivatives

$$\begin{aligned}
u_x &= U_\xi \xi_x + U_\eta \eta_x = U_\xi + U_\eta, \\
u_{xx} &= (U_{\xi\xi} + U_{\xi\eta}) + (U_{\eta\xi} + U_{\eta\eta}) = U_{\xi\xi} + 2U_{\xi\eta} + U_{\eta\eta}, \\
u_t &= U_\xi(-c) + U_\eta(c) = c(-U_\xi + U_\eta), \\
u_{tt} &= -c(U_{\xi\xi}(-c) + U_{\xi\eta}(c)) + c(U_{\eta\xi}(-c) + U_{\eta\eta}(c)) \\
&= c^2 U_{\xi\xi} - 2c^2 U_{\xi\eta} + c^2 U_{\eta\eta}.
\end{aligned}$$

In terms of the new variables, the wave equation transforms to

$$u_{tt} - c^2 u_{xx} = 0 = 4c^2 U_{\xi\eta},$$

or

$$U_{\xi\eta} = 0. \tag{3.5}$$

This is the *canonical form* of the wave equation (see problems 7–12 of Section 1.1). Now follow a line of reasoning developed by d'Alembert in the 1740's. Write equation 3.5 as

$$(U_\eta)_\xi = 0.$$

This says that U_η is independent of ξ and hence is a function of η only, say

$$U_\eta = w(\eta).$$

Integrate this with respect to η:

$$U(\xi,\eta) = \int U_\eta \, d\eta = \int w(\eta)\, d\eta + F(\xi).$$

The constant of integration with respect to η may be a function of ξ, because such a function has partial derivative zero with respect to η. Now $\int w(\eta)\, d\eta$ is just another function of η, say $G(\eta)$. We conclude that $U(\xi,\eta)$ must look like

$$U(\xi,\eta) = F(\xi) + G(\eta),$$

in which F and G can be any functions of a single variable, having two continuous derivatives.

Because $\xi = x - ct$ and $\eta = x + ct$, we reach the important conclusions for functions defined for all real x and for $t \geq 0$:

1. Any continuous solution of $u_{tt} = c^2 u_{xx}$ must be a sum of a function of $x - ct$ and a function of $x + ct$,

$$u(x,t) = F(x-ct) + G(x+ct), \tag{3.6}$$

for some twice continuously differentiable functions F and G of a single variable.

2.Conversely, if F are any G are continuously differentiable functions of a single variable, then

$$u(x,t) = F(x-ct) + G(x+ct)$$

is a solution of $u_{tt} = c^2 u_{xx}$.

3.2.1 d'Alembert's Solution

The first problem we will solve using this observation is called the *Cauchy problem for the wave equation on the real line*:

$$\begin{aligned} u_{tt} &= c^2 u_{xx} \text{ for } -\infty < x < \infty, t > 0, \\ u(x,0) &= \varphi(x), u_t(x,0) = \psi(x) \text{ for } -\infty < x < \infty. \end{aligned} \tag{3.7}$$

This problem is posed on the entire real line, so there are no end points and no boundary conditions. We will write a solution for this problem in terms of the initial position and velocity functions. Begin with the fact that $u(x,t)$ must look like

$$u(x,t) = F(x-ct) + G(x+ct).$$

The idea is to use φ and ψ to choose F and G to produce a solution satisfying given initial conditions. First,

$$u(x,0) = F(x) + G(x) = \varphi(x) \tag{3.8}$$

and, by chain rule differentiations,

$$u_t(x,0) = -cF'(x) + cG'(x) = \psi(x). \tag{3.9}$$

Integrate equation 3.9 and rearrange terms to obtain

$$-F(x) + G(x) = \frac{1}{c}\int_0^x \psi(s)\,ds - F(0) + G(0).$$

Add this to equation 3.8 to obtain

$$2G(x) = \varphi(x) + \frac{1}{c}\int_0^x \psi(s)\,ds - F(0) + G(0).$$

Solve this for $G(x)$:

$$G(x) = \frac{1}{2}\varphi(x) + \frac{1}{2c}\int_0^x \psi(s)\,ds - \frac{1}{2}F(0) + \frac{1}{2}G(0). \tag{3.10}$$

But then, from equation 3.8,

$$\begin{aligned} F(x) &= \varphi(x) - G(x) \\ &= \frac{1}{2}\varphi(x) - \frac{1}{2c}\int_0^x \psi(s)\,ds + \frac{1}{2}F(0) - \frac{1}{2}G(0). \end{aligned} \tag{3.11}$$

Finally, use equations 3.10 and 3.11 to write

$$\begin{aligned} u(x,t) &= F(x-ct) + G(x+ct) \\ &= \frac{1}{2}\varphi(x-ct) - \frac{1}{2c}\int_0^{x-ct} \psi(s)\,ds + \frac{1}{2}(F(0) - G(0)) \\ &+ \frac{1}{2}\varphi(x+ct) + \frac{1}{2c}\int_0^{x+ct} \psi(s)\,ds - \frac{1}{2}(F(0) - G(0)). \end{aligned}$$

Upon making cancelations and combining the two integrals, we obtain *d'Alembert's solution of the Cauchy problem for the wave equation on the real line*:

$$u(x,t) = \frac{1}{2}(\varphi(x-ct) + \varphi(x+ct)) + \frac{1}{2c}\int_{x-ct}^{x+ct} \psi(s)\,ds. \tag{3.12}$$

Example 3.7 We will solve the Cauchy problem

$$\begin{aligned} u_{tt} &= 9u_{xx} \text{ for } -\infty < x < \infty, t > 0, \\ u(x,0) &= \cos(x), u_t(x,0) = \sin(2x) \text{ for } -\infty < x < \infty. \end{aligned}$$

With $\varphi(x) = \cos(x)$ and $\psi(x) = \sin(2x)$, and $c = 3$, the solution 3.12 is

$$\begin{aligned} u(x,t) &= \frac{1}{2}(\cos(x-3t) + \cos(x+3t)) \\ &+ \frac{1}{6}\int_{x-3t}^{x+3t} \sin(2s)\,ds \\ &= \frac{1}{2}(\cos(x-3t) + \cos(x+3t) \\ &- \frac{1}{12}(\cos(2(x+3t)) - \cos(2(x-3t))). \end{aligned}$$

Using trigonometric identities, this solution can also be written

$$u(x,t) = \cos(x)\cos(3t) + \frac{1}{6}\sin(2x)\sin(6t).$$

From the d'Alembert solution, we know that the Cauchy problem 3.7 has a solution, and that this solution is unique. We will also show that the solution depends continuously on the initial data, so this Cauchy problem is well posed.

Theorem 3.1 (Continuous Dependence on Initial Data) Let $u_1(x,t)$ be the solution of problem 3.7 satisfying

$$u(x,0) = \varphi_1(x), u_t(x,0) = \psi_1(x).$$

Let $u_2(x,t)$ be the solution satisfying

$$u(x,0) = \varphi_2(x), u_t(x,0) = \psi_2(x).$$

Let $\epsilon > 0$ and $T > 0$. Then there exists a positive number δ such that, if

$$|\varphi_1(x) - \varphi_2(x)| < \delta \text{ and } |\psi_1(x) - \psi_2(x)| < \delta,$$

for all x and for $0 \le t \le T$, then

$$|u_1(x,t) - u_2(x,t)| < \epsilon$$

for all x and for $0 \le t \le T$.

Proof We know that

$$u_1(x,t) = \frac{1}{2}(\varphi_1(x-ct) + \varphi_1(x+ct)) + \int_{x-ct}^{x+ct} \psi_1(s)\,ds$$

and

$$u_2(x,t) = \frac{1}{2}(\varphi_2(x-ct) + \varphi_2(x+ct)) + \int_{x-ct}^{x+ct} \psi_2(s)\,ds.$$

If δ is a positive number so that the initial position and velocity functions satisfy the inequalities of the theorem, then

$$\begin{aligned}
&|u_1(x,t) - u_2(x,t)| \\
&\le \frac{1}{2}|\varphi_1(x-ct) - \varphi_2(x-ct)| + \frac{1}{2}|\varphi_1(x+ct) - \varphi_2(x+ct)| \\
&+ \frac{1}{2c}\int_{x-ct}^{x+ct} |\psi_1(s) - \psi_2(s)|\, ds \\
&< \frac{1}{2}\delta + \frac{1}{2}\delta + \frac{1}{2c}(2cT) = (1+T)\delta.
\end{aligned}$$

Given $\epsilon > 0$, we will therefore have

$$|u_1(x,t) - u_2(x,t)| < \epsilon$$

if δ is chosen as any positive number satisfying

$$(1+T)\delta < \epsilon,$$

or

$$0 < \delta < \frac{\epsilon}{1+T}.$$

This proves the theorem.

3.2.1.1 Forward and Backward Waves

We can write any solution of the wave equation as

$$u(x,t) = F(x-ct) + B(x+ct).$$

This provides a valuable interpretation of solutions. Think of $F(x-ct)$ as the wave $F(x)$ moving to the right (a *forward wave*) along the $x-$axis with velocity c, while $B(x+ct)$ is the wave $B(x)$ moving to the left (a *backward wave*), along the $x-$axis with velocity c. At any time t, the sum of these waves is the total wave profile $u(x,t)$. This allows us to envision the solution wave as the superposition of two waves moving in opposite directions along the line.

Given the initial position and velocity functions φ and ψ, we can write explicit formulas for these forward and backward waves. The solution can be written

$$\begin{aligned}
u(x,t) &= \frac{1}{2}(\varphi(x-ct) + \varphi(x+ct)) + \frac{1}{2c}\int_{x-ct}^{x+ct} \psi(s)\, ds \\
&= \frac{1}{2}\left(\varphi(x-ct) - \frac{1}{c}\int_0^{x-ct} \psi(\xi)\, d\xi\right) \\
&+ \frac{1}{2}\left(\varphi(x+ct) + \frac{1}{c}\int_0^{x+ct} \psi(\xi)\, d\xi\right) \\
&= F(x-ct) + B(x+ct), \qquad (3.13)
\end{aligned}$$

where

$$F(s) = \frac{1}{2}\varphi(s) - \frac{1}{2c}\int_0^s \psi(\xi)\,d\xi \tag{3.14}$$

and

$$B(s) = \frac{1}{2}\varphi(s) + \frac{1}{2c}\int_0^s \psi(\xi)\,d\xi. \tag{3.15}$$

Example 3.8 Look at the solution of problem 3.7 with $c = 1, \psi(x) = 0$ and

$$\varphi(x) = \begin{cases} x^2\sin(x) & \text{for } -\pi \le x \le \pi, \\ 0 & \text{for } |x| > \pi. \end{cases}$$

From equations 3.14 and 3.15, the forward and backward waves are $F(x-t)$ and $B(x+t)$, where

$$F(s) = B(s) = \begin{cases} \frac{1}{2}s^2\sin(s) & \text{for } |s| \le \pi, \\ 0 & \text{for } |s| > \pi. \end{cases}$$

The forward wave is $F(x-t)$, and the backward wave is $B(x+t)$, while the solution is the superposition

$$u(x,t) = F(x-t) + B(x+t).$$

Figure 3.27 is a graph of the initial position function, which is $u(x,0) = F(x,0) + B(x,0)$. Figure 3.28 shows the forward wave moving to the right for times $t = 0, 2, 7$, whereas Figure 3.29 shows the backward wave moving to the left at these times.

Figures 3.30–3.34 show the solution wave $u(x,t)$ as a sum of the forward and backward waves at times $t = 1/2, 1, 2, 5/2$, and 4. By the last time, the waves have separated and are moving without further interaction in opposite directions along the line. This occurs whenever the initial position function is nonzero only on a bounded interval, as is the case here.

Problems for Section 3.2.1

In each of problems 1–9, solve the Cauchy problem 3.7 on the real line for the given initial position and velocity functions and value of c.

1. $\varphi(x) = \cos(3x), \psi(x) = x, c = 7.$
2. $\varphi(x) = x^2, \psi(x) = \sin(2x), c = 4.$
3. $\varphi(x) = e^{-|x|}, \psi(x) = \sin^2(x), c = 3.$
4. $\varphi(x) = \cosh(x), \psi(x) = 2x, c = 2$
5. $\varphi(x) = \cos(x) - \sin(x), \psi(x) = \sin(x), c = 2.$

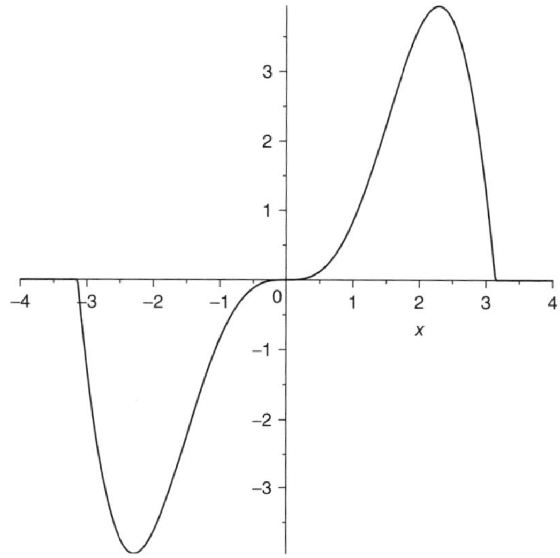

Figure 3.27: Initial position $u(x, 0)$ in example 3.8.

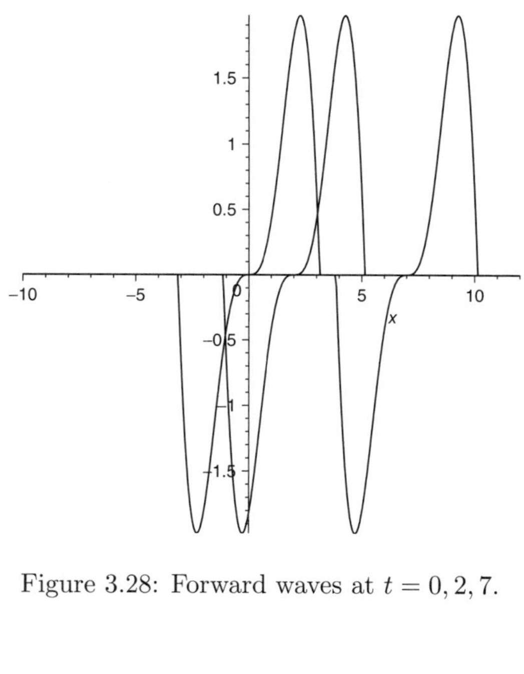

Figure 3.28: Forward waves at $t = 0, 2, 7$.

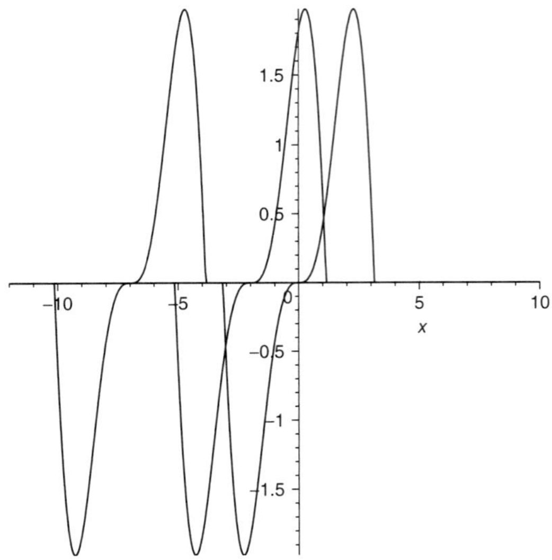

Figure 3.29: Backward waves at $t = 0, 2, 7$.

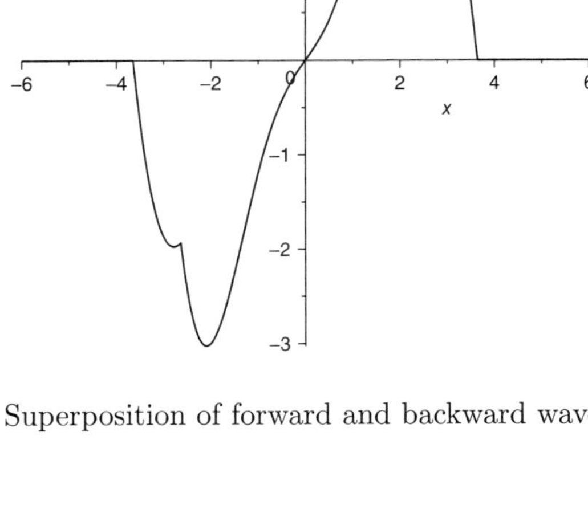

Figure 3.30: Superposition of forward and backward waves at $t = 1/2$.

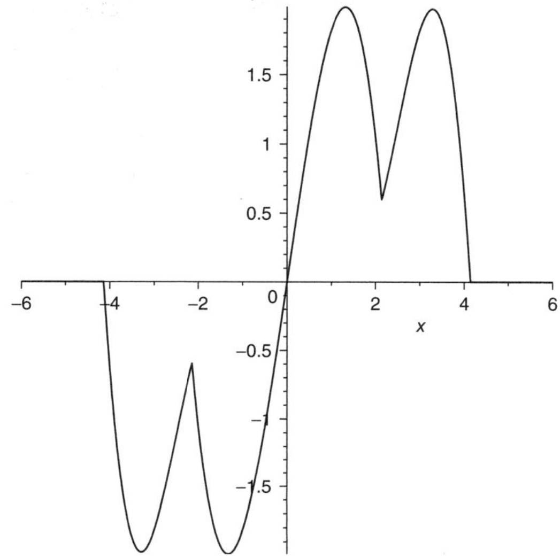

Figure 3.31: Superposition at time $t = 1$.

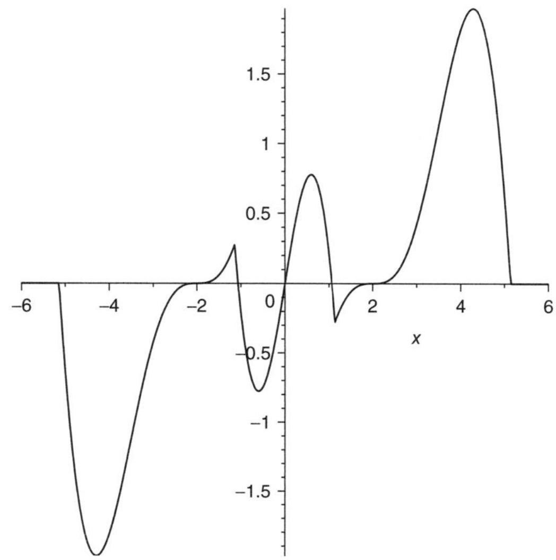

Figure 3.32: Superposition at $t = 2$.

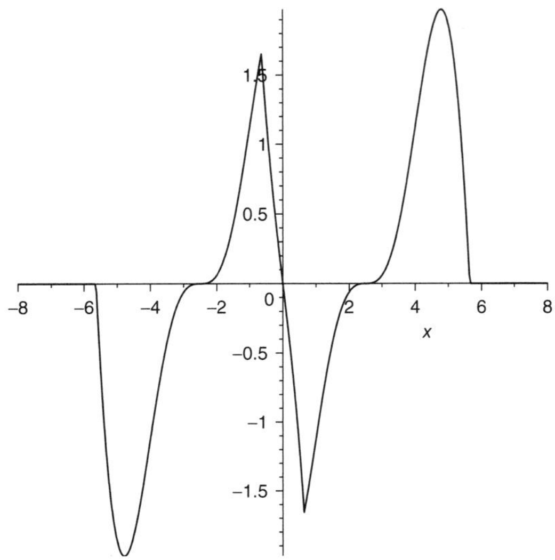

Figure 3.33: Superposition at $t = 5/2$.

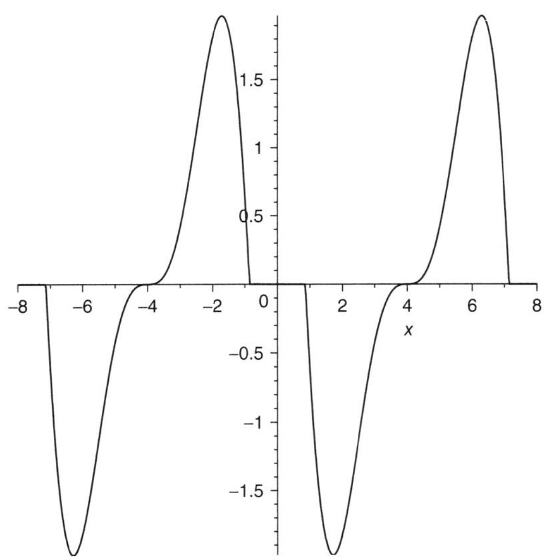

Figure 3.34: Superposition at $t = 4$.

6. $\varphi(x) = 2 + x, \psi(x) = e^x, c = 1.$

7. $\varphi(x) = \cos(x), \psi(x) = xe^{-x}, c = 4.$

8. $\varphi(x) = \sin(3x), \psi(x) = \cos(x), c = 1.$

9. $\varphi(x) = x^2, \psi(x) = x\cos(x), c = 3.$

10. Let w be a solution of

$$u_{tt} = c^2 u_{xx} \text{ for } -\infty < x < \infty, t > 0,$$
$$u(x,0) = \varphi(x), u_t(x,0) = 0 \text{ for } -\infty < x < \infty.$$

Let v be a solution of

$$u_{tt} = c^2 u_{xx} \text{ for } -\infty < x < \infty, t > 0,$$
$$u(x,0) = 0, u_t(x,0) = \psi(x) \text{ for } -\infty < x < \infty.$$

Show that $w + v$ is a solution of

$$u_{tt} = c^2 u_{xx} \text{ for } -\infty < x < \infty, t > 0,$$
$$u(x,0) = \varphi(x), u_t(x,0) = \psi(x) \text{ for } -\infty < x < \infty.$$

11. Write the solution of the Cauchy problem with $c = 1$, $\varphi(x) = \sin(x)$ and $\psi(x) = 0$. Then write the solution with $\varphi(x) = \sin(x) + \epsilon$ and $\psi(x) = 0$, in which ϵ is any positive number. Show that these solutions differ in magnitude by ϵ for all x and $t \geq 0$.

12. Write the solution of the Cauchy problem with $c = 1$, $\varphi(x) = \cos(x)$ and $\psi(x) = x$. Then write the solution with $\varphi(x) = \cos(x)+\epsilon$ and $\psi(x) = x+\epsilon$, with ϵ any positive number. Show that these solutions differ in magnitude by no more than $\epsilon(1 + T)$ for all x and $0 \leq t \leq T$, for every positive number T.

13. Let $u(x,t)$ be a solution of the problem

$$u_{tt} = u_{xx} + f(x,t) \text{ for } -\infty < x < \infty, t > 0,$$
$$u(x,0) = u_t(x,0) = 0 \text{ for } t > 0.$$

Let $w(x,t,T)$ satisfy

$$w_{xx} - w_{tt} = 0 \text{ for } -\infty < x < \infty, t > T,$$
$$w(x,T,T) = 0, w_t(x,T,T) = f(x,T).$$

Show that

$$u(x,t) = \int_0^t w(x,t,T)\,dT.$$

This conclusion is known as *Duhamel's principle for the wave equation.*

14. Verify that

$$u(x,t) = \frac{1}{2c}\int_0^t \int_{x-c(t-\tau)}^{x+c(t-\tau)} f(\xi,\tau)\,d\xi\,d\tau$$

is a solution of the problem

$$\begin{aligned} u_{tt} &= c^2 u_{xx} + f(x,t) \text{ for } -\infty < x < \infty, t > 0, \\ u(x,0) &= u_t(x,0) = 0 \text{ for } -\infty < x < \infty, \\ u(0,t) &= 0 \text{ for } t > 0. \end{aligned}$$

In each of problems 15–20, write the solution of the Cauchy problem with $c = 1$, the given $\varphi(x)$, and $\psi(x) = 0$, as a sum of forward and backward waves. Graph the progression of these waves and their sum for different values of t.

15. $\varphi(x) = e^{-x^2}$

16. $\varphi(x) = \begin{cases} 1 - |x| & \text{for } -1 \le x \le 1, \\ 0 & \text{for } |x| > 1. \end{cases}$

17. $\varphi(x) = \begin{cases} x\cos(x) & \text{for } -\pi/2 \le x \le \pi/2, \\ 0 & \text{for } |x| > \pi/2. \end{cases}$

18. $\varphi(x) = \begin{cases} 1 - x^2 & \text{for } |x| \le 1, \\ 0 & \text{for } |x > 1. \end{cases}$

19. $\varphi(x) = \begin{cases} x^3 - x^2 - 2 & \text{for } -1 \le x \le 2, \\ 0 & \text{for } x < -1 \text{ and for } x > 2. \end{cases}$

20. $\varphi(x) = \begin{cases} x^2 + x & \text{for } -1 \le x \le 1, \\ 0 & \text{for } |x| > 1. \end{cases}$

3.2.2 The Cauchy Problem on a Half-Line

We will use the solution of the Cauchy problem on the line to solve the Cauchy problem for the half line:

$$\begin{aligned} u_{tt} &= c^2 u_{xx} \text{ for } x > 0, t > 0, \\ u(x,0) &= \varphi(x), u_t(x,0) = \psi(x) \text{ for } x > 0, \\ u(0,t) &= 0 \text{ for } t > 0. \end{aligned} \tag{3.16}$$

This is the *Cauchy problem for the wave equation for the half-line.* Because this half line has an end point $x = 0$, a boundary condition is specified there. We have taken the left end of the string to be fixed.

We will use a device called the *method of images*, in which the plane is folded over the $t-$axis to extend the half line problem to a problem on the whole line, which we have solved. This is achieved by extending the initial position and velocity functions to odd functions defined for all x.

$$\Phi(x) = \begin{cases} \varphi(x) & \text{if } x \geq 0, \\ -\varphi(-x) & \text{if } x < 0 \end{cases}$$

and

$$\Psi(x) = \begin{cases} \psi(x) & \text{if } x \geq 0, \\ -\psi(-x) & \text{if } x < 0. \end{cases}$$

On the half line $x \geq 0$, $\Phi(x) = \varphi(x)$ and $\Psi(x) = \psi(x)$. Now consider the Cauchy problem for the entire line, with initial position $u(x,0) = \Phi(x)$ and initial velocity $u_t(x,0) = \Psi(x)$. We have solved this problem:

$$u(x,t) = \frac{1}{2}\left(\Phi(x-ct) + \Phi(x+ct)\right) + \frac{1}{2c}\int_{x-ct}^{x+ct} \Psi(s)\,ds. \tag{3.17}$$

This function satisfies the wave equation for all x, hence also for $x > 0$. Further, if $x \geq 0$,

$$u(x,0) = \frac{1}{2}(\Phi(x) + \Phi(x)) = \frac{1}{2}(\varphi(x) + \varphi(x)) = \varphi(x)$$

and

$$\begin{aligned} u_t(x,0) &= \frac{1}{2}\left(-c\Phi'(x) + c\Phi'(x)\right) + \frac{1}{2c}\left(c\Psi(x) - (-c)\Psi(x)\right) \\ &= \Psi(x) = \psi(x). \end{aligned}$$

Therefore, for $x \geq 0$, $u(x,t)$ as defined by equation 3.17 is also a solution of this Cauchy problem on the half line. Keep in mind, however, that this solution is written in terms of extensions of the initial position and velocity functions of the original problem. This was necessary because $\varphi(x)$ and $\psi(x)$ are defined only for $x \geq 0$, while, in d'Alembert's solution, $x - ct$ can be negative even if $x \geq 0$.

Example 3.9 Solve

$$\begin{aligned} u_{tt} &= u_{xx} \text{ for } x > 0, t > 0, \\ u(0,t) &= 0 \text{ for } t > 0, \\ u(x,0) &= \varphi(x) = 1 - e^{-x}, u_t(x,0) = \psi(x) = \cos(x) \text{ for } x \geq 0. \end{aligned}$$

Equation 3.17 gives the solution in terms of the odd extensions of $\varphi(x)$ and $\psi(x)$. These extensions are

$$\Phi(x) = \begin{cases} \varphi(x) = 1 - e^{-x} & \text{if } x \geq 0, \\ -\varphi(-x) = -1 + e^{x} & \text{if } x < 0. \end{cases}$$

and

$$\Psi(x) = \begin{cases} \psi(x) = \cos(x) & \text{if } x \geq 0, \\ -\psi(-x) = -\cos(x) & \text{if } x < 0. \end{cases}$$

The solution is

$$u(x,t) = \frac{1}{2}\left(\Phi(x-t) + \Phi(x+t)\right) + \frac{1}{2}\int_{x-t}^{x+t} \Psi(s)\, ds.$$

We will write $u(x,t)$ explicitly in terms of x and t. This requires two cases.

Case 1. If $x - t \geq 0$, then

$$\begin{aligned} u(x,t) &= \frac{1}{2}(\Phi(x-ct) + \Phi(x+ct)) + \frac{1}{2}\int_{x-ct}^{x+ct} \Psi(s)\, ds \\ &= \frac{1}{2}\left(1 - e^{-(x-t)} + 1 - e^{-(x+t)}\right) + \frac{1}{2}\int_{x-t}^{x+t} \cos(s)\, ds \\ &= 1 - \frac{1}{2}e^{-x}\left(e^{t} + e^{-t}\right) + \frac{1}{2}(\sin(x+t) - \sin(x-t)) \\ &= 1 - e^{-x}\cosh(t) + \cos(t)\sin(t). \end{aligned}$$

Case 2. If $x - t < 0$, then

$$\begin{aligned} u(x,t) &= \frac{1}{2}\left(\Phi(x-t) + \Phi(x+t)\right) + \frac{1}{2}\int_{x-t}^{x+t} \Psi(s)\, ds \\ &= \frac{1}{2}\left(-1 + e^{x-t} + 1 - e^{-(x+t)}\right) \\ &+ \frac{1}{2}\int_{x-t}^{0} -\cos(s)\, ds + \int_{0}^{x+t} \cos(s)\, ds \\ &= \frac{1}{2}\left(e^{x}e^{-t} - e^{-x}e^{-t}\right) - \left[\frac{1}{2}\sin(s)\right]_{x-t}^{0} + \left[\frac{1}{2}\sin(s)\right]_{0}^{x+t}. \end{aligned}$$

After some manipulation, this yields

$$u(x,t) = e^{-t}\sinh(x) + \sin(x)\cos(t).$$

In summary,

$$u(x,t) = \begin{cases} 1 - e^{-x}\cosh(t) + \cos(x)\sin(t) & \text{for } x - t \geq 0, \\ e^{-t}\sinh(x) + \sin(x)\cos(t) & \text{for } x - t < 0. \end{cases}$$

Problems for Section 3.2.2

In each of problems 1–10, write the solution of the problem 3.16 for the given initial position and velocity functions.

1. $\varphi(x) = x^2, \psi(x) = \sin(x), c = 1.$
2. $\varphi(x) = x^2, \psi(x) = x, c = 4.$
3. $\varphi(x) = 1 - e^x, \psi(x) = x^2, c = 2.$
4. $\varphi(x) = 1 - \cos(x), \psi(x) = e^{-x}, c = 4.$
5. $\varphi(x) = x\sin(x), \psi(x) = x^2, c = 2.$
6. $\varphi(x) = \sinh^2(x), \psi(x) = x, c = 7.$
7. $\varphi(x) = x^3, \psi(x) = e^{-x}, c = 3.$
8. $\varphi(x) = x^2 - x, \psi(x) = x - 1, c = 3.$
9. $\varphi(x) = \cosh(x) - 1, \psi(x) = \sin(x), c = 5.$
10. $\varphi(x) = x^3 + x, \psi(x) = \cos(2x), c = 6.$

3.2.3 Characteristic Triangles and Quadrilaterals

Characteristics, in combination with d'Alembert's solution, provide a way of solving several important types of problems involving the wave function.

Recall that the characteristics of the wave equation $u_{tt} = c^2 u_{xx}$ are the straight lines

$$x - ct = k_1, x + ct = k_2$$

having slope $1/c$ and $-1/c$, respectively, in the $x, t-$plane. There are exactly two characteristics through any point $P_0 : (x_0, t_0)$ with $t_0 > 0$. As shown in Figure 3.35, these are the characteristics $x - ct = x_0 - ct_0$ and $x + ct = x_0 + ct_0$, intersecting the $x-$axis at $(x_0 - ct_0, 0)$ and $(x_0 + ct_0, 0)$. These two points, together with (x_0, t_0), are the vertices of the *characteristic triangle* at (x_0, t_0). The base of this triangle is the interval $[x_0 - ct_0, x_0 + ct_0]$ on the $x-$axis.

If we are considering a Cauchy problem on the entire real line, then we know that the solution at P_0 is

$$u(x_0, t_0) = \frac{1}{2}(\varphi(x_0 - ct_0) + \varphi(x_0 + ct_0)) + \frac{1}{2c}\int_{x_0 - ct_0}^{x_0 + ct_0} \psi(s)\, ds.$$

This solution value depends on two things: the values of φ at the base vertices $x_0 - ct_0$ and $x_0 + ct_0$ of the characteristic triangle at (x_0, t_0), and the value of ψ on the entire base of this triangle (because of the integration).

We may also write $u(x_0, t_0)$ as a sum of forward and backward waves as

$$u(x_0, t_0) = F(x_0 - ct_0) + B(x_0 + ct_0).$$

Along the left slant side of the characteristic triangle, $x - ct = x_0 - ct_0$, so a disturbance initially at $(x_0 - ct_0, 0)$ is carried forward with constant velocity c along this characteristic. Along the right side of the characteristic triangle,

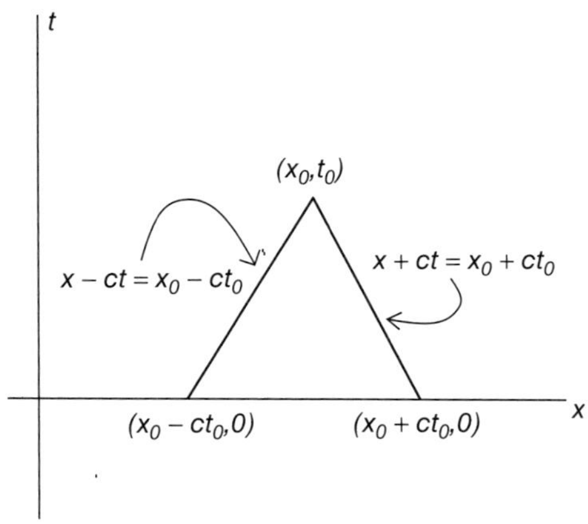

Figure 3.35: Characteristic triangle at (x_0, t_0).

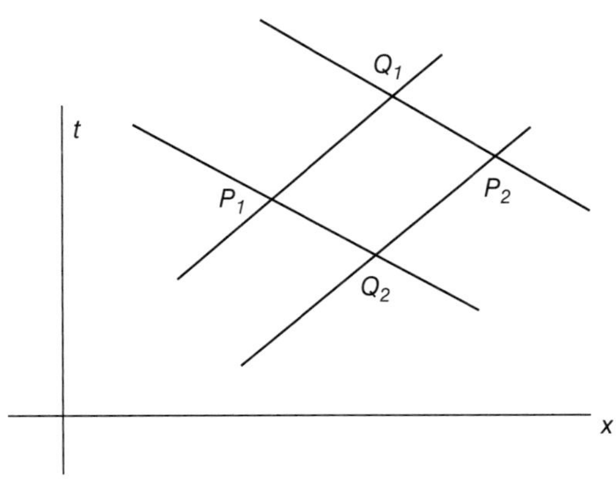

Figure 3.36: Characteristic quadrilateral.

$x + ct = x_0 + ct_0$, so a disturbance initially at $(x_0 + ct_0, 0)$ is carried back with constant velocity c. This says that *a solution of the wave equation propagates disturbances with constant velocity along the characteristics.*

In addition to characteristic triangles, we will use characteristic quadrilaterals. Figure 3.36 shows two pairs of parallel characteristics. These intersect at four points, which are the vertices of a *characteristic quadrilateral*. Denote opposite pairs of these points of intersection as P_1, P_2 and Q_1, Q_2. We will show that, for any solution of the wave equation $u_{tt} = c^2 u_{xx}$, regardless of initial and boundary conditions, the sum of the values of u at one pair of opposite vertices equals the sum of the values of u at the other pair:

$$u(P_1) + u(P_2) = u(Q_1) + u(Q_2). \tag{3.18}$$

This has the important consequence that, if the value of a solution is known at any three vertices of a characteristic quadrilateral, then it is also known at the fourth vertex. To verify this, use the fact that any solution must have the form

$$u(x,t) = F(x - ct) + G(x + ct)$$

for some functions F and G of a single variable. Label the characteristics

$$x + ct = K_1, x + ct = K_2, x - ct = K_3, x - ct = K_4,$$

as in Figure 3.37. The points of intersection are

$$P_1 = \left(\frac{1}{2}(K_1 + K_3), \frac{1}{2c}(K_1 - K_3)\right), P_2 = \left(\frac{1}{2}(K_2 + K_4), \frac{1}{2c}(K_2 - K_4)\right),$$
$$Q_1 = \left(\frac{1}{2}(K_2 + K_3), \frac{1}{2c}(K_2 - K_3)\right), Q_2 = \left(\frac{1}{2}(K_1 + K_4), \frac{1}{2c}(K_1 - K_4)\right).$$

Now compute $u(x,t)$ at each of these points. At P_1, $x - ct = K_3$ and $x + ct = K_1$, so

$$u(P_1) = F(K_3) + G(K_1).$$

Similarly,

$$u(P_2) = F(K_4) + G(K_2), u(Q_1) = F(K_3) + G(K_2),$$
$$\text{and } u(Q_2) = F(K_4) + G(K_1).$$

Then,

$$u(P_1) + u(P_2) = F(K_3) + F(K_4) + G(K_1) + G(K_2) = u(Q_1) + u(Q_2).$$

The power of characteristic triangles and quadrilaterals in solving wave propagation problems with moving ends and forcing terms will be seen in the next three sections.

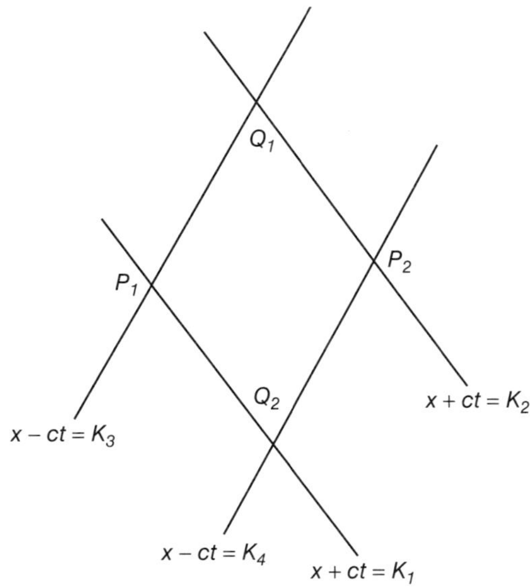

Figure 3.37: $u(P_1) + u(P_2) = u(Q_1) + u(Q_2)$.

Problems for Section 3.2.3

1. Suppose the initial displacement and velocity functions in the Cauchy problem on the real line vanish outside of some interval of finite length. Specifically, suppose, for some $a > 0$,

$$\varphi(x) = \psi(x) = 0 \text{ if } |x| > a.$$

Prove that, for any $t > 0$, the solution must vanish for x outside of the interval $[-a - ct, a + ct]$.

2. For the Cauchy problem on the real line, with $c = 1$ suppose there is an interval $[a, b]$ such that $\varphi(x)$ and $\psi(x)$ both vanish for x outside of this interval. Assume that the solution of the problem is continuous with continuous first and second partial derivatives. Show that

$$\int_{-\infty}^{\infty} \left[u_x^2 + u_t^2\right]_{t=T} dx = \int_{-\infty}^{\infty} \left[u_x^2 + u_t^2\right]_{t=0} dx$$

for every $T \geq 0$. *Hint*: First show that

$$(3u_t u_x)_x = (u_x^2 + u_t^2)_t.$$

Form a double integral over the rectangular region

$$-R - 2T \leq x \leq R + 2T, 0 \leq t \leq T.$$

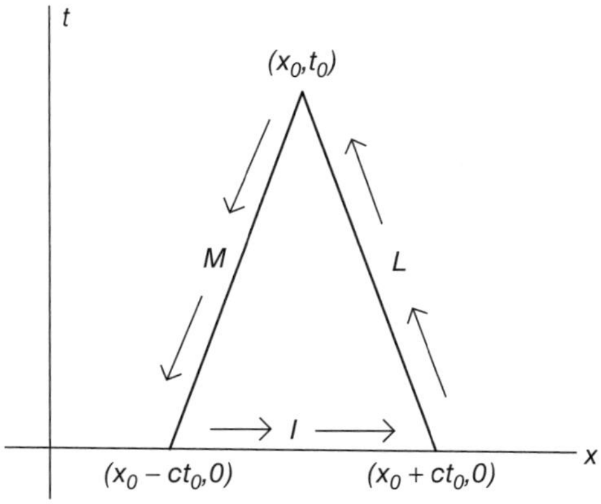

Figure 3.38: Characteristic triangle Δ at (x_0, t_0).

Apply Green's theorem to this integral.

The integral

$$\int_{\alpha}^{\beta} \left[u_x^2 + u_t^2\right]_{t=T} dx$$

is called the *energy* of $u(x,t)$ in the interval $[\alpha, \beta]$ at time T. The conclusion of this problem is a conservation of energy principle, stating that the total energy of the solution over the entire line is constant over time.

3.2.4 A Cauchy Problem with a Forcing Term

We will use the characteristic triangle to solve a Cauchy problem on the real line, with a forcing term:

$$\begin{aligned} u_{tt} &= c^2 u_{xx} + P(x,t) \text{ for } -\infty < x < \infty, t > 0, \\ u(x,0) &= \varphi(x), u_t(x,0) = \psi(x) \text{ for } -\infty < x < \infty. \end{aligned} \tag{3.19}$$

We may think of $P(x,t)$ as the magnitude of a damping or driving force.

Suppose we want the solution at $P_0 : (x_0, t_0)$. The characteristic triangle at P_0 has vertices $(x_0, t_0), (x_0 - ct_0, 0)$ and $(x_0 + ct_0, 0)$. Let Δ denote this triangle and all the points it encloses. The boundary C of Δ consists of three line segments, denoted I, L, and M in Figure 3.38. We will perform a calculation which will help us determine $u(x_0, t_0)$. Look at the double integral of $-P(x,t)$

over the solid triangle Δ:

$$-\iint_{\Delta} P(x,t)\,dA = \iint (c^2 u_{xx} - u_{tt})\,dA$$
$$= \iint_{\Delta} \left(\frac{\partial}{\partial x}(c^2 u_x) - \frac{\partial}{\partial t}(u_t) \right) dA$$
$$= \oint_C u_t\,dx + c^2 u_x\,dt,$$

with the line integral obtained by applying Green's theorem to the double integral. Evaluate this line integral as the sum of the line integrals counterclockwise over the three segments making up C.

On I, $t = 0$ so $dt = 0$ and x varies from $x_0 - ct_0$ to $x_0 + ct_0$:

$$\int_I (u_t\,dx + c^2 u_x)\,dt = \int_{x_0 - ct_0}^{x_0 + ct_0} u_t(x,0)\,dx = \int_{x_0 - ct_0}^{x_0 + ct_0} \psi(s)\,ds.$$

On L, $x + ct = x_0 + ct_0$, so $dx = -c\,dt$ and

$$\int_L u_t\,dx + c^2 u_x\,dt = \int_L \left(-cu_t\,dt + c^2 u_x \left(\frac{-1}{c} \right) dx \right)$$
$$= -c \int_L u_t\,dt + u_x\,dx = -c \int_L du$$
$$= -c(u(x_0, t_0) - u(x_0 + ct_0, 0)).$$

On M, $x - ct = x_0 - ct_0$, so $dx = c\,dt$ and

$$\int_M u_t\,dx + c^2 u_x\,dt = \int_M cu_t\,dt + c^2 u_x \left(\frac{1}{c} \right) dx$$
$$= \int_M c(u_t\,dt + u_x\,dx) = c \int_M du$$
$$= c(u(x_0 - ct_0) - u(x_0, t_0)).$$

Then

$$-\iint_{\Delta} P(x,t)\,dx\,dt = \int_{x_0 - ct_0}^{x_0 + ct_0} \psi(s)\,dx$$
$$+ c(u(x_0 - ct_0, 0) - 2u(x_0, t_0) + u(x_0 + ct_0, 0))$$
$$= c(\varphi(x_0 + ct_0) + \varphi(x_0 - ct_0) - 2cu(x_0, t_0) + \int_{x_0 - ct_0}^{x_0 + ct_0} \psi(s)\,ds.$$

Solve this equation for $u(x_0, t_0)$ to obtain

$$u(x_0, t_0) = \frac{1}{2}(\varphi(x_0 + ct_0) + \varphi(x_0 - ct_0)) + \frac{1}{2c} \int_{x_0 - ct_0}^{x_0 + ct_0} \psi(s)\,ds$$
$$+ \frac{1}{2c} \iint_{\Delta} P(x,t)\,dA.$$

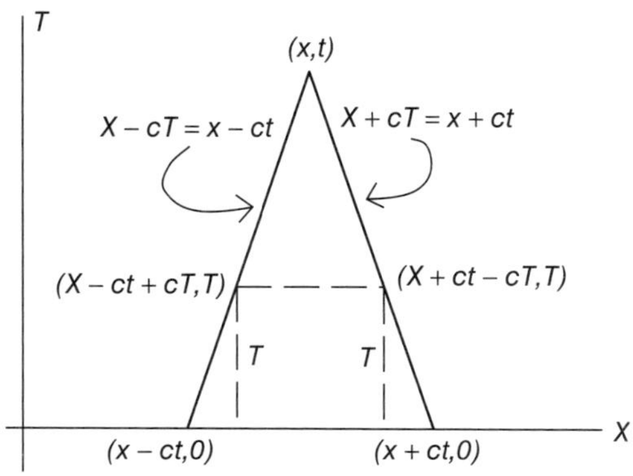

Figure 3.39: Limits of integration for $\int\int_{\Delta} P(X,T)\,dX\,dT$.

In this derivation we have used x_0 and t_0 for coordinates of the point at which the solution is computed. This allowed us to use x and t for the variables in the x, t − −plane. However, this is awkward if (x_0, t_0) is an arbitrary point. We will therefore rewrite the solution at an arbitrary point (x, t), using X and T as the variables in the space/time plane. In this notation,

$$u(x,t) = \frac{1}{x}(\varphi(x-ct) + \varphi(x+ct)) + \frac{1}{2c}\int_{x-ct}^{x+ct} \psi(s)\,ds + \frac{1}{2c}\iint_{\Delta} P(X,T)\,dX\,dT. \tag{3.20}$$

for all real x and for $t \geq 0$.

This is the solution of problem 3.19. In this notation Δ is the characteristic triangle having vertices $(x,t), (x-ct, 0)$ and $(x+ct, 0)$ in the $X, T-$ plane. Two sides of Δ are on the characteristics $X - cT = x - ct$ and $X + cT = x + ct$, and the base of Δ is the interval $[x - ct, x + ct]$ on the $X-$ axis.

It is usually easiest to carry out the $X-$integration first in the double integral. The limits of integration can be read from Figure 3.39. For any $T > 0$, X extends from $x - ct + cT$ to $x + ct - cT$, and T from 0 to t. These are obtained from the intersections of the horizontal line at height T with the left and right sides of the characteristic triangle at (x,t), respectively. Then

$$\iint_{\Delta} P(X,T)\,dX\,dT = \int_0^t \left(\int_{x-ct+cT}^{x+ct-cT} P(X,T)\,dX\right) dT. \tag{3.21}$$

Example 3.10 We will solve the problem

$$u_{tt} = 4u_{xx} + x\cos(t) \text{ for } -\infty < x < \infty, t > 0,$$
$$u(x,0) = e^{-x}, u_t(x,0) = \sin(x) \text{ for } -\infty < x < \infty.$$

The solution is given by equation 3.20. First evaluate the double integral, using equation 3.21 (and Figure 3.39). With $c = 2$,

$$\begin{aligned}
\iint_{\Delta} X\cos(T)\,dX\,dT &= \int_0^t \left(\int_{x-2t+2T}^{x+2t-2T} X\,dX\right)\cos(T)\,dT \\
&= \int_0^t \left[\frac{1}{2}X^2\right]_{x-2t+2T}^{x+2t-2T} \cos(T)\,dT \\
&= \int_0^t \frac{1}{2}\left((x+2t-2T)^2 - (x-2t+2T)^2\right)\cos(T)\,dT \\
&= \int_0^t 4x(t-T)\cos(T)\,dT \\
&= 4x(1-\cos(t)).
\end{aligned}$$

The solution is

$$\begin{aligned}
u(x,t) &= \frac{1}{2}\left(e^{-(x-2t)} + e^{-(x+2t)}\right) + \frac{1}{4}\int_{x-2t}^{x+2t} \sin(s)\,ds + x(1-\cos(t)) \\
&= e^{-x}\cosh(t) - \frac{1}{4}(\cos(x+2t) - \cos(x-2t)) + x(1-\cos(t)).
\end{aligned}$$

With some manipulation, we can also write

$$u(x,t) = e^{-x}\cosh(t) + \frac{1}{2}\sin(x)\sin(2t) + x(1-\cos(x)).$$

Problems for Section 3.2.4

In each of problems 1–10, solve problem 3.19 for the given information.

1. $\varphi(x) = x, \psi(x) = e^{-x}, P(x,t) = x + t, c = 4.$
2. $\varphi(x) = \sin(x), \psi(x) = 2x, P(x,t) = 2xt, c = 2.$
3. $\varphi(x) = x^2 - x, \psi(x) = \cos(2x), P(x,t) = t\cos(x), c = 8.$
4. $\varphi(x) = x^2, \psi(x) = xe^{-x}, P(x,t) = x\sin(t), c = 4.$
5. $\varphi(x) = \cosh(x), \psi(x) = 1, P(x,t) = 3xt^2, c = 1.$
6. $\varphi(x) = 1 + x, \psi(x) = \sin(x), P(x,t) = x - \cos(t), c = 7.$
7. $\varphi(x) = \cos(2x), \psi(x) = 1 - \cos(x), P(x,t) = t^2, c = 2.$
8. $\varphi(x) = x^3, \psi(x) = \sin^2(x), P(x,t) = e^{-x}\cos(t), c = 1.$
9. $\varphi(x) = x\sin(x), \psi(x) = e^{-x}, P(x,t) = xt, c = 2.$
10. $\varphi(x) = 1 - x^2, \psi(x) = x\sin(x), P(x,t) = t\sin(x), c = 6.$

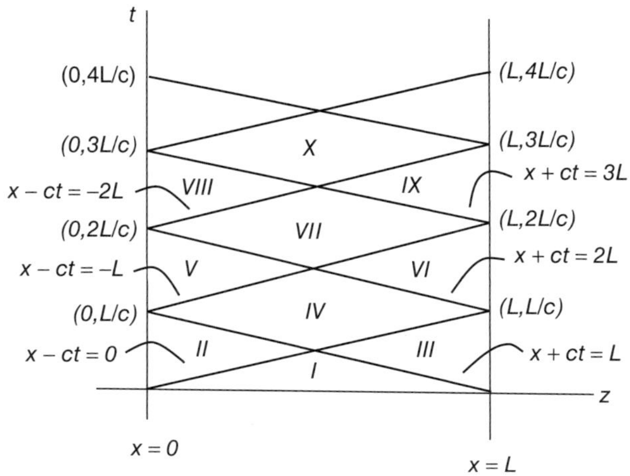

Figure 3.40: Characteristics partitioning the strip $0 \leq x \leq L, t \geq 0$.

3.2.5 String with Moving Ends

As an example of the power of characteristic quadrilaterals, we will solve a problem of a string of finite length, with a given initial position and velocity, but with ends moved by given control functions. The initial-boundary value model is

$$\begin{aligned} u_{tt} &= c^2 u_{xx} \text{ for } 0 < x < L, t > 0, \\ u(x,0) &= \varphi(x), u_t(x,0) = \psi(x) \text{ for } 0 < x < L, \\ u(0,t) &= a(t), u(L,t) = b(t) \text{ for } t > 0. \end{aligned} \tag{3.22}$$

We can separate the variables in the wave equation, leading to

$$X'' + \lambda X = 0, T'' + \lambda c^2 T = 0.$$

However, look at the boundary conditions. These require that

$$X(0)T(t) = a(t), \text{ or } T(t) = \frac{a(t)}{X(0)}$$

and

$$X(L)T(t) = b(t), \text{ or } T(t) = \frac{a(t)}{X(0)}.$$

Then $T(t)$ would be a constant multiple of $a(t)$ and also of $b(t)$. This would make $a(t)$ and $b(t)$ constant multiples of each other, and there is no rationale for this.

We can use characteristics quadrilaterals to solve this problem. Figure 3.40 shows the $x, t-$plane, and the vertical lines $x = 0$ and $x = L$. Think of these

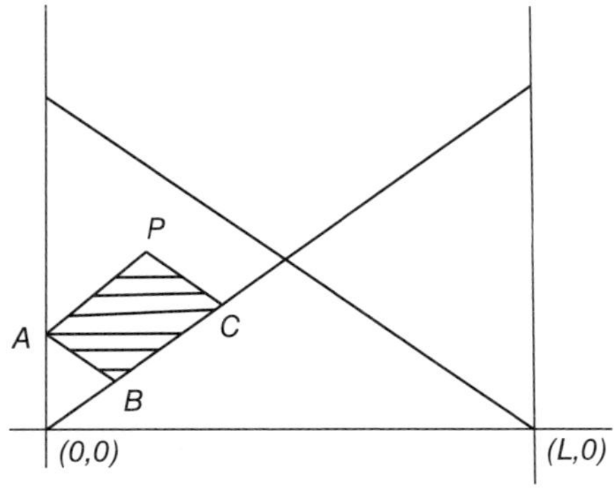

Figure 3.41: Characteristic quadrilateral in region II.

lines as vertical walls or barriers. The characteristic $x - ct = 0$, having positive slope $1/c$ and emanating from the origin, can be thought of as a beam of light that strikes the wall $x = L$ at $(L, L/c)$. From here it reflects upward to the left along the characteristic $x + ct = 2L$, and hits the wall $x = 0$ at $(0, 2L/c)$. This beam reflects upward to the right along the characteristic $x - ct = -2L$, which strikes the wall $x = L$ at $(L, 3L/c)$, then reflects back upward to the left, and so on.

Similarly, the characteristic emanating from $(L, 0)$, and aimed upward to the left, has equation $x + ct = L$. It hits the wall $x = 0$ at $(0, L/c)$, reflects upward to the right along the characteristic $x - ct = -L$ to hit the wall $x = L$ at $(L, 2L/c)$, reflects upward to the left along the characteristic $x + ct = 3L$ to hit the wall $x = 0$ at $(0, 3L/c)$, and so on.

These characteristics and the base and vertical walls subdivide the strip $S : 0 \leq x \leq L, t \geq 0$ into regions, labeled $I, II, III, IV, V, VI, \cdots$. These regions include their boundary line segments. We will use characteristic quadrilaterals to write a solution for $u(x, t)$ in each region, starting at I and working side to side and up through S. In doing this, keep in mind that we know the value of the solution on the vertical and bottom sides of S, because

$$u(0, t) = a(t), u(L, t) = b(t), \text{ and } u(x, 0) = \varphi(x).$$

First look at region I. A person standing at $P : (x, t)$ in this region does not feel any effects of the boundary conditions, and the solution here is given by d'Alembert's formula:

$$\begin{aligned} u(x, t) &= \frac{1}{2}(\varphi(x - ct) + \varphi(x + ct)) \\ &\quad + \frac{1}{2c} \int_{x-ct}^{x+ct} \psi(s)\, ds \text{ for } (x, t) \text{ in region I.} \end{aligned}$$

Next suppose P is in region II, as in Figure 3.41. Using the characteristics through P, construct a characteristic quadrilateral at P having two vertices

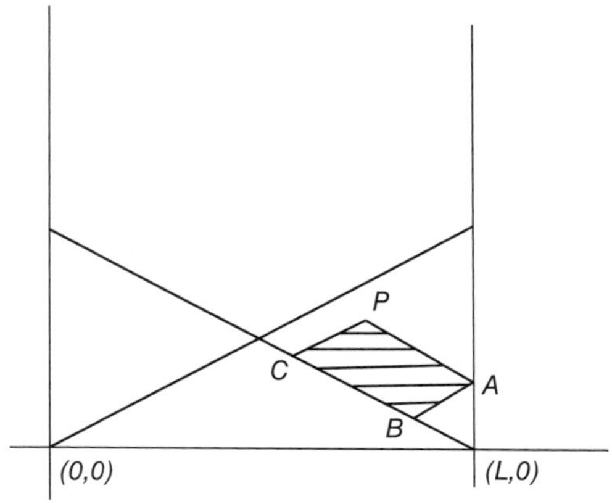

Figure 3.42: Characteristic quadrilateral in region III.

B and C on a boundary characteristic of region I, and one vertex A on the left boundary of S. $u(B)$ and $u(C)$ are known because we know the solution at points of region I. And $u(A)$ is known from the boundary condition $u(0, t) = a(t)$. From equation 3.18, we know $u(x, t)$:

$$u(x, t) = u(A) + u(C) - u(B).$$

This strategy produces $u(x, t)$ for all (x, t) in region II, including its boundary.

Next turn to region III. If P is in this region, construct the characteristic quadrilateral of Figure 3.42, using pieces of characteristics as sides, with two vertices, B and C on a boundary line of region I, and one vertex A on the right boundary line of the strip. We know $u(B)$ and $u(C)$ from the solution in region I, and, from the boundary condition $u(L, t) = b(t)$, we know $u(A)$ as well. Therefore, we know $u(x, t)$.

From here, work upwards in the strip S. Suppose P is in region IV (Figure 3.43). Construct a characteristic quadrilateral with one vertex at P, a vertex A on part of the boundary of region II, a vertex B on part of the boundary of region III, and a vertex C on part of the boundary of region I. We know $u(A)$, $u(B)$ and $u(C)$, hence also $u(x, t)$.

From here, go on to region V, then VI, VII, and so on.

The method involves a good deal of elementary computation (mostly algebra), because at each stage we must find equations of characteristics through specific points, then the coordinates of points of intersection of pairs of characteristics, then the values of $u(x, t)$ at these points.

Example 3.11 Consider the problem

$$u_{tt} = 9u_{xx} \text{ for } 0 < x < 4, t > 0,$$
$$u(x, 0) = \varphi(x) = x\sin(\pi x), u_t(x, 0) = \psi(x) = x(x-4) \text{ for } 0 < x < 4,$$
$$u(0, t) = u(4, t) = 9\pi t^2 \text{ for } t > 0.$$

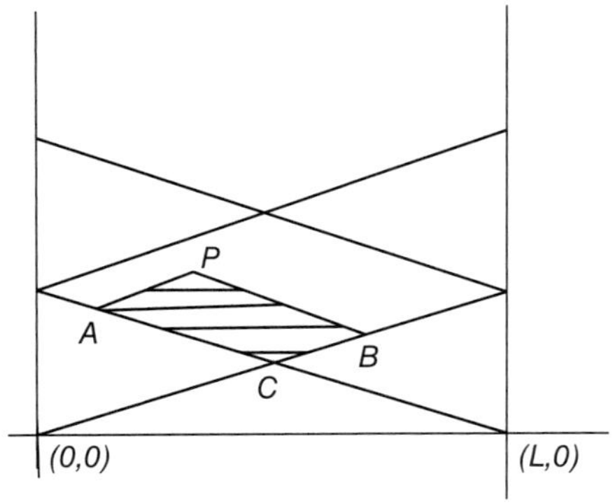

Figure 3.43: Characteristic quadrilateral in region IV.

We will determine $u(x,t)$ at arbitrary points in regions I, II and III. In carrying out the calculations of the vertices of the needed characteristic quadrilaterals, use the notaxtional device introduced in the preceding section (Figure 3.39). Let (x,t) be the point at which we are computing the solution and (X,T), a variable point in computing lines and their intersections.

With this convention, the region of interest for this example is $0 \le X \le 4$, $T \ge 0$. We also have $c = 3$.

In region I, we can use the d'Alembert solution:

$$\begin{aligned} &u(x,t)\\ &= \frac{1}{2}\left((x-3t)\sin(\pi(x-3t)) + (x+3t)\sin(\pi(x+3t))\right) + \frac{1}{6}\int_{x-3t}^{x+3t} s(s-4)\,ds \\ &= \frac{1}{2}\left((x-3t)\sin(\pi(x-3t)) + (x+3t)\sin(\pi(x+3t))\right) + x^2t + 3t^3 - 4xt. \end{aligned}$$

Next let (x,t) be in region II. Figure 3.44 shows a characteristic quadrilateral constructed with one vertex on the line $X = 0$ and one side on part of the boundary of region I. The characteristics and vertices are labeled in the diagram. Then

$$u(x,t) = u\left(0, t - \frac{x}{3}\right) + u\left(\frac{3t+x}{2}, \frac{3t+x}{6}\right) - u\left(\frac{3t-x}{2}, \frac{3t-x}{6}\right).$$

We can compute $u(0, t - x/3)$ from the boundary condition at $X = 0$, and

$$u\left(\frac{3t+2}{2}, \frac{3t+x}{6}\right) \text{ and } u\left(\frac{3t-x}{2}, \frac{3t-x}{6}\right)$$

from the solution in region I, which we know. This is a straightforward but

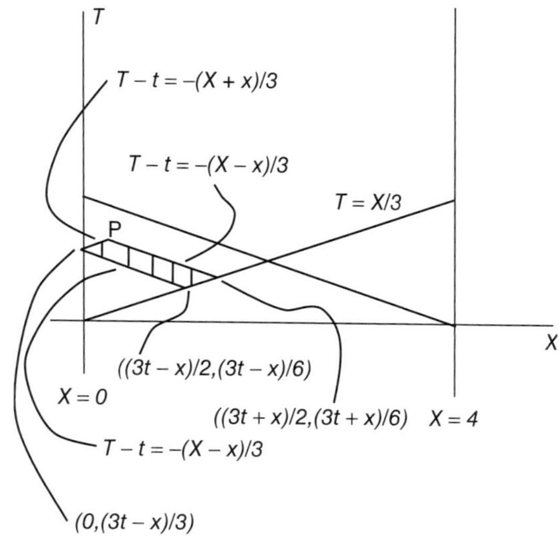

Figure 3.44: Characteristic quadrilateral in region I in example 3.11.

lengthy calculation. We obtain, for region II,

$$u(x,t) = 9\pi\left(\frac{x-3t}{3}\right)^2 + \frac{1}{2}u(x+3t)\sin(\pi(x+3t))$$
$$-\frac{1}{2}(x-3t)\sin(\pi(x-3t)) + 3xt^2 + \frac{1}{9}x^3 - 4xt.$$

Now move on to region III. From the characteristic quadrilateral labeled in Figure 3.45,

$$u(x,t) = u\left(\frac{4-3t+x}{2}, \frac{4+3t-x}{6}\right)$$
$$+ u\left(4, \frac{-4+3t+x}{3}\right) - u\left(\frac{12-x-3t}{2}, \frac{x+3t-4}{6}\right).$$

We know $u(4,(-4+3t+x)/3)$ from the boundary condition at $X = 4$, and we know the other two values of u from the solution in region I. Again, a computation yields

$$u(x,t) = \frac{1}{2}(x-3t)\sin(\pi(x-3t)) - \frac{1}{2}(8-x-3t)\sin(\pi(8-x-3t))$$
$$+ 9\pi\left(\frac{-4+x+3t}{3}\right)^2 + \frac{64}{9} + 12t^2$$
$$+ 4xt + \frac{4}{3}x^2 - 3xt^2 - \frac{1}{9}x^3 - 16t - \frac{16}{3}x$$

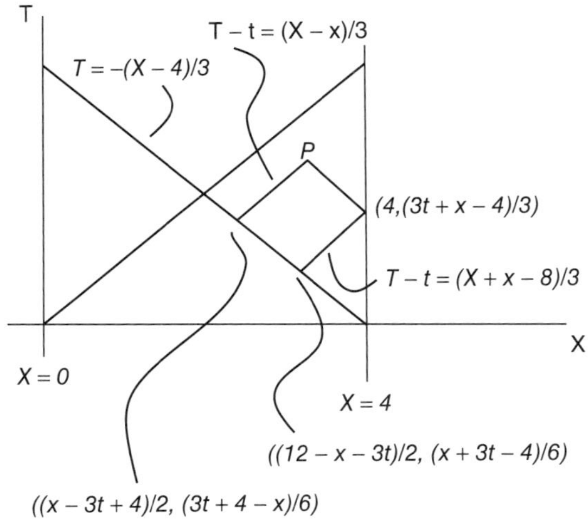

Figure 3.45: Characteristic quadrilateral in region III in example 3.11.

for (x, t) in region III.

In developing the details of this method of solution, we have glossed over an issue of compatibility of the initial and boundary conditions at $(0, 0)$ and at $(L, 0)$. To illustrate the idea, consider $u(0, 0)$. On one hand, this is $u(x, 0)$ at $x = 0$, so we would expect that $u(0, 0) = \varphi(0)$. But $u(0, 0)$ is also $u(0, t)$ at $t = 0$, we we also expect $u(0, 0) = a(0)$. This requires that $\varphi(0) = a(0)$. Extending this reasoning to the other end of the bar and including the initial velocity condition, the *compatibility conditions* are

$$a(0) = \varphi(0), a'(0) = \psi(0), a''(0) = c^2\varphi''(0)$$

and, at the right end of the medium,

$$b(0) = \varphi(L), b'(0) = \psi(L), \text{ and } b''(0) = c^2\varphi''(L).$$

Problems for Section 3.2.5

In each of problems 1–5, obtain expressions for $u(x, t)$ in regions I, II and III, for problem 3.22.

1. $c = 1, L = 2, \varphi(x) = x(2 - x), \psi(x) = 0, a(t) = b(t) = -t^2$.

2. $c = 2, L = 3, \varphi(x) = \sin(\pi x), \psi(x) = 0, a(t) = b(t) = t^3$.

3. $c = 3, L = 2, \varphi(x) = x^2(2 - x), \psi(x) = 0, a(t) = 18t^2, b(t) = -36t^2$.

4. $c = 4, L = 1, \varphi(x) = x\sin(\pi x), \psi(x) = 4x, a(t) = 16\pi t^2, b(t) = -16\pi t^2 + 4t$.

5. $c = 1, L = 2, \varphi(x) = x(2-x)^2, \psi(x) = x^2, a(t) = -4t^2, b(t) = 2t^2 + 4t$.

6. Obtain expressions for $u(x,t)$ with (x,t) in regions I, II and III and IV, for problem 3.22 if

$$c = 2, L = 1, \varphi(x) = x\cos(\pi x) + x, \psi(x) = x^2, a(t) = 0, b(t) = 2\pi t^2 + t.$$

7. Derive the compatibility conditions given at the end of this section.

3.3 The Wave Equation in Higher Dimensions

This section is an introduction to wave problems in higher dimensions. We will discuss two types of results. First, we will use double Fourier series to solve a problem for a bounded medium in two dimensions. This is similar to the solution of a two-dimension diffusion problem in Chapter 2. Following this, we will develop the Kirchhoff-Poisson integral formula for a solution in 3−space.

3.3.1 Vibrations in a Membrane with Fixed Frame

Imagine a flat membrane or plate being acted on in some way, with boundary fixed on a frame (as, for example, with a drum). A point of the membrane (other than on the frame) is free to oscillate in a vertical plane, but does not move horizontally.

To be specific, suppose the frame is a rectangle. At time zero, the membrane is lifted to some initial position and released from rest. The initial-boundary value problem for the wave function describing the motion is

$$\begin{aligned} u_{tt} &= c^2(u_{xx} + u_{yy}) \text{ for } 0 < x < L, 0 < y < K, t > 0, \\ u(x,0,t) &= u(x,K,t) = 0 \text{ for } 0 < x < L, t > 0, \\ u(0,y,t) &= u(L,y,t) = 0 \text{ for } 0 < y < K, t > 0, \\ u(x,y,0) &= \varphi(x,y), u_t(x,y,0) = 0 \text{ for } 0 < x < L, 0 < y < K. \end{aligned} \tag{3.23}$$

The variables can be separated. Let $u(x,y,t) = X(x)Y(y)T(t)$ to obtain

$$XYT'' = c^2(X''YT + XY''T).$$

Then

$$\frac{T''}{c^2T} - \frac{Y''}{Y} = \frac{X''}{X}.$$

Because x, y, and t are independent, both sides of this equation must equal some constant λ, so

$$\frac{T''}{c^2T} - \frac{Y''}{Y} = \frac{X''}{X} = -\lambda.$$

Then

$$X'' + \lambda X = 0 \text{ and } \frac{T''}{c^2T} + \lambda = \frac{Y''}{Y}.$$

In the second equation, y and t are independent, so for some constant μ,

$$\frac{T''}{c^2T} + \lambda = \frac{Y''}{Y} = -\mu.$$

Then

$$Y'' + \mu Y = 0$$

and

$$T'' + (\lambda + \mu)c^2T = 0.$$

From the boundary conditions, $X(0) = X(L) = Y(0) = Y(K) = 0$. Now the problems for X and Y are

$$X'' + \lambda X = 0; X(0) = X(L) = 0$$

and

$$Y'' + \mu Y = 0; Y(0) = Y(K) = 0,$$

with eigenvalues and eigenfunctions

$$\lambda_n = \frac{n^2\pi^2}{L^2}, X_n(x) = \sin\left(\frac{n\pi x}{L}\right) \text{ for } n = 1, 2, 3, \cdots$$

and

$$\mu_m = \frac{m^2\pi^2}{L^2}, Y_m(y) = \sin\left(\frac{m\pi y}{K}\right) \text{ for } m = 1, 2, 3, \cdots.$$

Here n and m independently vary over the positive integers.

For T, observe that, because of the zero initial velocity,

$$X(x)Y(y)T'(0) = 0$$

so $T'(0) = 0$. The problem for T is

$$T'' + \left(\frac{n^2\pi^2}{L^2} + \frac{m^2\pi^2}{K^2}\right)c^2T = 0; T'(0) = 0.$$

This has solutions that are constant multiples of

$$T_{nm}(t) = \cos(\alpha_{nm}ct),$$

where

$$\alpha_{nm} = \sqrt{\frac{n^2\pi^2}{L^2} + \frac{m^2\pi^2}{K^2}}.$$

For each positive integer n and m, we now have a function

$$u_{nm}(x, y, t) = b_{nm}\sin\left(\frac{n\pi x}{L}\right)\sin\left(\frac{m\pi y}{K}\right)\cos(\alpha_{nm}ct),$$

which satisfies the two-dimensional wave equation and the boundary conditions, as well as the condition that the initial velocity is zero.

A superposition of these functions is a double sum

$$u(x,y,t)=\sum_{n=1}^{\infty}\sum_{m=1}^{\infty} b_{nm}\sin\left(\frac{n\pi x}{L}\right)\sin\left(\frac{m\pi y}{K}\right)\cos(\alpha_{nm}ct).$$

To satisfy the initial position condition, choose the coefficients b_{nm} so that

$$u(x,y,0)=\sum_{n=1}^{\infty}\sum_{m=1}^{\infty} b_{nm}\sin\left(\frac{n\pi x}{L}\right)\sin\left(\frac{m\pi y}{K}\right)=\varphi(x,y).$$

This is a double Fourier sine series expansion of $\varphi(x,y)$ on the rectangle $0\le x\le L, 0\le y\le K$. We encountered such a problem in connection with temperature distributions in flat plates, and found that we should choose

$$b_{nm}=\frac{4}{LK}\int_0^K\int_0^L \varphi(\xi,\zeta)\sin\left(\frac{n\pi\xi}{L}\right)\sin\left(\frac{m\pi\zeta}{K}\right)d\xi\,d\zeta.$$

With these coefficients, $u(x,y,t)$ is the solution.

Example 3.12 We will solve the problem for the case that $L=K=\pi$ and $\varphi(x,y)=xy(\pi-x)(\pi-y)$. All that is required is to compute the coefficients:

$$\begin{aligned}b_{nm}&=\frac{4}{\pi^2}\int_0^{\pi}\int_0^{\pi}\xi(\pi-\xi)\sin(n\xi)\zeta(\pi-\zeta)\sin(m\zeta)\,d\xi\,d\zeta\\&=\frac{16}{n^3m^3\pi^2}((-1)^n-1)((-1)^m-1).\end{aligned}$$

The solution is

$$u(x,y,t)=$$
$$\frac{16}{\pi^2}\sum_{n=1}^{\infty}\sum_{m=1}^{\infty}\frac{((-1)^n-1)((-1)^m-1)}{n^3m^3}\sin(nx)\sin(my)\cos(\alpha_{nm}ct),$$

where

$$\alpha_{nm}=\sqrt{n^2+m^2}.$$

Problems for Section 3.3.1

1. Solve problem 3.23 if $c=3, L=3, K=6$ and $\varphi(x,y)=\sin(\pi x/3)y(6-y)$.
2. Solve problem 3.23 if $c=2, L=\pi, K=2\pi$ and $\varphi(x,y)=x^2(\pi-x)y^2(2\pi-y)$.

3. Outline a procedure for solving the two-dimensional wave equation with a fixed rectangular frame, if the membrane is not initially displaced, but an initial velocity is imposed on the membrane by requiring that $u_t(x,y,0) = \psi(x,y)$.

4. Solve the initial-value problem of problem 3 if $c = 4, L = K = 1, \psi(x,y) = \sin(\pi x)\sin(\pi y)$.

5. Solve the problem if there is both an initial displacement and an initial velocity, if $c = 1, L = K = 1$, $\varphi(x,y) = x(x-1)^2 y(y-1)$ and $\psi(x,y) = xy^2$.

6. Solve the problem if there is both an initial displacement and an initial velocity if $c = 1, L = 1, K = \pi$, $\varphi(x,y) = y\cos(\pi x/2)$ and $\psi(x,y) = x+y$.

3.3.2 The Poisson Integral Solution

We will solve the Cauchy problem for the wave equation in three dimensions:

$$\begin{aligned} u_{tt} &= u_{xx} + u_{yy} + u_{zz} \\ u(x,y,z,0) &= \varphi(x,y,z), u_t(x,y,z,0) = \psi(x,y,z) \end{aligned} \tag{3.24}$$

in which x, y, and z can independently assume all real values. We have taken $c = 1$. Assuming familiarity with vector calculus and the divergence theorem of Gauss, we will derive an integral formula for the solution. This integral formula has an important consequence for waves in 3−space.

We can solve problem 3.24 by adding the solution of the problem with zero initial velocity to the solution of the problem with no initial displacement. We will use this observation, together with another which is perhaps more subtle, to solve the problem. Suppose we look at problem 3.24 with zero initial displacement and initial velocity $\eta(x,y,z)$. Call this problem 3.25.

$$\begin{aligned} u_{tt} &= u_{xx} + u_{yy} + u_{zz}, \\ u(x,y,z,0) &= 0, u_t(x,y,z,0) = \eta(x,y,z). \end{aligned} \tag{3.25}$$

We claim that, if v is a solution of problem 3.25, then v_t is a solution of problem 3.24 with initial velocity $\eta(x,y,z)$ and zero initial velocity.

Theorem 3.2 Let $v(x,y,z)$ be the solution of problem 3.25. Then $v_t(x,y,z)$ is the solution of problem 3.24 with initial position $\eta(x,y,z)$, and zero initial velocity.

Proof Let $u = v_t$. We must first show that u satisfies the wave equation. Compute

$$\begin{aligned} u_{tt} &= (v_t)_{tt} = (v_{tt})_t = (v_{xx} + v_{yy} + v_{zz})_t \\ &= (v_t)_{xx} + (v_t)_{yy} + (v_t)_{zz} = u_{xx} + u_{yy} + u_{zz}. \end{aligned}$$

Now look at the initial conditions. Immediately

$$u(x,y,z,0) = v_t(x,y,z,0) = \eta(x,y,z).$$

Further,

$$\begin{aligned} u_t(x,y,z,0) &= v_{tt}(x,y,z,0) \\ &= \left[v_{xx} + v_{yy} + v_{zz}\right]_{t=0} = 0 \end{aligned}$$

because, in problem 3.25, $v(x,y,z,0) = 0$ for all (x,y,z). This proves the theorem.

Using Theorem 3.2, we can solve problem 3.24 if we can solve problem 3.25. For, if u_ψ is the solution of 3.25 with initial velocity $\psi(x,y,z)$, and u_φ is the solution of 3.25 with initial velocity φ, then

$$u(x,y,z,t) = \frac{\partial u_\varphi}{\partial t} + u_\psi. \tag{3.26}$$

What makes this result useful is that there is an integral formula for the solution of problem 3.25.

Theorem 3.3 (Kirchhoff's Integral Solution of Problem 3.25) Let ψ be continuous with continuous first and second partial derivatives for all (x,y,z). Then, for all real x,y,z, and $t > 0$, the solution of problem 3.25 is

$$u(x,y,z,t) = \frac{1}{4\pi t}\iint_{S(x,y,z;t)} \psi(X,Y,Z)\, d\sigma_t, \tag{3.27}$$

in which the integral is a surface integral over the sphere $S(x,y,z;t)$ of radius t about (x,y,z), and (X,Y,Z) is the variable of integration on $S(x,y,z;t)$.

$S(x,y,z;t)$ consists of all points (X,Y,Z) with

$$(X-x)^2 + (Y-y)^2 + (Z-z)^2 = t^2.$$

The integral in equation 3.27 is known as *Kirchhoff's integral* for the wave equation.

Proof We will first show that u satisfies the wave equation. It is convenient to use boldface for vectors, with $\mathbf{x} = (x,y,z)$ denoting an arbitrary point of 3−space, and $\mathbf{X} = (X,Y,Z)$ an integration variable. In this notation, $u(x,y,z,t) = u(\mathbf{x},t)$.

Denote $S(x,y,z;t)$ as S_t. Let U be the sphere of radius 1 about the origin, and let $d\sigma$ be the differential element of surface area on U, while $d\sigma_t$ is the differential element of surface area of S_t. Then

$$d\sigma_t = t^2 d\sigma.$$

Let $\mathbf{n}_t$ be the unit outer normal vector on S_t, and $\mathbf{n}$ the unit outer normal on U.

With this notation, equation 3.27 becomes

$$u(\mathbf{x},t) = \frac{1}{4\pi t}\iint_{S_t} \psi(\mathbf{X})\, d\sigma_t. \tag{3.28}$$

Transform this surface integral over S_t to one over U, in which the variable is $\tilde{\mathbf{x}}$, by putting

$$\mathbf{X} = \mathbf{x} + t\tilde{\mathbf{x}}.$$

We obtain

$$u(\mathbf{x},t) = \frac{t}{4\pi}\iint_{U} \psi(\mathbf{x}+t\tilde{\mathbf{x}})\, d\sigma. \tag{3.29}$$

Now compute

$$\begin{aligned} u_{xx}+u_{yy}+u_{zz} &= \frac{t}{4\pi}\iint [\psi_{xx}+\psi_{yy}+\psi_{zz}]_{\mathbf{x}+t\tilde{\mathbf{x}}}\, d\sigma \\ &= \frac{1}{4\pi t}\iint_{S_t} (\psi_{xx}(\mathbf{X})+\psi_{yy}(\mathbf{X})+\psi_{zz}(\mathbf{X}))\, d\sigma_t, \end{aligned} \tag{3.30}$$

and

$$\begin{aligned} u_t(\mathbf{x},t) &= \frac{1}{4\pi}\iint_U \psi(\mathbf{x}+t\tilde{\mathbf{x}})\, d\sigma + \frac{t}{4\pi}\iint_U \nabla\psi(\mathbf{x}+t\tilde{\mathbf{x}})\cdot\mathbf{n}\, d\sigma \\ &= \frac{1}{t}u(\mathbf{x},t) + \frac{1}{4\pi t}\nabla\psi(\mathbf{X})\cdot\mathbf{n}_t\, d\sigma_t, \end{aligned} \tag{3.31}$$

in which ∇ is the gradient operator. Apply Gauss's divergence theorem to the last integral to write it as

$$\iint_{S_t} \nabla\psi(\mathbf{X})\cdot\mathbf{n}_t\, d\sigma_t = \iiint_{B_t} \operatorname{div}(\nabla\psi)\, dV,$$

where B_t is the solid ball of radius t about (x,y,z). By a straightforward calculation,

$$\operatorname{div}\,(\nabla\psi) = \psi_{xx}+\psi_{yy}+\psi_{zz}.$$

Now equation 3.31 becomes

$$u_t = \frac{1}{t}u(\mathbf{x},t) + \frac{1}{4\pi t}\iiint_{B_t} (\psi_{xx}+\psi_{yy}+\psi_{zz})\, dV.$$

Denoting this triple integral as I, the last equation is

$$u_t(\mathbf{x},t) = \frac{1}{t}u(\mathbf{x},t) + \frac{1}{4\pi t}I. \tag{3.32}$$

Differentiate this equation with respect to t to obtain

$$\begin{aligned} u_{tt}(\mathbf{x},t) &= -\frac{1}{t^2}u(\mathbf{x},t) + \frac{1}{t}u_t(\mathbf{x},t) + \frac{1}{4\pi t}I_t - \frac{1}{4\pi t^2}I \\ &= \frac{1}{t}\left(-\frac{1}{t}u(\mathbf{x},t) + u_t(\mathbf{x},t) - \frac{1}{4\pi t}I\right) + \frac{1}{4\pi t}I_t \\ &= \frac{1}{4\pi t}I_t., \end{aligned} \tag{3.33}$$

with the term in large parentheses vanishing by equation 3.32. Now,

$$I_t = \iint_{S_t} (\psi_{xx}(\mathbf{X}) + \psi_{yy}(\mathbf{X}) + \psi_{zz}(\mathbf{X}))\, d\sigma_t. \tag{3.34}$$

By equations 3.33, 3.34 and 3.30,

$$\begin{aligned} u_{tt} &= \frac{1}{4\pi t}\iint_{S_t} (\psi_{xx}(\mathbf{X}) + \psi_{yy}(\mathbf{X}) + \psi_{zz}(\mathbf{X}))\, d\sigma_t \\ &= u_{xx} + u_{yy} + u_{zz}. \end{aligned}$$

Therefore the function defined by equation 3.27 satisfies the wave equation.

There remains to show that $u(\mathbf{x},t)$ satisfies the initial conditions. First,

$$u(\mathbf{x},0) = 0$$

by putting $t = 0$ into equation 3.29. Finally, from the first line of equation 3.31,

$$\begin{aligned} u_t(\mathbf{x},t) &= \frac{1}{4\pi}\iint_U \psi(\mathbf{x})\, d\sigma \\ &= \psi(\mathbf{x})\frac{1}{4\pi}\iint_U d\sigma = \psi(\mathbf{x}), \end{aligned}$$

because

$$\iint_U d\sigma = \text{ area of } U = 4\pi.$$

This completes the proof of the theorem.

Corollary 3.1 (Poisson's Formula) The solution of problem 3.24 is

$$u(\mathbf{x},t) = \frac{1}{4\pi}\frac{\partial}{\partial t}\left[\frac{1}{t}\iint_{S_t} \varphi(\mathbf{X})\, d\sigma_t\right] + \frac{1}{4\pi t}\iint_{S_t} \psi(\mathbf{X})\, d\sigma_t. \tag{3.35}$$

This conclusion, which is known as *Poisson's formula*, follows immediately from equation 3.26 and Theorem 3.3.

Poisson's integral formula suggests a principle named for the Dutch natural philosopher Christian Huygens (1629–1695), who was a contemporary of Newton and Leibniz. The integrals in the solution for $u(x, y, z, t)$ are surface integrals over the sphere of radius t about (x, y, z), and these depend on the values of the initial position and velocity functions only at points on this surface. For this reason S_t is called the *domain of influence* of the solution.

Suppose an initial disturbance vanishes outside a solid ball B_R of radius R about the origin, and imagine standing at (x, y, z) outside this ball. Let ρ be the distance from the origin to (x, y, z). S_t intersects B_R exactly when

$$\rho - R \leq t \leq \rho + R.$$

This means that the disturbance is felt first at time $\rho - R$, but after time $\rho + R$ the effect has passed, because the initial disturbance vanishes outside B_R and the domain of influence no longer intersects any nonzero data. This is called *Huygens's principle*, and it applies to solutions of the wave equation in 3 space dimensions.

If we think of the disturbance as a sound wave (say a whistle blown by a person at the origin), this means that we first hear the sound at time $\rho - R$, continuing to experience it until time $\rho + R$, after which time the sound vanishes.

Problems for Section 3.3.2

1. Suppose in problem 3.24 that the initial position function is identically zero, while the initial velocity is constant. Show that the solution is independent of $\mathbf{x}$. Does this make sense from a physical point of view, thinking of the solution as a sound wave?

3.3.3 Hadamard's Method of Descent

French mathematician Jacques Hadamard (1865–1963) developed a method that can be used to solve the Cauchy problem for the wave equation in two dimensions, using the Poisson integral solution for three dimensions. Because the method descends from three to two dimensions, it is called the *method of descent*.

Letting $c = 1$, the problem in two dimensions is

$$u_{tt} = u_{xx} + u_{yy},$$
$$u(x, y, 0) = \varphi(x, y), u_t(x, y, 0) = \psi(x, y).$$

The idea is to think of the $x, y-$plane as consisting of all points $(x, y, 0)$ in $3-$space, The Cauchy problem in the plane can be thought of as a Cauchy problem in $3-$space, in which the initial position and velocity functions are independent of z. This problem in R^3 has the solution

$$u(x, y, t) = \frac{1}{4\pi}\frac{\partial}{\partial t}\left[\frac{1}{t}\iint_{S_t} \varphi(\xi, \eta)\, d\sigma_t\right] + \frac{1}{4\pi t}\iint_{S_t} \psi(\xi, \eta)\, d\sigma_t.$$

Because φ and ψ are functions of just x and y, these surface integrals can be projected to double integrals over a region of the plane. Let $\mathbf{n}_t$ denote the unit outer normal to S_t and consider a typical surface element on S_t having area $d\sigma$. This surface element projects onto a $d\xi$ by $d\eta$ rectangle centered at $(x, y, 0)$ in the $(\xi, \eta, 0)$ plane. The area of this rectangle is related to the area of the surface element by

$$d\xi\, d\eta = \mathbf{n}_t \cdot \mathbf{k} = \frac{1}{t}\sqrt{t^2 - (\xi - x)^2 - (\eta - y)^2}\, d\sigma.$$

Here we have used the fact that the sphere of radius t about $(x, y, 0)$ has equation

$$(\xi - x)^2 + (\eta - y)^2 + \zeta^2 = t^2.$$

The fact that the sphere has both an upper and a lower hemisphere introduces a factor of 2 and we have

$$\iint_{S_t} \varphi(\xi, \eta)\, d\sigma_t = 2t \iint_{D_t} \frac{\varphi(\xi, \eta)}{\sqrt{t^2 - (\xi - x)^2 - (\eta - y)^2}}\, d\xi\, d\eta,$$

in which D_t is the disk of radius t about (x, y) in the plane, and consists of all (ξ, η) with

$$(\xi - x)^2 + (\eta - y)^2 \le t^2.$$

A similar reduction holds for the integral of ψ over S_t. This yields an integral solution for the problem in the plane:

$$\begin{aligned} u(x, y, t) = {} & \frac{1}{2\pi}\frac{\partial}{\partial t}\left[\iint_{D_t} \frac{\varphi(\xi, \eta)}{\sqrt{t^2 - (\xi - x)^2 - (\eta - y)^2}}\right] d\xi\, d\eta \\ & + \frac{1}{2\pi}\iint_{D_t} \frac{\psi(\xi, \eta)}{\sqrt{t^2 - (\xi - x)^2 - (\eta - y)^2}}\, d\xi\, d\eta. \end{aligned}$$

There is a significant difference between the solutions in the plane and in 3−space. The domain of influence of the initial data at a point in 3− space at time t is the surface of a sphere of radius t about the point, and Huygens's principle holds. In the plane, however, the domain of influence of (ξ, η) at time t is the *solid disk* of radius t about the point, not just its circular boundary. A disturbance, say in a small disk D about the origin, will eventually be felt at (x, y) and then be felt at all later times, because once D_t intersects D at some time, it will do so at all later times. Huygens's principle does not hold in the plane.

If Hadamard's method is applied to descend from the plane to the line, one can obtain d'Alembert's solution. This is problem 1.

Problems for Section 3.3.3

1. Use the method of descent to derive d'Alembert's formula from the integral obtained for the plane.

2. Using d'Alembert's solution, does Huygens's principle hold for the line?

Chapter 4

Dirichlet and Neumann Problems

4.1 Laplace's Equation and Harmonic Functions

In two dimensions, the heat equation is

$$u_t = k(u_{xx} + u_{yy}).$$

In the limit as $t \to \infty$, and in the absence of sources or sinks of energy, the effects of the initial conditions are no longer felt in the solution and we think of u_t as zero to obtain *Laplace's equation*

$$u_{xx} + u_{yy} = 0.$$

Laplace's equation is also known as the *potential equation* and also as the *steady state heat equation*.

In three space variables, Laplace's equation is

$$u_{xx} + u_{yy} + u_{zz} = 0.$$

In vector calculus, the del symbol ∇ denotes the gradient operator, where

$$\nabla^2 u(x, y) = u_{xx} + u_{yy}$$

and, in three dimensions,

$$\nabla^2 u(x, y, z) = u_{xx} + u_{yy} + u_{zz}.$$

Now Laplace's equation can be written

$$\nabla^2 u = 0.$$

∇^2 is read del squared. The del symbol ∇ is also called *nabla* after the Greek word *nevel*, for an ancient Hebrew harp, which has the shape of an inverted upper case Greek delta Δ.

A function satisfying Laplace's equation in two or more dimensions is called a *harmonic function*. Sums and constant multiples of harmonic functions are again harmonic. If u is harmonic at each point of some region D, then we say that u is *harmonic on* D.

In this chapter we explore properties of harmonic functions, methods of solution of boundary value problems involving Laplace's equation, and existence questions for these boundary value problems.

Harmonic functions occur in many contexts. In complex analysis, if $f(z) = u(x,y) + iv(x,y)$ is a differentiable function of a complex variable $z = x + iy$ defined over some region D of the complex plane, then both u and v are harmonic on D.

For example,

$$f(z) = z^4 = (x+iy)^4 = x^4 - 6x^2y^2 + y^4 + i(4x^3y - 4xy^3).$$

Both of the functions

$$u(x,y) = x^4 - 6x^2y^2 + y^4 \text{ and } v(x,y) = 4x^3y - 4xy^3$$

are harmonic on the entire plane.

4.1.1 Laplace's Equation in Polar Coordinates

Depending on the region of interest, we may want to use polar coordinates (r,θ) instead of rectangular coordinates (x,y). As preparation for this, we will derive Laplace's equation in polar coordinates. Suppose u is harmonic on a region D of the plane, and let

$$x = r\cos(\theta), y = r\sin(\theta)$$

to obtain a new function

$$U(r,\theta) = u(r\cos(\theta), r\sin(\theta))$$

of r and θ. By the chain rule,

$$\begin{aligned}
U_r &= u_x\cos(\theta) + u_y\sin(\theta),\\
U_{rr} &= u_{xx}\cos^2(\theta) + 2u_{xy}\cos(\theta)\sin(\theta) + u_{yy}\sin^2(\theta),\\
U_\theta &= -u_x r\sin(\theta) + u_y r\cos(\theta),\\
U_{\theta\theta} &= -u_x r\cos(\theta) - u_y r\sin(\theta)\\
&\quad - r\sin(\theta)[u_{xx}(-r\sin(\theta)) + u_{xy} r\cos(\theta)]\\
&\quad + r\cos(\theta)[u_{yx}(-r\sin(\theta)) + u_{yy} r\cos(\theta)]\\
&= -rU_r + r^2[u_{xx}\sin^2(\theta) - 2u_{xy}\cos(\theta)\sin(\theta) + u_{yy}\cos^2(\theta)].
\end{aligned}$$

Observe that

$$U_{rr} + \frac{1}{r^2}U_{\theta\theta} = u_{xx} + u_{yy} - \frac{1}{r}U_r,$$

so Laplace's equation in polar coordinates is

$$U_{rr} + \frac{1}{r}U_r + \frac{1}{r^2}U_{\theta\theta} = 0. \tag{4.1}$$

It is easy to produce examples of harmonic functions in polar coordinates. To begin with a simple case, look for functions $U(r)$ that are independent of θ. Then $U_\theta = 0$ and Laplace's equation has no $\theta-$ term.

$$U_{rr} + \frac{1}{r}U_r = 0,$$

or

$$rU_{rr} + U_r = 0.$$

Because U is a function of just r, write this as

$$rU'' + U' = (rU')' = 0.$$

Integrate to get

$$rU' = c$$

so

$$U' = \frac{c}{r}.$$

A second integration yields

$$U(r) = c\ln(r) + k$$

with c and k arbitrary constants. This function is harmonic for $r > 0$.

We can find harmonic functions $U(r, \theta)$ of both r and θ by separating variables in equation 4.1. Put $U(r, \theta) = R(r)\Theta(\theta)$ to obtain

$$R''\Theta + \frac{1}{r}R'\Theta + \frac{1}{r^2}\Theta'' = 0.$$

Write this as

$$\frac{r^2R'' + rR'}{R} = -\frac{\Theta''}{\Theta} = \lambda.$$

Then

$$r^2R'' + rR' - \lambda R = 0$$

and

$$\Theta'' + \lambda\Theta = 0.$$

Because (r, θ) and $(r, -\theta)$ are polar coordinates of the same point, we require that

$$\Theta(\pi) = \Theta(-\pi). \tag{4.2}$$

This is a *periodicity condition*. Now consider cases on λ.

Case 1. $\lambda = 0$. Then $\Theta(\theta) = a\theta + b$. To have

$$\Theta(\pi) = a\pi + b = \Theta(-\pi) = -a\pi + b,$$

we need $a = 0$, so this case yields only the constant solution for $\Theta(\theta)$.

Case 2. Suppose $\lambda < 0$, say $\lambda = -k^2$ with $k > 0$. Now

$$\Theta'' - k^2\Theta = 0,$$

with solutions

$$\Theta(\theta) = ae^{k\theta} + be^{-k\theta}.$$

This function can satisfy conditions 4.2 only if $a = b = 0$, so this case does not produce any nontrivial solutions.

Case 3. Suppose $\lambda > 0$, say $\lambda = k^2$ with $k > 0$. Then

$$\Theta'' + k^2\Theta = 0,$$

with solutions

$$\Theta(\theta) = a\cos(k\theta) + b\sin(k\theta).$$

This function will satisfy condition 4.2 exactly when k is an integer. Because $k > 0$, then $\lambda = k^2$ for $k = 1, 2, 3, \cdots$.

Now consider the equation for $R(r)$. This is an Euler equation, with solutions of the form $R(r) = r^\alpha$. Substituting this into the differential equation with $\lambda = k^2$, we must have

$$\alpha(\alpha - 1)r^2 r^{\alpha-2} + \alpha r r^{\alpha-1} - \lambda r^\alpha = 0.$$

Divide by r^α to obtain

$$\alpha(\alpha - 1) + \alpha - k^2 = 0.$$

Then $\alpha^2 = k^2$, so $\alpha = \pm k$ and solutions for R are

$$R(r) = cr^k + dr^{-k}.$$

This means that

$$U(r, \theta) = (cr^k + dr^{-k})(a\cos(k\theta) + b\sin(k\theta))$$

in which k can be any nonnegative integer (we repeat these solutions if k is a negative integer).

In summary, we have obtained the harmonic functions

$$1 \text{ and } \ln(r),$$

which are independent of θ, with $\ln(r)$ defined for $r > 0$, as well as θ−dependent harmonic functions

$$r^k\cos(k\theta), r^k\sin(k\theta) \text{ for } r \geq 0$$

and

$$r^{-k}\cos(k\theta), r^{-k}\sin(\theta) \text{ for } r > 0,$$

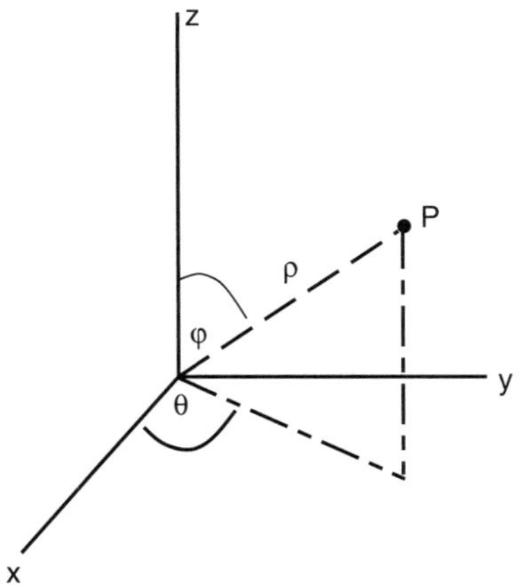

Figure 4.1: Spherical coordinates (ρ, θ, φ).

with k any nonnegative integer and θ any real number. These functions will be used to solve Dirichlet and Neumann problems for disks.

4.1.2 Laplace's Equation in 3 Dimensions

Spherical coordinates (ρ, θ, φ) are related to rectangular coordinates by

$$x = \rho\cos(\varphi)\cos(\theta), y = \rho\cos(\varphi)\sin(\theta), z = \rho\sin(\varphi).$$

Figure 4.1 shows the significance of the coordinates ρ, θ and φ. ρ is the distance from the origin to (x, y, z), θ is the polar angle, and φ is the azimuthal angle.

Laplace's equation in spherical coordinates is

$$u_{\rho\rho} + \frac{2}{\rho}u_\rho + \frac{1}{\rho^2\sin^2(\varphi)}u_{\theta\theta} + \frac{1}{\rho^2}u_{\varphi\varphi} + \frac{\cot(\varphi)}{\rho^2}u_\varphi = 0.$$

One solution that depends only on ρ is

$$U(\rho) = \frac{1}{\rho} = \frac{1}{\sqrt{x^2 + y^2 + z^2}}.$$

This is defined for all of 3−space with the origin removed. Other solutions can be produced using separation of variables.

Problems for Section 4.1

1. Write $f(z) = z^3 - 4z^2 + z - 1$ in the form $u(x, y) + iv(x, y)$, with $u(x, y)$ and $v(x, y)$ real-valued. Show that $u(x, y)$ and $v(x, y)$ are harmonic for all x and y.

2. Suppose u is harmonic on a region Ω of points in the plane. Let a and b be given numbers, and let Ω^* consist of all points $(x+a, y+b)$ for (x, y) in Ω. Ω^* is a translation of Ω in the plane. Define

$$v(x^*, y^*) = u(x^* - a, y^* - b)$$

for all (x^*, y^*) in Ω^*. Show that v is harmonic on Ω^*. This shows that translations of harmonic functions are harmonic.

3. Let θ be a real number and let

$$x^* = \cos(\theta)x + \sin(\theta)y,$$
$$y^* = -\sin(\theta)x + \cos(\theta)y.$$

This defines a rotation through the angle θ radians in the plane. The inverse transformation is

$$x = x^* \cos(\theta) - y^* \sin(\theta), y = x^* \sin(\theta) + y^* \cos(\theta).$$

Suppose u is harmonic on Ω, and let Ω^* be obtained by applying this rotation to each point of Ω. Let

$$w(x^*, y^*) = u(x^* \cos(\theta) - y^* \sin(\theta), x^* \sin(\theta) + y^* \cos(\theta))$$

for each (x^*, y^*) in Ω^*. Show that w is harmonic on Ω^*. This means that rotations take harmonic functions to harmonic functions.

4. Let B be the open ball of radius a about the origin in 3–space. B consists of all points at distance less than a from the origin. Suppose $u(x, y, z)$ is harmonic on the region exterior to B, consisting of all points at distance strictly greater than a from the origin.

An *inversion* I of B is defined to be the mapping

$$I(\mathbf{x}) = \frac{a^2}{|\mathbf{x}|^2}\mathbf{x}$$

for $\mathbf{x}$ in B and $\mathbf{x} \neq \mathbf{0}$.

(a) Show that $I(\mathbf{x})$ is on a line from the origin through $\mathbf{x}$, and that the product of the distance from the origin to $\mathbf{x}$ and from the origin to $I(\mathbf{x})$ is a^2, for each nonzero $\mathbf{x}$ in B.

(b) Prove that every point $\mathbf{y}$ outside of B (so $|\mathbf{y}| > a$) is the image of exactly one nonzero $\mathbf{x}$ in $B(\mathbf{0}, a)$. This is why I is called an inversion. It inverts the open ball about the origin, sending each of its points (except the origin) to a unique image outside the ball, in such a way that these correspondences are one to one.

(c) Let $w(\mathbf{x}) = u(I(\mathbf{x}))$ for each $\mathbf{x}$ in B. Show that w is harmonic on the set B with the origin removed (the punctured ball of radius a about the origin).

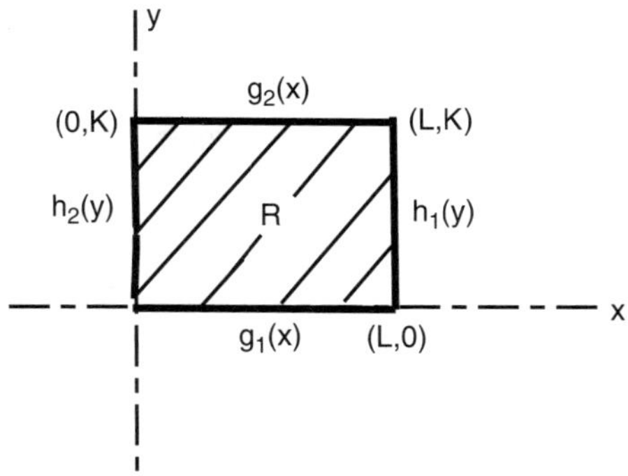

Figure 4.2: Boundary data for the Dirichlet problem for a rectangle.

4.2 The Dirichlet Problem for a Rectangle

Let D be a set of points in the plane, bounded by a curve C. A *Dirichlet problem* for D consists of finding a function that is harmonic on D and assumes given values on C. This is the boundary value problem

$$\nabla^2 u(x,y) = 0 \text{ for } (x,y) \in D,$$
$$u(x,y) = g(x,y) \text{ for } (x,y) \text{ on } C,$$

with $g(x,y)$ a given function.

The Dirichlet problem derives its name from German mathematician Johann Peter Gustav Lejeune Dirichlet (1805–1859), who did pioneering work in number theory, function theory, differential equations, and Fourier analysis.

There is no general method for solving Dirichlet problems for arbitrary regions. Indeed, for some fairly innocent looking regions D and data functions $g(x,y)$ defined on the boundary, this problem may have no solution. This issue is addressed in Section 4.7.

We will use separation of variables to solve Dirichlet problems for the case that the region is a rectangle. Let R be the rectangle consisting of all (x,y) with $0 \le x \le L, 0 \le y \le K$. The boundary of R consists of four line segments and function values are specified on each side. This Dirichlet problem is

$$\nabla^2 u(x,y) = 0 \text{ for } 0 < x < L, 0 < y < K,$$
$$u(x,0) = g_1(x), u(x,K) = g_2(x) \text{ for } 0 < x < L,$$
$$u(L,y) = h_1(y), u(0,y) = h_2(y) \text{ for } 0 < y < K,$$

with the functions $g_1(x), g_2(x), h_1(y)$, and $h_2(y)$ prescribing the values $u(x,y)$ is to have on the boundary sides. This information is displayed in Figure 4.2.

If we try a separation of variables, we get a rude surprise. Put $u(x,y) = X(x)Y(y)$ into $\nabla^2 u(x,y) = 0$:

$$X''Y + XY'' = 0.$$

Then

$$\frac{X''}{X} = -\frac{Y''}{Y} = -\lambda,$$

because x and y are independent, so

$$X'' + \lambda X = 0, Y'' - \lambda Y = 0.$$

Now everything falls apart. On the lower side of R, we need

$$u(x, 0) = X(x)Y(0) = g_1(x),$$

so $X(x)$ must be a constant multiple of $g_1(x)$. On the upper side,

$$u(x, L) = X(x)Y(L) = g_2(x),$$

so $X(x)$ must also be a multiple of $g_2(x)$. Together, these require that $g_1(x)$ and $g_2(x)$ must be constant multiples of each other, even though these functions have no connection to one another and can be chosen independently. The same kind of difficulty occurs for $Y(y)$ upon considering the conditions given on the vertical sides.

There is a way around this difficulty. Form four Dirichlet problems, P_1, P_2, P_3, and P_4. In each problem, one side has the boundary data from the original problem, whereas the other three sides have zero boundary data. This is displayed in Figure 4.3.

Suppose these problems have solutions $u_1(x, y), \cdots, u_4(x, y)$, respectively. We claim that

$$u(x, y) = \sum_{j=1}^{4} u_j(x, y)$$

is the solution of the original problem.

Clearly u is harmonic on the rectangle, because each u_j is. For the boundary conditions, verify by substitution that

$$\begin{aligned}
u(x, 0) &= u_1(x, 0) + u_2(x, 0) + u_3(x, 0) + u_4(x, 0)\\
&= g_1(x) + 0 + 0 + 0 = g_1(x),\\
u(x, K) &= u_1(x, K) + u_2(x, K) + u_3(x, K) + u_4(x, K)\\
&= 0 + 0 + g_2(x) + 0 = g_2(x),\\
u(L, y) &= u_1(L, y) + u_2(L, y) + u_3(L, y) + u_4(L, y)\\
&= 0 + h_1(y) + 0 + 0 = h_1(y),\\
&\text{and}\\
u(0, y) &= u_1(0, y) + u_2(0, y) + u_3(0, y) + u_4(0, y)\\
&= 0 + 0 + 0 + h_2(y) = h_2(y).
\end{aligned}$$

There remains to demonstrate that we can solve a Dirichlet problem for a rectangle if there is only nonzero data on one side. Example 4.1 illustrates the

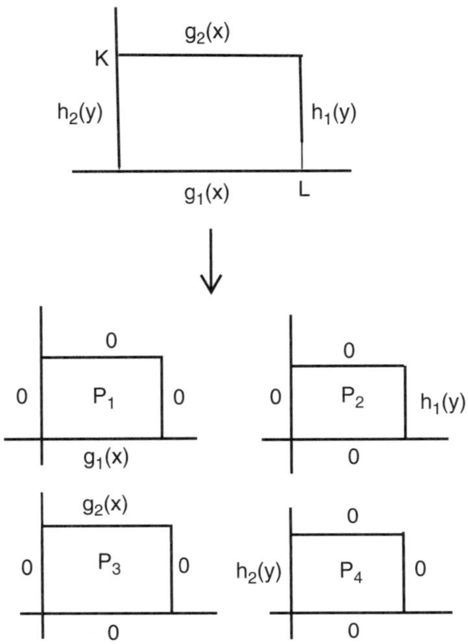

Figure 4.3: Partitioning a Dirichlet problem for a rectangle.

idea if the nonzero data are given only on the top side. The same strategy works for any side.

Example 4.1 Let R be the rectangle consisting of all (x, y) with $0 \leq x \leq 3, 0 \leq y \leq 7$. We will solve the Dirichlet problem

$$\begin{aligned}
\nabla^2 u(x,y) &= 0 \text{ for } (x,y) \text{ in } R, \\
u(x,0) &= 0 \text{ for } 0 < x < 3, \\
u(0,y) &= u(3,y) = 0 \text{ for } 0 < y < 7, \\
u(x,7) &= f(x) = x \sinh(3-x) \text{ for } 0 < x < 3.
\end{aligned}$$

Let $u(x, y) = X(x)Y(y)$ in Laplace's equation to obtain

$$X'' + \lambda X = 0, Y'' - \lambda Y = 0.$$

Use the three sides with zero boundary data to obtain conditions on $X(x)$ and $Y(y)$. First,

$$u(0, y) = X(0)Y(y) = 0$$

so $X(0) = 0$. Similarly, $u(3, y) = X(3)Y(y) = 0$ implies that $X(3) = 0$, and $u(x, 0) = X(x)Y(0) = 0$ implies that $Y(0) = 0$. The problems for $X(x)$ and

$Y(y)$ are

$$X'' + \lambda X = 0; X(0) = X(3) = 0,$$
$$Y'' - \lambda Y = 0; Y(0) = 0.$$

The problem for X has eigenvalues and eigenfunctions

$$\lambda_n = \frac{n^2\pi^2}{9}, X_n(x) = \sin\left(\frac{n\pi x}{3}\right)$$

for $n = 1, 2, \cdots$. Now the problem for Y is

$$Y''(y) - \frac{n^2\pi^2}{9}Y = 0; Y(0) = 0.$$

This differential equation has general solution

$$Y(y) = ae^{n\pi y/3} + be^{-n\pi y/3}.$$

Because $Y(0) = a + b = 0$, $b = -a$, so solutions for Y are constant multiples of

$$Y_n(y) = \sinh\left(\frac{n\pi y}{3}\right).$$

For $n = 1, 2, \cdots$, we now have functions

$$u_n(x, y) = c_n \sin\left(\frac{n\pi x}{3}\right) \sinh\left(\frac{n\pi y}{3}\right),$$

which are harmonic on the rectangle, and satisfy the homogeneous boundary conditions on the left, right and lower sides. To satisfy $u(x, 7) = f(x)$, let

$$u(x, y) = \sum_{n=1}^{\infty} c_n \sin\left(\frac{n\pi x}{3}\right) \sinh\left(\frac{n\pi y}{3}\right)$$

and choose the coefficients so that

$$u(x, 7) = \sum_{n=1}^{\infty} c_n \sin\left(\frac{n\pi x}{3}\right) \sinh\left(\frac{7n\pi}{3}\right) = x\sinh(3 - x).$$

This is a Fourier sine expansion of $x\sinh(3-x)$ on $[0, 3]$, so we must choose the entire coefficient to be the Fourier sine coefficient:

$$c_n \sinh\left(\frac{7n\pi}{3}\right) = \frac{2}{3}\int_0^3 \xi \sinh(3 - \xi) \sin\left(\frac{n\pi\xi}{3}\right) d\xi.$$

Upon carrying out the integration, we obtain

$$c_n = \frac{36n\pi}{(n^2\pi^2 + 9)^2 \sinh(7n\pi/3)} \left(\cosh(3) - (-1)^n\right).$$

The solution is

$$u(x,y) =$$
$$\sum_{n=1}^{\infty} \frac{36n\pi}{(n^2\pi^2+9)^2 \sinh(7n\pi/3)} (\cosh(3) - (-1)^n) \sin\left(\frac{n\pi x}{3}\right) \sinh\left(\frac{n\pi y}{3}\right).$$

Problems for Section 4.2

In each of problems 1–5, solve the Dirichlet problem for the indicated rectangle and boundary conditions.

1.

$$u(0,y) = u(1,y) = 0 \text{ for } 0 < y < \pi,$$
$$u(x,0) = \sin(\pi x), u(x,\pi) = 0 \text{ for } 0 < x < 1.$$

2.

$$u(0,y) = y(2-y), u(3,y) = 0 \text{ for } 0 < y < 2,$$
$$u(x,0) = u(x,2) = 0 \text{ for } 0 < x < 3.$$

3.

$$u(0,y) = u(1,y) = 0 \text{ for } 0 < y < 4,$$
$$u(x,0) = 0, u(x,4) = x\cos(\pi x/2) \text{ for } 0 < x < 1.$$

4.

$$u(0,y) = \sin(y), u(\pi,y) = 0 \text{ for } 0 < y < \pi,$$
$$u(x,0) = x(\pi - x), u(x,\pi) = 0 \text{ for } 0 < x < \pi.$$

5.

$$u(0,y) = 0, u(2,y) = \sin(y) \text{ for } 0 < y < \pi,$$
$$u(x,0) = 0, u(x,\pi) = x\sin(\pi x) \text{ for } 0 < x < 2.$$

Apply separation of variables to solve each of the following mixed boundary value problems (in which boundary data is a mixture of function values, and values of partial derivatives of the function).

6.

$$\nabla^2 u = 0 \text{ for } 0 < x < L, 0 < y < K,$$
$$u(x,0) = u_y(x,K) = 0 \text{ for } 0 < x < L,$$
$$u(0,y) = 0, u(L,y) = g(y) \text{ for } 0 < y < K.$$

7.

$$\nabla^2 u = 0 \text{ for } 0 < x < L, 0 < y < K,$$
$$u(x,0) = 0, u(x,K) = f(x) \text{ for } 0 < x < L,$$
$$u(0,y) = u_x(L,y) = 0 \text{ for } 0 < y < K.$$

4.3 The Dirichlet Problem for a Disk

As with rectangles, we can solve Dirichlet problems for disks. Let D be a disk of radius ρ about the origin. Using polar coordinates, we will solve the Dirichet problem

$$\nabla^2 u(r,\theta) = u_{rr} + \frac{1}{r}u_r + \frac{1}{r^2}u_{\theta\theta} = 0 \text{ for } 0 \le r < \rho, -\pi \le \theta \le \pi,$$
$$u(\rho,\theta) = f(\theta) \text{ for } -\pi \le \theta \le \pi.$$

We have already done a lot of the work. Recall from Section 4.1 that the functions

$$1, r^n\cos(n\theta), r^n\sin(n\theta)$$

are harmonic for $n = 1.2, \cdots$. We will attempt to write a solution of this Dirichlet problem as a superposition of these harmonic functions:

$$u(r,\theta) = \frac{1}{2}a_0 + \sum_{n=1}^{\infty}\left(a_n r^n\cos(n\theta) + b_n r^n\sin(n\theta)\right). \tag{4.3}$$

The harmonic functions $r^{-n}\cos(n\theta)$ and $r^{-n}\sin(n\theta)$ are not used in this superposition because they are undefined at $r = 0$, the center of the disk.

$u(r,\theta)$ is harmonic over the entire plane. To satisfy the boundary condition, we need

$$u(\rho,\theta) = \frac{1}{2}a_0 + \sum_{n=1}^{\infty}\left(a_n\rho^n\cos(n\theta) + b_n\rho^n\sin(n\theta)\right) = f(\theta).$$

This is a Fourier series expansion of $f(\theta)$ on $[-\pi,\pi]$. The coefficients are

$$a_0 = \frac{1}{\pi}\int_{-\pi}^{\pi} f(\xi)\,d\xi,$$
$$a_n\rho^n = \frac{1}{\pi}\int_{-\pi}^{\pi} f(\xi)\cos(n\xi)\,d\xi,$$
$$b_n\rho^n = \frac{1}{\pi}\int_{-\pi}^{\pi} f(\xi)\sin(n\xi)\,d\xi.$$

for $n = 1, 2, \cdots$.

Then

$$a_0 = \frac{1}{\pi}\int_{-\pi}^{\pi} f(\xi)\,d\xi,$$
$$a_n = \frac{1}{\rho^n\pi}\int_{-\pi}^{\pi} f(\xi)\cos(n\xi)\,d\xi.$$
$$b_n = \frac{1}{\rho^n\pi}\int_{-\pi}^{\pi} f(\xi)\sin(n\xi)\,d\xi.$$

Example 4.2 Solve

$$\nabla^2 u(r,\theta) = 0 \text{ for } 0 \le r < 3, -\pi < \theta < \pi,$$
$$u(3,\theta) = f(\theta) = |\cos(\theta/2)| \text{ for } -\pi \le \theta \le \pi.$$

The solution is equation 4.3 with $\rho = 3$. Compute

$$a_0 = \frac{1}{\pi}\int_{-\pi}^{\pi} |\cos(\xi/2)|\,d\xi = \frac{4}{\pi},$$
$$a_n = \frac{1}{3^n\pi}\int_{-\pi}^{\pi} |\cos(\xi/2)|\cos(n\xi)\,d\xi = \frac{4(-1)^{n+1}}{3^n\pi(4n^2-1)},$$
$$b_n = \frac{1}{3^n\pi}\int_{-\pi}^{\pi} |\cos(\xi/2)|\sin(n\xi)\,d\xi = 0.$$

The solution is

$$u(r,\theta) = \frac{2}{\pi} + \frac{4}{\pi}\sum_{n=1}^{\infty}\frac{(-1)^{n+1}}{3^n(4n^2-1)}r^n\cos(n\theta)$$

for $0 \le r < 3, -\pi \le \theta \le \theta$.

Sometimes a Dirichlet problem is posed on a disk, but articulated in rectangular coordinates. In this case we may proceed by converting the problem to polar coordinates.

Example 4.3 Solve the Dirichlet problem

$$\nabla^2 v(x,y) = 0 \text{ for } x^2 + y^2 < 12,$$
$$v(x,y) = x^2 - y \text{ for } x^2 + y^2 = 12.$$

This problem is posed on the disk of radius $\rho = 2\sqrt{3}$ about the origin. With

$$x = r\cos(\theta) \text{ and } y = r\sin(\theta),$$

the boundary function translates to

$$x^2 - y = \rho^2\cos^2(\theta) - \rho\sin(\theta)$$
$$= 12\cos^2(\theta) - 2\sqrt{3}\sin(\theta).$$

The problem in polar coordinates is

$$\nabla^2 u(r,\theta) = u_{rr} + \frac{1}{r}u_r + \frac{1}{r^2}u_{\theta\theta} = 0 \text{ for } r^2 < 12, -\pi \leq \theta \leq \pi,$$
$$u(2\sqrt{3},\theta) = f(\theta) = 12\cos^2(\theta) - 2\sqrt{3}\sin(\theta).$$

By equation 4.3, the solution is

$$u(r,\theta) = \frac{1}{2}a_0 + \sum_{n=1}^{\infty}(a_n r^n \cos(n\theta) + b_n r^n \sin(n\theta))$$

in which

$$a_0 = \frac{1}{\pi}\int_{-\pi}^{\pi} f(\xi)\,d\xi = 12,$$

$$a_n = \frac{1}{\pi}\frac{1}{(2\sqrt{3})^n}\int_{-\pi}^{\pi} f(\xi)\cos(n\xi)\,d\xi = \begin{cases} 0 & \text{for } n \neq 2, \\ 1/2 & \text{for } n = 2, \end{cases}$$

and

$$b_n = \frac{1}{\pi}\frac{1}{(2\sqrt{3})^n}\int_{-\pi}^{\pi} f(\xi)\sin(n\xi)\,d\xi = \begin{cases} 0 & \text{for } n = 2,3,\cdots, \\ -1 & \text{for } n = 1. \end{cases}$$

The solution is therefore

$$u(r,\theta) = 6 + \frac{1}{2}r^2\cos(2\theta) - r\sin(\theta).$$

This is a finite sum because only a_2 and b_1 are nonzero.

If we want the solution in rectangular coordinates, use the identity

$$\cos(2\theta) = \cos^2(\theta) - \sin^2(\theta)$$

to write

$$v(x,y) = 6 + \frac{1}{2}(x^2 - y^2) - y.$$

To check this calculation, on the boundary of the disk, $x^2 + y^2 = 12$, so $y^2 = 12 - x^2$ and

$$v(x,y) = 6 + \frac{1}{2}(x^2 - (12 - x^2)) - y = x^2 - y,$$

as required.

4.3.1 Poisson's Integral Solution

We will write an integral solution of the Dirichlet problem for a disk about the origin, starting with a disk of radius 1. Insert the integrals for the Fourier coefficients of $f(\theta)$ into equation 4.3, with $\rho = 1$, and rearrange terms:

$$\begin{aligned}u(r,\theta) &= \frac{1}{2\pi}\int_{-\pi}^{\pi} f(\xi)\,d\xi \\ &+ \sum_{n=1}^{\infty}\frac{1}{\pi}r^n\left[\int_{-\pi}^{\pi} f(\xi)\cos(n\xi)\,d\xi\cos(n\theta) + \int_{-\pi}^{\pi} f(\xi)\sin(n\xi)\,d\xi\sin(n\theta)\right] \\ &= \frac{1}{2\pi}\int_{-\pi}^{\pi}\left[1 + 2\sum_{n=1}^{\infty} r^n(\cos(n\xi)\cos(n\theta) + \sin(n\xi)\sin(n\theta))\right] f(\xi)\,d\xi \\ &= \frac{1}{2\pi}\int_{-\pi}^{\pi}\left[1 + 2\sum_{n=1}^{\infty} r^n\cos(n(\theta-\xi))\right] f(\xi)\,d\xi.\end{aligned}$$

The quantity

$$P(r;\zeta) = \frac{1}{2\pi}\left[1 + 2\sum_{n=1}^{\infty} r^n\cos(n\zeta)\right]$$

is called the *Poisson kernel*, and in terms of this kernel function, the solution is

$$u(r,\theta) = \int_{-\pi}^{\pi} P(r;\theta-\xi)f(\xi)\,d\xi. \tag{4.4}$$

for $0 \le r < 1$ and $-\pi \le \theta \le \pi$. This is reminiscent of the Dirichlet kernel used in the proof of the Fourier convergence theorem.

We will sum the infinite series in the Poisson kernel to write it in simpler form. Begin with Euler's formula

$$e^{i\zeta} = \cos(\zeta) + i\sin(\zeta).$$

Let $z = re^{i\zeta}$. This is the polar form of a complex number z having polar coordinates (r,ζ). By Euler's formula,

$$z^n = r^n e^{in\zeta} = r^n\cos(n\zeta) + ir^n\sin(n\zeta),$$

so $r^n\cos(n\zeta)$ is the real part of z^n:

$$r^n\cos(n\zeta) = \operatorname{Re}(z^n).$$

Then

$$1 + 2\sum_{n=1}^{\infty} r^n\cos(n\zeta) = \operatorname{Re}\left(1 + 2\sum_{n=1}^{\infty} z^n\right).$$

What we gain from this manipulation is that the series $\sum_{n=1}^{\infty} z^n$ is a geometric series whose sum is known:

$$\sum_{n=1}^{\infty} z^n = \frac{z}{1-z} \text{ if } |z| = r < 1.$$

This enables us to write

$$\begin{aligned} &1 + 2\sum_{n=1}^{\infty} r^n \cos(n\zeta) \\ &= \operatorname{Re}\left(1 + 2\frac{z}{1-z}\right) = \operatorname{Re}\left(\frac{1+z}{1-z}\right) \\ &= \operatorname{Re}\left(\frac{1+re^{i\zeta}}{1-re^{i\zeta}}\right). \end{aligned}$$

All that remains is to determine this real part. Multiply the numerator and denominator by $1 - re^{-i\zeta}$, which is the complex conjugate of the denominator, to obtain

$$\begin{aligned} \frac{1+re^{i\zeta}}{1-re^{i\zeta}} &= \left(\frac{1+re^{i\zeta}}{1-re^{i\zeta}}\right)\left(\frac{1-re^{-i\zeta}}{1-re^{-i\zeta}}\right) \\ &= \frac{1-r^2+r(e^{i\zeta}-e^{-i\zeta})}{1+r^2-r(e^{i\zeta}+e^{-i\zeta})} \\ &= \frac{1-r^2+2ir\sin(\zeta)}{1+r^2-2r\cos(\zeta)}. \end{aligned}$$

From this, we immediately read the real part of this quantity as

$$\frac{1-r^2}{1+r^2-2r\cos(\zeta)}.$$

Summarizing the calculation, we have

$$1 + 2\sum_{n=1}^{\infty} r^n \cos(n\zeta) = \operatorname{Re}\left(\frac{1+re^{i\zeta}}{1-re^{-i\zeta}}\right) = \frac{1-r^2}{1+r^2-2r\cos(\zeta)}.$$

This gives the Poisson kernel as the quotient:

$$P(r;\zeta) = \frac{1}{2\pi}\frac{1-r^2}{1+r^2-2r\cos(\zeta)}.$$

Now equation 4.4 is

$$u(r,\theta) = \frac{1}{2\pi}\int_{-\pi}^{\pi} \frac{1-r^2}{1+r^2-2r\cos(\theta-\xi)} f(\xi)\, d\xi. \tag{4.5}$$

This is *Poisson's integral formula* for the solution of the Dirichlet problem for the unit disk. It is valid for $0 \leq r < 1$ and $-\pi \leq \theta \leq \pi$.

By a change of variables, we can adapt this expression to obtain a solution of the Dirichlet problem on the disk of radius ρ about the origin:

$$u(r,\theta) = \frac{1}{2\pi}\int_{-\pi}^{\pi} \frac{\rho^2 - r^2}{\rho^2 + r^2 - 2\rho r\cos(\theta-\xi)} f(\xi)\,d\xi. \tag{4.6}$$

Sometimes it is useful to rewrite this as

$$u(r,\theta) = \frac{1}{2\pi}\int_{0}^{2\pi} \frac{\rho^2 - r^2}{\rho^2 + r^2 - 2\rho r\cos(\theta-\xi)} f(\xi)\,d\xi, \tag{4.7}$$

using the fact that $f(\theta)$ is assumed to be periodic of period 2π. In fact, the integral can be taken between a and $a + 2\pi$ for any number a.

Equations 4.5–4.7 are all known as *Poisson's integral solution of the Dirichlet problem for a disk.*

Example 4.4 We will use Poisson's integral to rewrite the solution of the problem of example 4.2. There we had $\rho = 3$ and $u(\rho,\theta) = f(\theta) = |\cos(\theta/2|$. The solution is

$$u(r,\theta) = \frac{1}{2\pi}\int_{0}^{2\pi} \frac{9 - r^2}{9 + r^2 - 6r\cos(\theta-\xi)} |\cos(\xi/2)|d\xi.$$

Which is to be preferred, a series or an integral solution? This depends on what we want to do with the solution. If, for example, we want numerical values at particular points, there are techniques for approximating integrals to high degrees of accuracy. Series are not as well suited to such numerical approximations, except in special cases.

Problems for Section 4.3

In each of problems 1–3, use equation 4.3 to write a solution of the Dirichlet problem in polar coordinates, for the given disk and boundary function.

1. $\rho = 7, u(7,\theta) = \cos^2(\theta)$.

2. $\rho = 6, u(6,\theta) = \sin^3(\theta) + \cos^2(\theta)$.

3. $\rho = 4, u(4,\theta) = \sin^2(\theta)\cos^2(\theta)$.

In problems 4 and 5, solve the Dirichlet problem by converting it to polar coordinates and using equation 4.3.

4.

$$\nabla^2 u(x,y) = 0 \text{ for } x^2 + y^2 < 9,$$
$$u(x,y) = x^2 \text{ for } x^2 + y^2 = 9.$$

5.

$$\nabla^2 u(x, y) = 0 \text{ for } x^2 + y^2 < 16,$$
$$u(x, y) = x^2 y^2 \text{ for } x^2 + y^2 = 16.$$

6. Consider the Dirichlet problem for the disk $r < \rho$, with $u(\rho, \theta) = f(\theta)$. Suppose $f(\theta)$ is an odd function and is periodic of period 2π on $[-\pi, \pi]$. Show that the solution $u(r, \theta)$ is also an odd function of θ. If $f(\theta)$ is an even function of θ, is $u(r, \theta)$ even as a function of θ?

7. Solve the Dirichlet problem for an annulus (the region between two concentric circles). The problem may be stated:

$$\nabla^2 u(r, \theta) = 0 \text{ for } 0 < \rho_1 < \theta < \rho_2, -\pi \leq \theta \leq \pi,$$
$$u(\rho_1, \theta) = g(\theta), u(\rho_2, \theta) = f(\theta) \text{ for } -\pi \leq \theta \leq \pi.$$

Hint: It is assumed that $\rho_1 > 0$. In this case attempt a series similar to that of equation 4.3, except now include the terms $\ln(r)$, $r^{-n}\cos(n\theta)$ and $r^{-n}\sin(n\theta)$, because the origin is not in the annulus. Use the boundary conditions on both of the boundary circles to find the coefficients.

In each of problems 8–13, solve the Dirichlet problem for annulus, which is specified by giving boundary data on each of the boundary circles.

8. $u(1, \theta) = 1, u(2, \theta) = 2$.

9. $u(1, \theta) = 1, u(2, \theta) = \cos(\theta)$.

10. $u(1, \theta) = \sin(\theta), u(2, \theta) = \cos(\theta)$.

11. $u(1, \theta) = 1, u(2, \theta) = \cos^2(\theta)$.

12. $u(2, \theta) = \sin(2\theta), u(4, \theta) = \sin(4\theta)$.

13. $u(2, \theta) = \sin^2(\theta), u(4, \theta) = \cos^2(\theta)$.

14. Derive equation 4.6 from 4.5.

15. By differentiating under the integral, verify that the function defined by equation 4.6 is harmonic (in polar coordinates) on the disk $r < \rho$.

16. Choose $u(r, \theta) = r^n \sin(n\theta)$ in equation 4.5, with n a given positive integer. Evaluate $u(1/2, \pi/2)$ to derive the integral formula

$$\int_{-\pi}^{\pi} \frac{\sin(n\xi)}{5 - 4\sin(\xi)}\, d\xi = \frac{\pi}{3(2^{n-1})} \sin(n\pi/2).$$

17. Use equation 4.5 with $u(r, \theta) = r^n \cos(n\theta)$ to evaluate

$$\int_{-\pi}^{\pi} \frac{\cos(n\xi)}{5 - 4\cos(\xi)}\, d\xi$$

for n an arbitrary positive integer.

18. Derive integral formulas by choosing $u(r, \theta) = r^n \cos(n\theta)$, with n a given positive integer, and evaluating $u(1/2, \pi/2)$ in Poisson's formula.

4.4 Properties of Harmonic Functions

In this section we will explore some important properties of harmonic functions. Although this can be done in n variables, for the most part we will work in the plane and 3−space, assuming familiarity with vector calculus, line integrals, and the integral theorems of Gauss and Green. We will also need to know some topology of R^n, which we will review first.

4.4.1 Topology of R^n

Let R^n denote the space of ordered n−tuples $(x_1, x_2, \cdots, x_n)$ of real numbers. We think of R^2 as the plane, consisting of points (x, y), and R^3 as 3−space consisting of triples (x, y, z) of real numbers.

Objects in R^n are often called n−vectors, and we will denote them in boldface, for example

$$\mathbf{x} = (x_1, x_2, \cdots, x_n).$$

If

$$\mathbf{x} = (x_1, \cdots, x_n) \text{ and } \mathbf{y} = (y_1, \cdots, y_n)$$

the *distance* between $\mathbf{x}$ and $\mathbf{y}$ is defined by the metric

$$|\mathbf{x} - \mathbf{y}| = \sqrt{\sum_{j=1}^{n} (x_j - y_j)^2}.$$

In the plane, this is the usual

$$|\mathbf{x} - \mathbf{y}| = \sqrt{(x_1 - y_1)^2 + (x_2 - y_2)^2}.$$

Because the distance between $\mathbf{x}$ and $\mathbf{y}$ is the same as the distance between $\mathbf{y}$ and $\mathbf{x}$, this distance function is *symmetric*:

$$|\mathbf{x} - \mathbf{y}| = |\mathbf{y} - \mathbf{x}|.$$

Distance also satisfies a *triangle inequality*

$$|\mathbf{x} - \mathbf{z}| \leq |\mathbf{x} - \mathbf{y}| + |\mathbf{y} - \mathbf{z}|.$$

This says that the distance between $\mathbf{x}$ and $\mathbf{z}$ cannot exceed the distance between $\mathbf{x}$ and $\mathbf{y}$ plus the distance between $\mathbf{y}$ and $\mathbf{z}$. We all know this from our experience in walking from one place to another.

The inequality also follows from the fact that any side of a triangle has length less than or equal to the sum of the lengths of the other two sides (Figure 4.4).

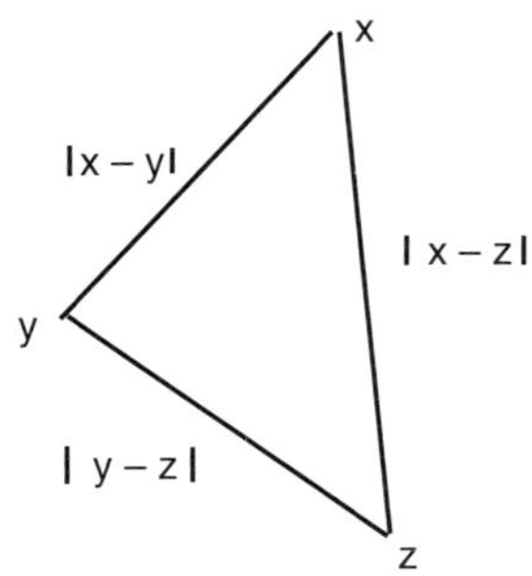

Figure 4.4: Triangle inequality in R^n.

If A is a set of points in R^n, we will write $\mathbf{x} \in A$ if $\mathbf{x}$ belongs to (is an element of) A.

Given a point $\mathbf{a} \in R^n$, the *open ball of radius* r about $\mathbf{a}$ is denoted $B(\mathbf{a}, r)$, and consists of all points $\mathbf{x} \in R^n$ at distance less than r from $\mathbf{a}$.

Sets are conveniently specified using set notation. For this open ball, write

$$B(\mathbf{a}, r) = \{\mathbf{x} \,|\, \mathbf{x} \in R^n \text{ and } |\mathbf{x} - \mathbf{a}| < r\}.$$

The left-most curly bracket announces that a set of points or objects of some kind is about to be defined. Following this bracket is some suitable name or symbol, say, for example, $\mathbf{x}$. Following this, the vertical line | announces the beginning of the rules that $\mathbf{x}$ must satisfy, or properties $\mathbf{x}$ must have, to be in the set. Finally, the right curly bracket indicates that the definition of this set is complete. In everyday terms, a set is a club, and the set notation lists the rules of membership.

Stated more generally, the set of all objects having some property $\mathcal{P}$ can be denoted

$$\{x \,|x \text{ has property } \mathcal{P}\}.$$

There is nothing special about use of the symbol x in defining this set. Any "reasonable" symbol can be used. For example, the set just defined can also be specified by

$$\{y \,|y \text{ has property } \mathcal{P}\}.$$

We will write $A \subset B$, read A is a subset of B, if every object in A is also in B. This does not preclude the possibility that the sets are the same, consisting of exactly the same objects. Note that $A \subset A$ for every set A because it is trivially true that every element of A is also in A. If both $A \subset B$ and $B \subset A$, then $A = B$ and the sets are identical, containing the same elements, because every object in A is in B, and, conversely, every object in B is in A.

Set difference is defined as follows. If B and A are sets, then $B - A$ denotes the set of elements of B that are not in A. If A and B have no elements in common, then of course $B - A = B$. We may think of this as scooping A out of B.

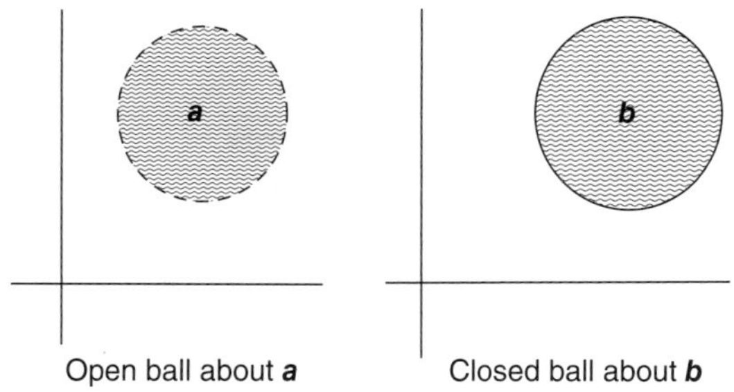

Figure 4.5: Open and closed balls in the plane.

If $B \subset A$, then $B - A = \varnothing$, the *empty set* having no elements, because $B \subset A$ implies that there is no element of B that is not in A.

The *closed ball of radius* r *about* $\mathbf{b}$ is denoted $\overline{B(\mathbf{b}, r)}$, and consists of all $\mathbf{x} \in R^n$ at distance less than or equal to r from $\mathbf{b}$:

$$\overline{B(\mathbf{b}, r)} = \{\mathbf{x} | \mathbf{x} \in R^n \text{ and } |\mathbf{x} - \mathbf{b}| \leq r\}.$$

In the plane, an open ball $B(\mathbf{a}, r)$ consists of all points interior to the circle of radius r about $\mathbf{a}$, while $\overline{B(\mathbf{b}, r)}$ consists of all points in the open ball about $\mathbf{b}$, together with those on the boundary circle (see Figure 4.5). The exclusion of points on the bounding circle for the open ball is often indicated by drawing this circle in dashed lines.

In R^3, circles are replaced by spheres, with the open ball about a point consisting of all points interior to some sphere about the point, while the closed ball contains these points as well as those on the surface of the sphere.

An open ball about a point is also known as a *neighborhood* of the point.

Now let A be a set of points in R^n. A point $\mathbf{a} \in A$ is called an *interior point* of A if there is some neighborhood of $\mathbf{a}$, all of whose points belong to A. In the plane, this means that we can draw a circle about $\mathbf{a}$ (perhaps of very small radius) enclosing only points of A. In 3−space this means that we can find a sphere about $\mathbf{a}$ enclosing only points of A. Thus a point is interior to A if, while standing at the point, there is some distance we can reach out in any direction and encounter only points of A.

The *interior* of A consists of all interior points of A. All interior points of A are in A, so the interior of A is a subset of A. However, there may be points in A that are not interior points.

A point $\mathbf{b}$ in R^n is called a *boundary point* of A if *every* neighborhood of $\mathbf{b}$ contains at least one point in A and at least one point not in A. In this sense, boundary points are on the edge of A. A person standing on a boundary point can reach out and touch a point not in A and a point in A, no matter how close the reach is restricted to be. A boundary point of A may or may not belong to A.

The *boundary* of A consists of all the boundary points of A. The boundary of A is denoted ∂A. If A has no boundary points, then $\partial A = \varnothing$.

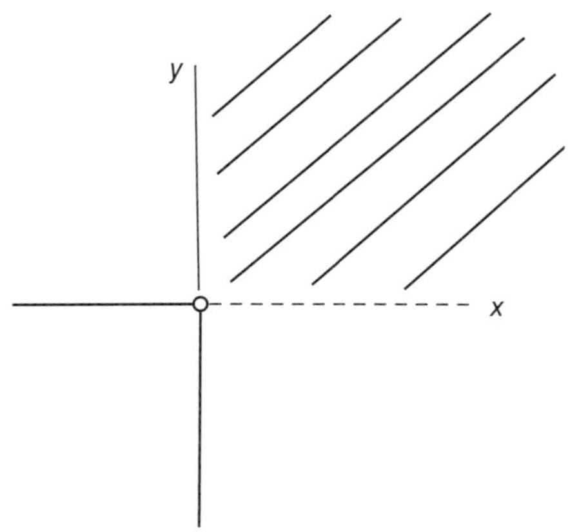

Figure 4.6: Points (x, y) with $x \geq 0$ and $y > 0$.

Example 4.5 In the plane, let

$$A = \{(x, y) \mid x \geq 0 \text{ and } y > 0\}.$$

A is depicted in Figure 4.6 and consists of all points in the plane having positive coordinates, together with those on the vertical axis above the origin, but not including the origin or points on the horizontal axis.

Every point $(x, 0)$ with $x \geq 0$ is a boundary point of A not contained in A. Points $(0, y)$ with $y > 0$ are also boundary points of A, but are contained in A. These are all the boundary points of A:

$$\partial A = \{(x, y) \mid x = 0 \text{ and } y \geq 0 \text{ or } x \geq 0 \text{ and } y = 0\}.$$

The small open dot at the origin in the diagram, and the dashed positive $x-$axis, are reminders that these points are not in A.

Every point (x, y) with $x > 0$ and $y > 0$ is an interior point of A, and these are all the interior points.

Example 4.6 Let $A = B(\mathbf{a}, r)$, an open ball of radius r about some point $\mathbf{a}$ in R^n. Then every point in A is an interior point of A. The boundary points are exactly those at distance r from $\mathbf{a}$, and none of these is in A.

Example 4.7 Let $A = B(\mathbf{a}, r) - \{\mathbf{a}\}$. This is the open ball of radius r about $\mathbf{a}$, with the center $\mathbf{a}$ removed. (Think of a circle or sphere with the center punched out). Every point of A is an interior point. However, the boundary points are now those at distance r from the center, together with the center $\mathbf{a}$. The center is a boundary point because every neighborhood of the center contains the center (which is a point not in A), as well as points in A.

A point can never be both an interior point and a boundary point of a set. However, each point in a set must be either an interior point or boundary point

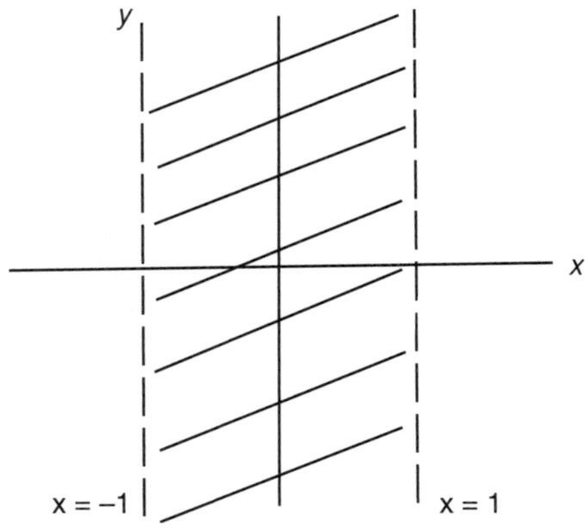

Figure 4.7: Vertical strip consisting of (x, y) with $|x| < 1$.

of this set. Keep in mind, however, that a set may have boundary points that do not belong to the set.

A set A is an *open set* if each of its points is an interior point of A. And A a *closed set* if A contains all of its boundary points.

In example 4.5, A is neither open nor closed. A is not open because it contains some of its boundary points, and these are not interior points. It is not closed because there are boundary points of A that are not in A. Although the language might suggest otherwise, the terms open and closed are not opposite concepts when referring to sets.

In example 4.6, any open ball about a point is open (all of its points are interior points, and it contains no boundary points).

In example 4.7, the punctured open ball (center removed) is open (all of its points are interior points) and not closed (it has boundary points, but none of these is contained in the set).

Example 4.8 R^n itself is open (all of its points are interior points), and also closed (it has no boundary points, and so vacuously contains all of them!).

Example 4.9 Let S consist of all points (x, y) in the plane with $-1 < x < 1$ and y arbitrary:

$$S = \{(x, y)| - 1 < x < 1\}.$$

This is the infinite vertical strip of Figure 4.7, consisting of all points strictly between the vertical lines $x = 1$ and $x = -1$.

Every point of S is an interior point. The boundary points are all the points on the vertical sides, specifically, points $(1, y)$ or $(-1, y)$. No boundary point of S belongs to S, and all points of S are interior points. S is open and not closed.

Let T consist of all points (x, y) with $-1 < x \leq 1$. This is the strip S together with all points on the right vertical side. The points $(1, y)$ are also boundary points of T, but are contained in T and are not interior points. Therefore T is not open. But T is not closed either, because some boundary points of T (those of the form $(-1, y)$) are not in T.

The *closure* of a set A consists of all points in A, together with all of the boundary points of A (if there are any). This set is denoted $\overline{A}$. For example, the closure of an open ball $B(\mathbf{a}, r)$ consists of all points in the open ball (distance less than r from the center), together with all points on the surface of the ball (at distance r from the center).

It is always true that $A \subset \overline{A}$, but these sets may not be equal. This occurs when there are boundary points of A that are not in A, which can happen, for example, if A is open.

One can prove that a set is closed exactly when the set is equal to its own closure. Further, different sets may have the same closure.

In example 4.9, $\overline{S} = \overline{T}$, with both closures consisting of all points (x, y) with $-1 \leq x \leq 1$. In this case, neither set equals its closure, because neither set is closed.

In example 4.5, the closure of A consists of all points (x, y) with $x \geq 0, y \geq 0$, formed by including all of the points of A and all its boundary points. This is a closed set, but does not equal the set itself, because some boundary points of A that are not in A are included in the closure.

A set A of points in R^n is *bounded* if it can be fit inside some open ball of sufficiently large radius. Put another way, A is bounded if, for some number R,

$$|\mathbf{x}| < R \text{ for all } \mathbf{x} \in A.$$

This says nothing about whether the set is open, closed, not open, or not closed. It simply says that A cannot contain points arbitrarily far from the origin. A set is *unbounded* if it is not bounded. Every open ball is a bounded set, and the closure of any bounded set is a bounded set. In the plane, the (open) upper half-plane, consisting of all (x, y) with $y > 0$, is unbounded.

There are two other properties of sets that we will use. A set is *compact* if it is closed and bounded. And a set A is *connected* if, given any two points $\mathbf{x}$ and $\mathbf{y}$ in A, there is a polygonal path (made up of a finite number of straight line segments) from one point to the other and lying entirely in A. This is not a general topological notion of connectedness, but is intuitively appealing and is suited to the analysis we will carry out.

Example 4.10 Let S consist of two disjoint solid disks in the plane. A typical such set is shown in Figure 4.8. S is not connected. No path can get from a point in one disk to a point in the other without going outside both disks.

Finally, a *domain* is a set of points in R^n that is both open and connected.

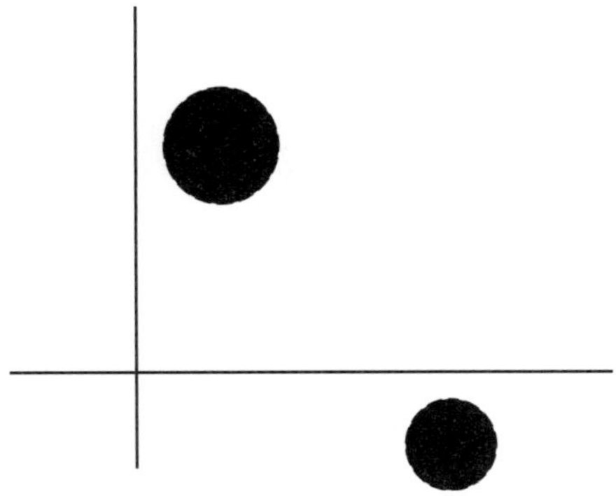

Figure 4.8: A non-connected set in the plane.

Problems for Section 4.4.1

In each of problems 1–9, determine all the interior point and boundary points of the set, as well as whether the set is open, closed, bounded, connected, and/or compact. Also determine if the set is a domain.

1.
$$S = \{(x, y)|0 \le x < 1 \text{ and } 0 < y \le 1\}.$$

2.
$$P = \{(x, y)|x < 0 \text{ and } 1 \le y \le 4\}.$$

3.
$$Q = \{(x, y)|1 < x^2 + y^2 < 4\}.$$

4.
$$W = \{(x, y)| -1 < x < 6 \text{ and } x < y\}.$$

5.
$$M = \{(x, y)|x \text{ and } y \text{ are rational numbers }\}.$$

6.
$$D = \{(x, y)|x \text{ and } y \text{ are irrational numbers }\}.$$

7.
$$P = \{(x, y)|x > 3 \text{ or } x \le -2\}.$$

8.
$$B = \{(x, y, z)|x^2 + y^2 + z^2 \ge 4\}.$$

9.
$$T = \{(x, y, z)|x + y + z = 7\}.$$

10. E consists of all (x, y, z) in R^3 that are within distance 7 of $(-3, -4, -3)$ or within distance 2 of $(11, 15, 12)$.

11. If A is any set of points in R^n, the *complement* of A is the set A^C consisting of all points of R^n that are not in A. Show that A and A^C have the same boundary points.

12. Let $A \subset R^n$. Show that A is open if and only if A^C is closed.

13. Show that R^n and the empty set $\varnothing$ are both open and closed. Can any other subset of R^n be both open and closed?

4.4.2 Representation Theorems

Functions that are harmonic on domains in R^n have representations as integrals. These are important in deriving and understanding further properties of harmonic functions and in solving Dirichlet problems. We restrict our attention to integral representations in the plane and 3−space. These provide some insight into what generalizations to R^n might look like.

If A is a set of points in R^2 or R^3, let $C^2(A)$ denote the set of all functions that are continuous, with continuous first and second partial derivatives, on A. A function in $C^2(A)$ is called a C^2 function on A.

4.4.2.1 A Representation Theorem in R^3

We begin with a representation theorem for C^2 functions on bounded domains in 3−space. This will require familiarity with surfaces, closed surfaces, outer normal vectors, normal derivatives, surface integrals and the divergence theorem of Gauss.

We will also need the following lemma from 19th century British amateur natural philosopher George Green and, independently, Ukrainian mathematician Michel Ostrogradsky. In the statement, recall that

$$\frac{\partial u}{\partial n} = \nabla u \cdot \mathbf{n},$$

where $\mathbf{n}$ is the unit outer normal to the surface.

Lemma 4.1 (Green's Second Identity in R^3) Let Ω be a bounded domain in R^3, and assume that $\partial\Omega$ is a piecewise smooth closed surface. Let u and v be C^2 functions on $\overline{\Omega}$. Then

$$\iiint_{\overline{\Omega}} (u\nabla^2 v - v\nabla^2 u)\, dV = \iint_{\partial\Omega} \left(u\frac{\partial v}{\partial n} - v\frac{\partial u}{\partial n} \right) d\sigma.$$

Proof Let $\mathbf{n}$ be the unit outer normal to $\partial\Omega$. Apply Gauss's divergence theorem to the surface integrals to obtain

$$\begin{aligned}
&\iint_{\partial\Omega} u\frac{\partial v}{\partial n}\,d\sigma - \iint_{\partial\Omega} v\frac{\partial u}{\partial n}\,d\sigma \\
&= \iint_{\partial\Omega} u\nabla v\cdot\mathbf{n}\,d\sigma - \iint_{\partial\Omega} v\nabla u\cdot\mathbf{n}\,d\sigma \\
&= \iiint_{\overline{\Omega}} \operatorname{div}(u\nabla v)\,dV - \iiint_{\overline{\Omega}} \operatorname{div}(v\nabla u)\,dV.
\end{aligned}$$

It is a straightforward calculation to verify that

$$\operatorname{div}(u\nabla v) - \operatorname{div}(v\nabla u) = u\nabla^2 v - v\nabla^2 u,$$

and this completes the proof.

We will use this lemma to derive a representation theorem for C^2 functions in 3−space, which we will then apply to harmonic functions.

Theorem 4.1 (Representation Theorem for C^2 Functions in R^3) Let Ω be a bounded domain in R^3 with piecewise smooth closed boundary surface $\partial\Omega$. Let $u \in C^2(\overline{\Omega})$. Then, at any $\mathbf{x} \in \Omega$,

$$\begin{aligned}
u(\mathbf{x}) = &\frac{1}{4\pi}\iint_{\partial\Omega}\left[\frac{1}{|\mathbf{y}-\mathbf{x}|}\frac{\partial u(\mathbf{y})}{\partial n} - u(\mathbf{y})\frac{\partial}{\partial n}\frac{1}{|\mathbf{y}-\mathbf{x}|}\right] d\sigma_{\mathbf{y}} \\
&- \frac{1}{4\pi}\iiint_{\overline{\Omega}} \frac{\nabla^2 u(\mathbf{y})}{|\mathbf{y}-\mathbf{x}|}\,dV_{\mathbf{y}}. \qquad (4.8)
\end{aligned}$$

The notations $d\sigma_{\mathbf{y}}$ and $dV_{\mathbf{y}}$ are reminders that the variable of integration is $\mathbf{y}$, while $\mathbf{x}$ is any point of Ω.

Proof For $\mathbf{y} \in \Omega$ with $\mathbf{y} \neq \mathbf{x}$, let

$$v(\mathbf{y}) = \frac{1}{|\mathbf{y}-\mathbf{x}|}.$$

We would like to apply Green's second identity to u and v, but v is not defined at $\mathbf{x}$. However, Ω is an open set, so $\mathbf{x}$ is an interior point of Ω and there is a neighborhood $B(\mathbf{x}, \epsilon)$ of $\mathbf{x}$ containing only points of Ω. Denote $B(\mathbf{x}, \epsilon) =$ B. By choosing ϵ small enough, we can ensure that $\overline{B}$ is wholly contained in Ω. Form Ω_ϵ by removing all points of $\overline{B}$ from Ω. It is easy to check that Ω_ϵ is also a domain (open and connected). Figure 4.9 suggests this construction.

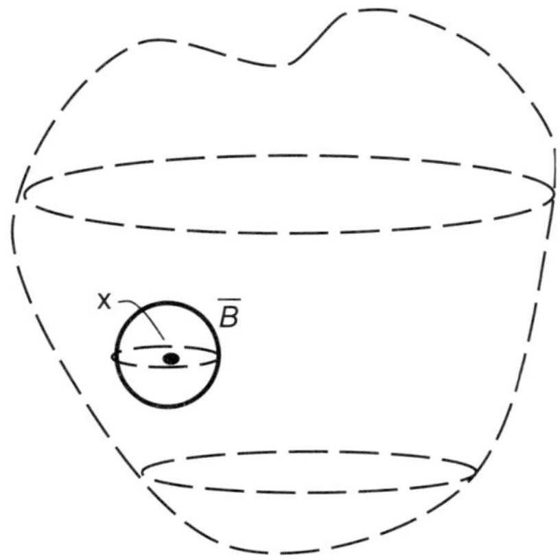

Figure 4.9: Ω_ϵ is formed by cutting $\overline{B}$ out of Ω.

Now, v is harmonic for $\mathbf{y} \neq \mathbf{x}$, so $\nabla^2 v = 0$ on Ω_ϵ. By Green's second identity,

$$-\iiint_{\overline{\Omega_\epsilon}} \frac{\nabla^2 u(\mathbf{y})}{|\mathbf{y}-\mathbf{x}|}\, dV_{\mathbf{y}}$$
$$= \iint_{\partial\Omega_\epsilon} \left(u(\mathbf{y}) \frac{\partial}{\partial n}\left(\frac{1}{|\mathbf{y}-\mathbf{x}|}\right) - \frac{1}{|\mathbf{y}-\mathbf{x}|}\frac{\partial u(\mathbf{y})}{\partial n}\right) d\sigma_{\mathbf{y}}.$$

$\partial\Omega_\epsilon$ consists of two disjoint parts: the boundary $\partial\Omega$ of Ω, and $S = S(\mathbf{x}, \epsilon)$, the sphere bounding B (Figure 4.9). The surface integral over $\partial\Omega_\epsilon$ is the sum of the surface integrals over $\partial\Omega$ and S, so the last equation can be written

$$-\iiint_{\overline{\Omega_\epsilon}} \frac{\nabla^2 u(\mathbf{y})}{|\mathbf{y}-\mathbf{x}|}\, dV_{\mathbf{y}}$$
$$= \iint_{\partial\Omega} \left(u(\mathbf{y}) \frac{\partial}{\partial n}\left(\frac{1}{|\mathbf{y}-\mathbf{x}|}\right) - \frac{1}{|\mathbf{y}-\mathbf{x}|}\frac{\partial u(\mathbf{y})}{\partial n}\right) d\sigma_{\mathbf{y}}$$
$$+ \iint_{S} \left(u(\mathbf{y}) \frac{\partial}{\partial n}\left(\frac{1}{|\mathbf{y}-\mathbf{x}|}\right) - \frac{1}{|\mathbf{y}-\mathbf{x}|}\frac{\partial u(\mathbf{y})}{\partial n}\right) d\sigma_{\mathbf{y}} \qquad (4.9)$$

We want to determine what happens to the terms of this equation in the limit as $\epsilon \to 0$. First, the convergent improper triple integral over Ω_ϵ approaches the triple integral over Ω:

$$\iiint_{\overline{\Omega_\epsilon}} \frac{\nabla^2 u(\mathbf{y})}{|\mathbf{y}-\mathbf{x}|}\, dV_{\mathbf{y}} \to \iiint_{\overline{\Omega}} \frac{\nabla^2 u(\mathbf{y})}{|\mathbf{y}-\mathbf{x}|}\, dV_{\mathbf{y}}.$$

The surface integral over $\partial\Omega$ on the right side of equation 4.9 is independent of ϵ, and, therefore, remains unchanged as $\epsilon \to 0$.

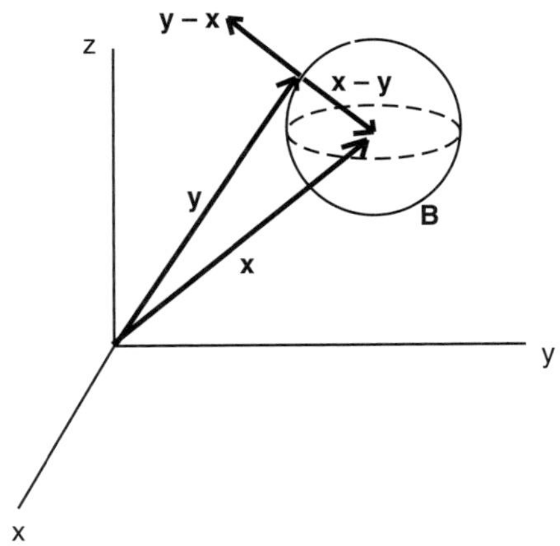

Figure 4.10: Exterior normal $\mathbf{y} - \mathbf{x}$ to B at $\mathbf{y}$.

Finally, we will show that the surface integral over S in equation 4.9 has limit $4\pi u(\mathbf{x})$. If $\mathbf{y} \in S$, then

$$\frac{1}{|\mathbf{y} - \mathbf{x}|} = \frac{1}{\epsilon}.$$

To compute the normal derivative of v on S, recall that the line from any point of a sphere to its center is normal to the sphere. This means that for any $\mathbf{y} \in S$, the vector $\mathbf{y} - \mathbf{x}$ is normal to S and points out of the ball B. We want the normal vector oriented away from Ω_ϵ, so choose a vector along $-(\mathbf{y} - \mathbf{x})$, toward $\mathbf{x}$, and into the ball (hence exterior to Ω_ϵ; Figure 4.10). To obtain a unit vector with this orientation, let

$$\mathbf{y} = y_1\mathbf{i} + y_2\mathbf{j} + y_3\mathbf{k} \text{ and } \mathbf{x} = x_1\mathbf{i} + x_2\mathbf{j} + x_3\mathbf{k}$$

and choose

$$\begin{aligned}\mathbf{n}(\mathbf{y}) &= -\frac{\mathbf{y} - \mathbf{x}}{|\mathbf{y} - \mathbf{x}|} \\ &= -\frac{(y_1 - x_1)\mathbf{i} + (y_2 - x_2)\mathbf{j} + (y_3 - x_3)\mathbf{k}}{\sqrt{(y_1 - x_1)^2 + (y_2 - x_2)^2 + (y_3 - x_3)^2}}.\end{aligned}$$

Now compute the normal derivative

$$\begin{aligned}
\frac{\partial}{\partial n}\left(\frac{1}{|\mathbf{y}-\mathbf{x}|}\right) &= \nabla\left(\frac{1}{|\mathbf{y}-\mathbf{x}|}\right)\cdot\mathbf{n}(\mathbf{y}) \\
&= -\frac{(y_1-x_1)\mathbf{i}+(y_2-x_2)\mathbf{j}+(y_3-x_3)\mathbf{k}}{((y_1-x_1)^2+(y_2-x_2)^2+(y_3-x_3)^2)^{3/2}}\cdot\mathbf{n}(\mathbf{y}) \\
&= \frac{(y_1-x_1)^2+(y_2-x_2)^2+(y_3-x_3)^2}{((y_1-x_1)^2+(y_2-x_2)^2+(y_3-x_3)^2)^2} \\
&= \frac{1}{(y_1-x_1)^2+(y_2-x_2)^2+(y_3-x_3)^2} \\
&= \frac{1}{\epsilon^2}.
\end{aligned}$$

Therefore

$$\begin{aligned}
&\iint_S \left(u(\mathbf{y})\frac{\partial}{\partial n}\left(\frac{1}{|\mathbf{y}-\mathbf{x}|}\right)-\frac{1}{|\mathbf{y}-\mathbf{x}|}\frac{\partial u(\mathbf{y})}{\partial n}\right)d\sigma_{\mathbf{y}} \\
&= \iint_S \left(\frac{1}{\epsilon^2}u(\mathbf{y})-\frac{1}{\epsilon}\frac{\partial u(\mathbf{y})}{\partial n}\right)d\sigma_{\mathbf{y}} \\
&= \frac{1}{\epsilon^2}\iint_S u(\mathbf{x})\,d\sigma_{\mathbf{y}}+\iint_S\left(\frac{1}{\epsilon^2}[u(\mathbf{y})-u(\mathbf{x})]-\frac{1}{\epsilon}\frac{\partial u(\mathbf{y})}{\partial n}\right)d\sigma_{\mathbf{y}}.
\end{aligned}$$

Because the integration is with respect to $\mathbf{y}$, $u(\mathbf{x})$ factors through the integral, and

$$\frac{1}{\epsilon^2}\iint_S u(\mathbf{x})\,d\sigma_{\mathbf{y}} = \frac{1}{\epsilon^2}u(\mathbf{x})(\text{area of } S) = 4\pi u(\mathbf{x}).$$

Finally,

$$\begin{aligned}
&\left|\iint_S\left(\frac{1}{\epsilon^2}[u(\mathbf{y})-u(\mathbf{x})]-\frac{1}{\epsilon}\frac{\partial u(\mathbf{y})}{\partial n}\right)d\sigma_y\right| \\
&= \frac{1}{\epsilon^2}(4\pi\epsilon^2)\max_{\mathbf{y}\in S}|u(\mathbf{y})-u(\mathbf{x})|+\frac{1}{\epsilon}(4\pi\epsilon^2)\max_{\mathbf{y}\in S}\left|\frac{\partial u(\mathbf{y})}{\partial n}\right| \to 0
\end{aligned}$$

as $\epsilon \to 0$ because by continuity $u(\mathbf{y}) \to u(\mathbf{x})$ as $\mathbf{y} \to \mathbf{x}$ and, for the last term, because

$$\left|\frac{\partial u(\mathbf{y})}{\partial n}\right|$$

is bounded for $\mathbf{y} \in S$.

Therefore, in the limit as $\epsilon \to 0$, equation 4.9 becomes

$$\begin{aligned}
&-\iiint_{\overline{\Omega}}\frac{\nabla^2 u(\mathbf{y})}{|\mathbf{y}-\mathbf{x}|}\,dV_{\mathbf{y}} \\
&= \iint_{\partial\Omega}\left(u(\mathbf{y})\frac{\partial}{\partial n}\left(\frac{1}{|\mathbf{y}-\mathbf{x}|}\right)-\frac{1}{|\mathbf{y}-\mathbf{x}|}\frac{\partial u(\mathbf{y})}{\partial n}\right)d\sigma_{\mathbf{y}}+4\pi u(\mathbf{x}). \qquad (4.10)
\end{aligned}$$

This is equivalent to the conclusion of the theorem, completing the proof.

This representation theorem does not assume that u is harmonic. If, however, $\nabla^2 u = 0$, then the triple integral in equation 4.10 vanishes and we are left with

$$u(\mathbf{x}) = \frac{1}{4\pi} \iint_{\partial\Omega} \left(\frac{1}{|\mathbf{y}-\mathbf{x}|} \frac{\partial u(\mathbf{y})}{\partial n} - u(\mathbf{y}) \frac{\partial}{\partial n} \frac{1}{|\mathbf{y}-\mathbf{x}|} \right) d\sigma_{\mathbf{y}} \tag{4.11}$$

for $\mathbf{x} \in \Omega$. The importance of this equation is that it relates the value $u(\mathbf{x})$ of a harmonic function u at $\mathbf{x} \in \Omega$ to an integral over the boundary, hence to a quantity that depends only on values of u on the boundary. This is exactly the information that is given in a Dirichlet problem.

4.4.2.2 A Representation Theorem in the Plane

Suppose Ω is a bounded domain in the plane, with boundary $\partial\Omega$ a piecewise smooth curve. The planar version of Green's second lemma looks the same as the lemma in 3−space, but the triple integral is replaced with a double integral, and the surface integral over the bounding surface with a line integral over the boundary curve.

For the representation theorem in 3−space, we used the function

$$\frac{1}{|\mathbf{y}-\mathbf{x}|} \text{ for } \mathbf{y} \neq \mathbf{x}.$$

In the plane, this function is replaced with

$$v(\mathbf{y}) = \ln\left(\frac{1}{|\mathbf{y}-\mathbf{x}|}\right),$$

which is harmonic as long as $\mathbf{y} \neq \mathbf{x}$. We obtain the following version of Theorem 4.1, adapted to domains in the plane.

Theorem 4.2 (Representation Theorem for $C^2(\overline{\Omega})$ Functions in R^2) Let Ω be a bounded domain in R^2 whose boundary $\partial\Omega$ is a piecewise smooth closed curve. Let $u \in C^2(\overline{\Omega})$. Then, for $\mathbf{x} \in \Omega$,

$$\begin{aligned} u(\mathbf{x}) = &\frac{1}{2\pi} \oint_{\partial\Omega} \left(\ln\left(\frac{1}{|\mathbf{y}-\mathbf{x}|}\right) \frac{\partial u(\mathbf{y})}{\partial n} - u(\mathbf{y}) \frac{\partial}{\partial n} \ln\left(\frac{1}{|\mathbf{y}-\mathbf{x}|}\right) \right) ds \\ &- \frac{1}{2\pi} \iint_{\overline{\Omega}} \nabla^2 u(\mathbf{y}) \ln\left(\frac{1}{|\mathbf{y}-\mathbf{x}|}\right) dA. \end{aligned}$$

As in the three-dimensional case, in the case that u is harmonic, this representation theorem gives $u(\mathbf{x})$ at points of Ω in terms of information on the boundary of Ω.

Problems for Section 4.4.2

1. In R^3, what does the representation theorem reduce to for the case that $u(\mathbf{x}) = 1$?
2. In R^3, what does the representation theorem say for the case $u(\mathbf{x}) = 1/|\mathbf{x}|$?
3. Prove Green's second lemma for a domain in the plane.
4. Write out the details of proof of the representation theorem for domains in the plane, at a level of rigor comparable to that given for the theorem in 3−space.

4.4.3 The Mean Value Property and the Maximum Principle

Harmonic functions enjoy a *mean value property*. In 3−space, this says that if u is harmonic on a bounded domain Ω, then, at any $\mathbf{x} \in \Omega$, $u(\mathbf{x})$ is the average of the values of u over any sphere centered at $\mathbf{x}$ and lying entirely in Ω. This is proved using the representation theorem for harmonic functions.

Theorem 4.3 (Mean Value Property for Harmonic Functions in R^3) Let u be harmonic on a bounded domain Ω in R^3. Let $\mathbf{x} \in \Omega$. Then

$$u(\mathbf{x}) = \frac{1}{4\pi\epsilon^2} \iint_{S(\mathbf{x},\epsilon)} u(\mathbf{y})\, d\sigma_y,$$

provided that ϵ is sufficiently small that $\overline{B(\mathbf{x},\epsilon)} \subset \Omega$.

In this statement, $4\pi\epsilon^2$ is the surface area of the sphere of radius ϵ centered at $\mathbf{x}$. The surface integral is in a sense a sum of the values of the function u over this sphere. The theorem therefore says that $u(\mathbf{x})$ is the sum of the values of u over the sphere centered at $\mathbf{x}$, divided by the area of this sphere. This number is interpreted as the average of the function values over this sphere.

Proof Denote $S = S(\mathbf{x},\epsilon)$. By the representation theorem,

$$u(\mathbf{x}) = \frac{1}{4\pi} \iint_S \left(\frac{1}{|\mathbf{y}-\mathbf{x}|} \frac{\partial u(\mathbf{y})}{\partial n} - u(\mathbf{y}) \frac{\partial}{\partial n} \frac{1}{|\mathbf{y}-\mathbf{x}|} \right) d\sigma_{\mathbf{y}}.$$

By a calculation done in the proof of the representation theorem,

$$\frac{1}{|\mathbf{y}-\mathbf{x}|} = \frac{1}{\epsilon} \text{ and } \frac{\partial}{\partial n} \frac{1}{|\mathbf{y}-\mathbf{x}|} = -\frac{1}{\epsilon^2}$$

for $\mathbf{y} \in S$. This normal derivative is negative because the exterior normal points out of the ball away from $\mathbf{x}$, opposite the orientation in the representation theorem. Then,

$$u(\mathbf{x}) = \frac{1}{4\pi} \iint_S \frac{1}{\epsilon} \frac{\partial u(\mathbf{y})}{\partial n}\, d\sigma_{\mathbf{y}} + \frac{1}{4\pi} \iint_S \frac{1}{\epsilon^2} u(\mathbf{y})\, d\sigma_{\mathbf{y}}.$$

In this equation, the first integral on the right is zero by the divergence theorem:

$$\iint_S \frac{\partial u(\mathbf{y})}{\partial n}\, d\sigma_{\mathbf{y}} = \iint_S \nabla u(\mathbf{y}) \cdot \mathbf{n}\, d\sigma_{\mathbf{y}}$$
$$= \iiint_B \operatorname{div}(\nabla u)\, dV_{\mathbf{y}} = \iiint_B \nabla^2 u\, dV_{\mathbf{y}} = 0.$$

Therefore

$$u(\mathbf{x}) = \frac{1}{4\pi\epsilon^2} \iint_S u(\mathbf{y})\, d\sigma_{\mathbf{y}}.$$

For two space dimensions, the mean value property takes the following form, with a sphere replaced by a circle, and a solid ball about a point by a disk.

Theorem 4.4 (Mean Value Property for Harmonic Functions in R^2) Let u be harmonic on a bounded domain Ω in R^2 and let $\mathbf{x} \in \Omega$. Then

$$u(\mathbf{x}) = \frac{1}{2\pi\epsilon} \oint_C u(\mathbf{y})\, ds_{\mathbf{y}},$$

in which C is a circle of radius ϵ about $\mathbf{x}$, and ϵ is sufficiently small that the circle and all points it encloses are in Ω.

The factor $2\pi\epsilon$ in the theorem is the length of C, so the conclusion states that the value of u at any point of Ω is the sum of its values on the circle, divided by the length of the circle. This number is interpreted as the average of the function over the circle.

The representation theorems also enable us to derive a maximum principle for harmonic functions (recall the weak maximum principle for solutions of the heat equation). Suppose u is harmonic on a bounded domain Ω, and continuous on the closure $\overline{\Omega}$. Because $\overline{\Omega}$ is compact, $u(\mathbf{x})$ achieves a maximum value on $\overline{\Omega}$. We will prove that if u is harmonic, this maximum must occur at a boundary point and, further, cannot also occur as an interior point of Ω. This is a stronger result than the weak maximum principle for the heat equation.

The proof makes use of a subtle fact from topology. Suppose Γ is a polygonal path in Ω, consisting of a finite number of line segments. Then there is a positive number ρ such that every point P of Γ is at distance at least ρ from each boundary point of Ω. This means that a polygonal path in a bounded domain cannot come arbitrarily close to the boundary of Ω (Figure 4.11). This is false if we replace polygonal paths with piecewise smooth paths. Such a path may indeed come arbitrarily close to the boundary of a domain in which it is contained.

Theorem 4.5 (Maximum Principle for Harmonic Functions) Let u be harmonic and nonconstant on a bounded domain Ω in R^n and continuous on the closure $\overline{\Omega}$. Then $u(\mathbf{x})$ achieves its maximum and minimum values on $\overline{\Omega}$ only at points of the boundary of Ω.

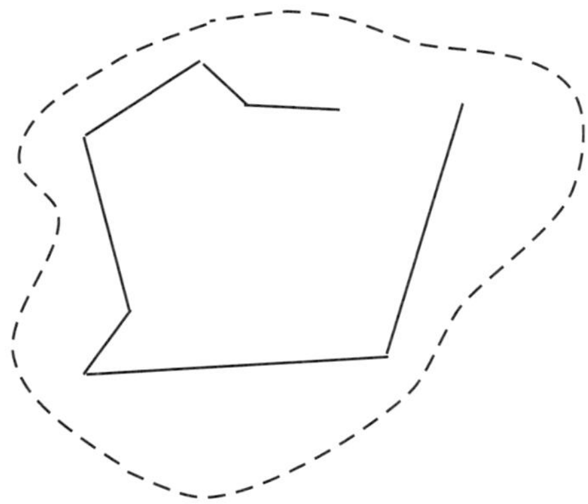

Figure 4.11: A polygonal path in a domain must remain at some distance from the boundary.

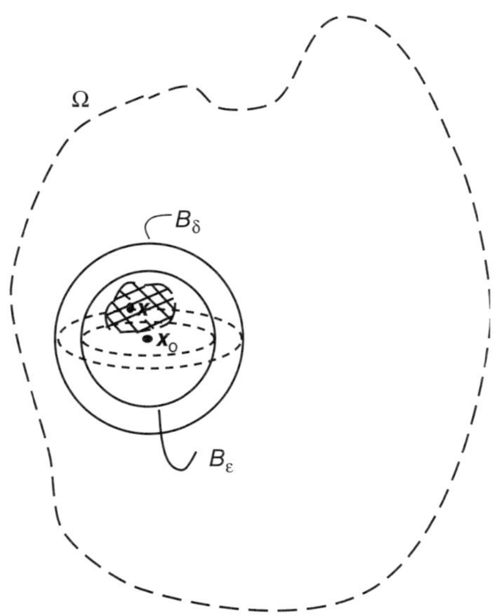

Figure 4.12: Patch on B_ϵ within B_δ within Ω.

Proof Let

$$M = \max_{\mathbf{x}\in\overline{\Omega}} u(\mathbf{x}).$$

We want to show that this maximum value can be achieved only at one or more points on the boundary of Ω.

Suppose instead that $u(\mathbf{x})$ achieves its maximum value on $\overline{\Omega}$ at a point $\mathbf{x}_0$ in Ω (hence at an interior point, since Ω is an open set). Let $B_\delta = B(\mathbf{x}_0, \delta)$, with δ chosen so that $\overline{B}_\delta \subset \Omega$. This can be done because $\mathbf{x}_0$ is an interior point of Ω. We will prove that $u(\mathbf{x}) = M$ for all $\mathbf{x} \in \overline{B_\delta}$.

To prove this, choose any ϵ with $0 < \epsilon < \delta$ and let B_ϵ be the open ball of radius ϵ about $\mathbf{x}_0$. If $u(\mathbf{x}) < M$ for any $\mathbf{x} \in S(\mathbf{x}_0, \epsilon)$, then by the continuity of u we would have $u(\mathbf{x}) < M$ on some open patch of points about $\mathbf{x}$ and lying on $S(\mathbf{x}_0, \epsilon)$ (Figure 4.12). This would make the average of the values of u over $S(\mathbf{x}_0, \epsilon)$ less than M, contradicting the fact that, by the mean value property, $u(\mathbf{x}_0) = M$ is the average of values of u over $S(\mathbf{x}_0, \epsilon)$. We conclude that $u(\mathbf{x}) = M$ at every point on $S(\mathbf{x}_0, \epsilon)$. But ϵ is any positive number with $0 < \epsilon < \delta$, so again by the continuity of u, $u(\mathbf{x}) = M$ at every point of B, and then $u(\mathbf{x}) = M$ at every point of $\overline{B_\delta}$.

Now we claim that $u(\mathbf{x}) = M$ at every point of Ω. To prove this, let $\mathbf{y} \in \Omega$ and $\mathbf{y} \neq \mathbf{x}_0$. We will show that $u(\mathbf{y}) = M$.

Because Ω is connected, there is a polygonal path Γ in Ω from $\mathbf{x}_0$ to $\mathbf{y}$, consisting of a finite number of line segments. Let ρ be a positive number such that every point of Γ is at least distance ρ from every point of $\partial\Omega$.

Move along Γ from $\mathbf{x}_0$ to $\mathbf{y}$, choosing a finite number of intermediary points $\mathbf{x}_1, \mathbf{x}_2, \cdots, \mathbf{x}_n$ along Γ in such a way that, for some positive numbers $\epsilon_0, \epsilon_1, \cdots, \epsilon_n$, each open ball $\overline{B(\mathbf{x}_j, \epsilon_j)}$ is entirely within Ω, and

$$\mathbf{x}_1 \in B(\mathbf{x}_0, \epsilon_0),$$
$$\mathbf{x}_j \in B(\mathbf{x}_{j-1}, \epsilon_{j-1}) \text{ for } j = 2, \cdots, n$$

and

$$\mathbf{y} \in B(\mathbf{x}_n, \epsilon_n).$$

This construction is illustrated in Figure 4.13.

This yields a sequence of points in Ω from $\mathbf{x}_0$ to $\mathbf{y}$, each contained in an open ball about its center $\mathbf{x}_j$, with every pair of consecutive open balls about some $\mathbf{x}_j$ containing the center $\mathbf{x}_{j-1}$ of the preceding open ball.

Walk from $\mathbf{x}_0$ to $\mathbf{y}$ along Γ. First, $u(\mathbf{x}) = M$ at every point of $B(\mathbf{x}_0, \epsilon_0)$, and $\mathbf{x}_1$ is in this open ball, so $u(\mathbf{x}_1) = M$. Then $u(\mathbf{x}) = M$ on all of $B(\mathbf{x}_1, \epsilon_1)$, by the reasoning previously applied on $B(\mathbf{x}_0, \epsilon)$.

But $\mathbf{x}_2$ is in this open ball about $\mathbf{x}_1$, so $u(\mathbf{x}_2) = M$. But then

$$u(\mathbf{x}) = M$$

on all of $B(\mathbf{x}_2, \epsilon_2)$. Continuing toward $\mathbf{y}$ along Γ, we find that $u(\mathbf{x}) = M$ on each successive open ball, hence at the center of the next open ball, and so on.

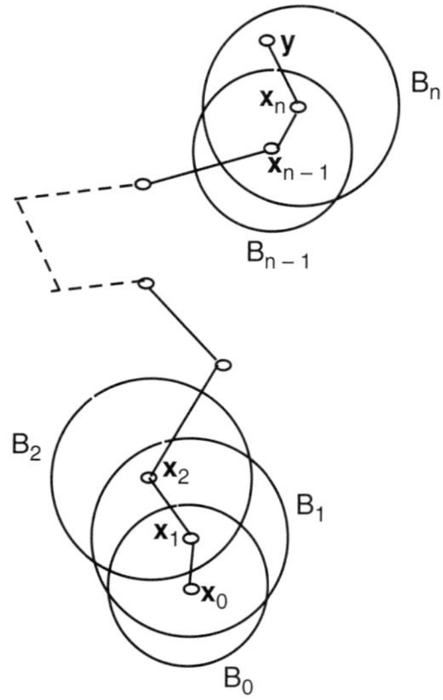

Figure 4.13: Chain of open balls connecting $\mathbf{x}_0$ to $\mathbf{y}$ along Γ in Ω.

After n steps, we conclude that $u(\mathbf{x}) = M$ on the last open ball, about $\mathbf{y}$, and hence

$$u(\mathbf{y}) = M$$

also. This proves that $u(\mathbf{x}) = M$ on all of Ω, and, by continuity, on $\overline{\Omega}$. This contradicts the assumption that u is not a constant function, and proves that u cannot assume a maximum value at a point of Ω. This maximum, therefore, can only be achieved at points on the boundary of Ω.

By applying this argument to $-u$, we can also show that, on $\overline{\Omega}$, u achieves its minimum value only at one or more boundary points of Ω.

One important consequence of the maximum principle is that a Dirichlet problem can have only one continuous solution on a bounded domain.

Theorem 4.6 (Uniqueness of Solutions of Dirichlet Problems) Let Ω be a bounded domain in R^n and let $f(\mathbf{x})$ be continuous on $\partial\Omega$. Then the Dirichlet problem

$$\nabla^2 u = 0 \text{ on } \Omega,$$
$$u(\mathbf{x}) = f(\mathbf{x}) \text{ for } \mathbf{x} \in \partial\Omega$$

can have only one solution that is continuous on Ω.

Proof Suppose u and v are continuous solutions. Let $h = u - v$. Then h is harmonic on Ω and $h = 0$ on $\partial\Omega$. Since h must achieve its maximum and minimum values for $\overline{\Omega}$ on $\partial\Omega$, then h must be identically zero on $\overline{\Omega}$. Therefore $u = v$.

Using the maximum property, it can also be proved that continuous solutions of the Dirichlet problem depend continuously on the boundary data. Small changes in the given boundary function of a Dirichlet problem cause small changes in the solution.

Theorem 4.7 (Continuous Dependence on Boundary Data) Let Ω be a bounded domain in R^n. Let f and g be continuous on $\partial\Omega$. Let w be the continuous solution of

$$\nabla^2 u = 0 \text{ on } \Omega, u = f \text{ on } \partial\Omega,$$

and let v be the continuous solution of

$$\nabla^2 u = 0 \text{ on } \Omega, u = g \text{ on } \partial\Omega.$$

Suppose ϵ is a positive number, and that

$$|f(\mathbf{x}) - g(\mathbf{x})| < \epsilon \text{ for } \mathbf{x} \in \partial\Omega.$$

Then

$$|w(\mathbf{x}) - v(\mathbf{x})| < \epsilon \text{ for } \mathbf{x} \in \overline{\Omega}.$$

We conclude this section with Harnack's inequality.

Suppose $u(r, \theta)$ is a solution of a Dirichlet problem on the disk $r < \rho$, with $u(\rho, \theta) = f(\theta)$ given on the boundary circle. By Poisson's integral formula 4.6,

$$u(r, \theta) = \frac{1}{2\pi} \int_{-\pi}^{\pi} \frac{\rho^2 - r^2}{\rho^2 + r^2 - 2r\rho\cos(\theta - \xi)} f(\xi)\, d\xi.$$

Now,

$$-1 \leq \cos(\theta - \xi) \leq 1$$

in the denominator. Replacing this cosine term first by -1, and then by 1, in Poisson's solution, we obtain

$$\frac{1}{2\pi} \int_{-\pi}^{\pi} \frac{\rho^2 - r^2}{\rho^2 + r^2 + 2r\rho} f(\xi)\, d\xi \leq u(r, \theta) \leq \frac{1}{2\pi} \int_{-\pi}^{\pi} \frac{\rho^2 - r^2}{\rho^2 + r^2 - 2r\rho} f(\xi)\, d\xi.$$

This inequality can be written as

$$\frac{(\rho - r)(\rho + r)}{(\rho + r)^2} \frac{1}{2\pi} \int_{-\pi}^{\pi} f(\xi)\, d\xi \leq u(r, \theta) \leq \frac{(\rho - r)(\rho + r)}{(\rho - r)^2} \frac{1}{2\pi} \int_{-\pi}^{\pi} f(\xi)\, d\xi.$$

Then

$$\frac{\rho - r}{\rho + r} \frac{1}{2\pi} \int_{-\pi}^{\pi} f(\xi)\, d\xi \leq u(r, \theta) \leq \frac{\rho + r}{\rho - r} \frac{1}{2\pi} \int_{-\pi}^{\pi} f(\xi)\, d\xi.$$

But, from Poisson's integral formula,

$$u(0,0) = \frac{1}{2\pi} \int_{-\pi}^{\pi} f(\xi)\, d\xi$$

We can, therefore, write the last inequalities as

$$\frac{\rho - r}{\rho + r} u(0,0) \leq u(r,\theta) \leq \frac{\rho + r}{\rho - r} u(0,0). \tag{4.12}$$

for $0 \leq r < \rho$ and $-\pi \leq \theta \leq \pi$. This is *Harnack's inequality*. It will be used to further explore further properties of harmonic functions, and will also play a role in an existence theorem for Dirichlet problems at the end of this chapter.

Problems for Section 4.4.3

1. Write out the details of a proof of the mean value property for bounded domains in the plane.

2. Let Ω be a domain in the plane bounded by a piecewise smooth closed curve C. Let $u, v \in C^2(\overline{\Omega})$. Prove Green's first identity:

$$\oint_C u \frac{\partial v}{\partial n}\, ds = \iint_{\overline{\Omega}} (u\nabla^2 v + \nabla u \cdot \nabla v)\, dA.$$

3. Use Green's first identity to prove Green's second identity for a bounded domain in the plane.

4. Let Ω be a bounded domain in the plane having a piecewise smooth curve as boundary. Let $u \in C^2(\overline{\Omega})$ and suppose that

$$\nabla^2 u = 0 \text{ on } \Omega,$$
$$u = 0 \text{ on } \partial\Omega.$$

   (a) Use Green's first identity to show that $u = 0$ on $\overline{\Omega}$.

   (b) Use the maximum principle to show that $u = 0$ on $\overline{\Omega}$.

5. Fill in the details of the following proof of the mean value theorem for harmonic functions on a bounded domain in the plane. Let $(x, y) \in \Omega$ and consider an open ball B of radius ϵ about (x, y) and lying entirely in Ω. If $(\xi, \eta) \in B$, write

$$\xi = x + r\cos(\theta), \eta = y + r\sin(\theta).$$

   Apply Green's first identity to show that

$$\int_0^{2\pi} u_r(x + r\cos(\theta), y + r\sin(\theta)) r\, d\theta = \iint_B r\nabla^2 u\, dr\, d\theta.$$

Conclude that the integral on the left is zero; therefore,

$$\frac{\partial}{\partial r}\int_0^{2\pi} u(x + r\cos(\theta), y + r\sin(\theta))\, d\theta = 0.$$

This means that

$$\int_0^{2\pi} u(x + r\cos(\theta), y + r\sin(\theta))\, d\theta$$

is independent of r, and, therefore, must equal the value of this integral at $r = 0$. Thus conclude that

$$\int_0^{2\pi} u(x, y)\, d\theta = \int_0^{2\pi} u(x + r\cos(\theta), y + r\sin(\theta))\, d\theta.$$

From this, show that

$$u(x, y) = \frac{1}{2\pi}\int_0^{2\pi} u(x + r\cos(\theta), y + r\sin(\theta))\, d\theta.$$

Finally, explain why this represents $u(x, y)$ as the average of function values on a circle of radius r about (x, y),

6. We will outline another proof of the maximum principle for harmonic functions. Fill in the details. Let u be harmonic on a bounded domain in the plane, having piecewise continuous closed boundary $\partial\Omega$. We want to show that the maximum value of $u(x, y)$ on $\overline{\Omega}$ is achieved only on $\partial\Omega$.

   First prove that, if v is continuous on $\overline{\Omega}$ and $v_{xx}+v_{yy} > 0$ on Ω, then $v(x, y)$ achieves its maximum value only on $\partial\Omega$. (Suppose that this maximum is achieved at an interior point $(x_0, y_0) \in \Omega$. Show that $v_{xx}(x_0, y_0) \leq 0$ and $v_{yy}(x_0, y_0) \leq 0$, obtaining a contradiction). Now let

$$M_{\partial\Omega} = \max_{(x,y)\in\partial\Omega} u(x, y)$$

   and let

$$v(x, y) = u(x, y) + \epsilon(x^2 + y^2),$$

   where ϵ is any positive number. Show that u must achieve its maximum M_v on $\overline{\Omega}$ at a point of $\partial\Omega$. Then

$$v(x, y) \leq M_v \leq M_{\partial\Omega} + \epsilon D,$$

   where D is the largest value of $x^2 + y^2$ for $(x, y) \in \overline{\Omega}$. Thus show that

$$u(x, y) \leq M_{\partial\omega} + \epsilon D.$$

   Finally, use the fact that ϵ can be arbitrarily close to 0.

7. Prove the *normal derivative lemma.* Let Ω be a domain in the plane and let u be continuous on $\overline{\Omega}$ and harmonic on Ω. Let the maximum of u on $\overline{\Omega}$ be achieved at a boundary point $\mathbf{x}_0$ and suppose that $\partial\Omega$ has a tangent at $\mathbf{x}_0$. Suppose also that the normal derivative

$$\left.\frac{\partial u}{\partial n}\right]_{\mathbf{x}=\mathbf{x}_0}$$

exists and that u is not a constant function. Then

$$\left.\frac{\partial u}{\partial n}\right]_{\mathbf{x}=\mathbf{x}_0} > 0.$$

Hint: Recall that

$$\left.\frac{\partial u}{\partial n}\right]_{\mathbf{x}=\mathbf{x}_0} = \lim_{h\to 0}\frac{1}{h}(u(\mathbf{x}_0) - u(\mathbf{x}_0 - h\mathbf{n})),$$

where $\mathbf{n}$ is the unit outer normal to $\partial\Omega$ at $\mathbf{x}_0$. It is enough to prove the lemma when $\Omega = B(\mathbf{x}_0, r)$. Define

$$v(\mathbf{x}) = \ln\left(\frac{r}{|\mathbf{x}|}\right)$$

for $\mathbf{x} \neq \mathbf{0}$. Then $w(\mathbf{x}) = u(\mathbf{x}) + hv(\mathbf{x})$ is harmonic for $r/2 < |\mathbf{x}| < r$ and continuous on the closure of this domain. Show that

$$\max_{|\mathbf{x}|=r/2} u(\mathbf{x}) < u(\mathbf{x}_0).$$

Hence show that $w(\mathbf{x}) < u(\mathbf{x}_0)$ for h sufficiently small. Conclude that w achieves a maximum at $\mathbf{x} = \mathbf{x}_0$. Conclude that

$$\left.\frac{\partial u}{\partial n}\right]_{\mathbf{x}=\mathbf{x}_0} \geq 0.$$

From this, complete the proof of the normal derivative lemma.

The lemma is valid for domains in R^n, with the concept of tangent line at $\mathbf{x}_0$ appropriately generalized (tangent plane if $n = 3$). For $n > 2$, the proof just outlined can be adapted by putting

$$v(\mathbf{x}) = |\mathbf{x}|^{2-n} - r^{2-n}.$$

8. Suppose u is harmonic and nonnegative on the entire plane. Prove that u must be a constant function. *Hint*: Notice that ρ can be any positive number in Harnack's inequality, and take the limit as $\rho \to \infty$.

   This result is a theorem of Liouville, and also follows from a theorem in complex analysis also named for Liouville (a bounded entire function must be constant).

9. Let u be harmonic on the entire plane, and suppose u is not a constant function. Show that u can have neither an upper bound or a lower bound over the entire plane.

4.5 The Neumann Problem

A Neumann problem is one in which we look for a function that is harmonic on a given domain, but, instead of being given function values on the boundary, we are given values of the normal derivative of the function on the boundary. This problem (in two dimensions) may be stated: find $u(,y)$ such that

$$\begin{aligned} \nabla^2 u &= 0 \text{ on } \Omega \\ \frac{\partial u}{\partial n} &= f(x,y) \text{ on } \partial\Omega \end{aligned} \tag{4.13}$$

In this section we will first look at the question of uniqueness of solutions of this problem, and derive a necessary condition for a solution to exist. Following this, we will solve the Neumann problem for a rectangle and a disk in the plane.

4.5.1 Uniqueness and Existence

Uniqueness of the solution of a Neumann problem is not an issue. If u is a solution, so is $u + c$ for any number c, because the added constant will vanish in the normal derivative condition on the boundary.

However, we can show that two continuous solutions of a Neumann problem can differ only by a constant.

Theorem 4.8 Let Ω be a bounded domain in R^n, having a closed, piecewise smooth boundary $\partial\Omega$. Let f be continuous on $\partial\Omega$. Let u and v be solutions of the Neumann problem 4.13. Then u and v differ by a constant on $\overline{\Omega}$.

We will prove the theorem for a domain in R^2, using Green's first identity.

Lemma 4.2 (Green's First Identity in R^2) Let Ω be a bounded domain in R^2, having a closed, piecewise smooth closed curve C as boundary. Let g and h be continuous with continuous first and second partial derivatives on $\overline{\Omega}$. Then

$$\oint_C g\frac{\partial h}{\partial n}\,ds = \iint_{\overline{\Omega}} (g\nabla^2 h + \nabla g \cdot \nabla h)\,dA.$$

Proof of the Lemma By Green's theorem,

$$\oint_C g\frac{\partial h}{\partial n}\,ds = \iint_{\overline{\Omega}} \operatorname{div}(g\nabla h)\,dA.$$

The rest of the proof of the lemma is just a computation of the integrand on the right:

$$\begin{aligned}\operatorname{div}(g\nabla h) &= \nabla \cdot \left(g\frac{\partial h}{\partial x}\mathbf{i} + g\frac{\partial h}{\partial y}\mathbf{j} \right) \\ &= \frac{\partial}{\partial x}\left(g\frac{\partial h}{\partial x} \right) + \frac{\partial}{\partial y}\left(g\frac{\partial h}{\partial y} \right) \\ &= g\left(\frac{\partial^2 h}{\partial x^2} + \frac{\partial^2 h}{\partial y^2} \right) + \frac{\partial g}{\partial x}\frac{\partial h}{\partial x} + \frac{\partial g}{\partial y}\frac{\partial h}{\partial y} \\ &= g\nabla^2 h + \nabla g \cdot \nabla h.\end{aligned}$$

This completes the proof of the lemma.

If $g = h$ in the lemma, then

$$\oint_C g\frac{\partial g}{\partial n}\,ds = \iint_{\overline{\Omega}} (g\nabla^2 g + |\nabla g|^2)\,dA. \tag{4.14}$$

Now we can prove the theorem.

Proof of Theorem 4.8 Let $u = v - w$. Then $\nabla^2 u = 0$ on Ω, and

$$\frac{\partial u}{\partial n} = \frac{\partial v}{\partial n} - \frac{\partial w}{\partial n} = f - f = 0 \text{ on } \partial\Omega.$$

Let $g = u$ into equation 4.14:

$$\begin{aligned}\oint_C u\frac{\partial u}{\partial n}\,ds &= 0 \\ &= \iint_{\overline{\Omega}} (u\nabla^2 u + |\nabla u|^2)\,dA \\ &= \iint_{\overline{\Omega}} |\nabla u|^2\,dA.\end{aligned}$$

But then

$$|\nabla u|^2 = u_x^2 + u_y^2 = 0$$

on $\overline{\Omega}$. Then

$$u_x(x, y) = u_y(x, y) = 0 \text{ on } \Omega$$

so $u(x, y)$ is constant on Ω. By continuity, $u(x, y)$ is constant on $\overline{\Omega}$.

A similar argument can be used to verify the conclusion of the theorem in R^3.

We will not attempt an existence theorem for solutions of a Neumann problem. However, it is easy to derive an important necessary condition for a solution to exist, for Neumann problems in the plane. Suppose u is a solution of problem 4.13. By equation 4.14, with $g = 1$ and $h = u$, we immediately have

$$\oint_{\partial\Omega} \frac{\partial u}{\partial n}\,ds = 0.$$

But $\partial u/\partial n = f$ on the boundary curve $\partial\Omega$, so existence of a solution requires that

$$\oint_{\partial\Omega} f\, ds = 0.$$

This means that, if this integral is not zero, a solution cannot exist. Vanishing of the integral of $\partial u/\partial n$ over the boundary of the region is a necessary condition for the Neumann problem 4.13 to have a solution.

For a domain in R^3, this necessary condition translates to

$$\iint_{\partial\Omega} f\, d\sigma = 0.$$

The surface integral of the given normal derivative over the boundary surface must be zero.

This condition has a physical interpretation. Think of Laplace's equation $\nabla^2 u = 0$ as the steady-state heat equation. Vanishing of this integral (in two or three dimensions) means that in the steady-state case the net flow of heat energy across the boundary of the domain must be zero. This is a conservation of energy condition, in the absence of sources or sinks of energy in the domain.

Problems for Section 4.5.1

1. Let Ω be a bounded domain in the plane and suppose $\partial\Omega$ is a piecewise smooth closed curve. Let q be continuous on $\overline{\Omega}$ and f on $\partial\Omega$. Prove that the problem

$$\nabla^2 u = q \text{ on } \Omega,$$
$$\frac{\partial u}{\partial n} = f \text{ on } \partial\Omega$$

can have a solution only if

$$\iint_{\overline{\Omega}} q(x,y)\, dA = \oint_{\partial\Omega} f\, ds.$$

This gives a necessary condition for this extended Neumann problem to have a solution.

2. Consider the boundary value problem

$$\nabla^2 u = 0 \text{ on } \Omega,$$
$$\frac{\partial u}{\partial n} + hu = f \text{ on } \partial\Omega.$$

Here it is assumed that Ω is a bounded domain in the plane, with piecewise smooth closed boundary curve $\partial\Omega$, while h and f are continuous on $\partial\Omega$, $h(\mathbf{x})$ is not identically zero, and $h(\mathbf{x}) \geq 0$ on $\partial\Omega$. Prove that this problem can have at most one solution.

3. Prove the conclusion of problem 2 for the 3−dimensional version of this problem.

4. Let f be continuous on the bounded domain Ω in R^2, and let g be continuous on the piecewise smooth curve bounding Ω. Let k be a real number. Prove that the problem

$$\nabla^2 u + ku = f \text{ on } \Omega,$$
$$\frac{\partial u}{\partial n} = g \text{ on } \partial\Omega$$

can have at most one solution if $k < 0$.

5. Let Ω be a bounded domain in R^2, whose boundary consists of two piecewise smooth closed curves C and K. Let f be continuous on Ω, and let g be continuous on C. Let k be a real number. Prove that the problem

$$\nabla^2 u = f \text{ on } \Omega,$$
$$u = g \text{ on } C,$$
$$\frac{\partial u}{\partial n} + ku = 0 \text{ on } K,$$

can have at most one solution.

4.5.2 Neumann Problem for a Rectangle

We will solve a Neumann problem for an L by K rectangle in the plane. Assume that the rectangle has one corner at the origin and two sides along the positive horizontal and vertical axes. In this case the normal derivative of u on each vertical side is $\pm\partial u/\partial x$, because the unit outer normal vector to these sides is either $\mathbf{i}$ or $-\mathbf{i}$. The normal derivative of u on each horizontal side is $\pm\partial u/\partial y$, because the unit outer normal vector to these sides is either $\mathbf{j}$ or $-\mathbf{j}$ (Figure 4.14).

As with the Dirichlet problem for a rectangle, a Neumann problem for a rectangle is often split into four subproblems, in each of which nonzero data (a normal derivative) is specified on only one side. As an illustration, the particular problem we will solve is

$$\nabla^2 u(x, y) = 0 \text{ for } 0 < x < L, 0 < y < K,$$
$$u_y(x, 0) = u_y(x, K) = 0 \text{ for } 0 < x < L,$$
$$u_x(0, y) = 0, u_x(L, y) = g(y) \text{ for } 0 < y < K.$$

For a solution to exist, the integral of the given normal derivative around the boundary of the rectangle must be zero. For the problem as stated, this is the requirement that

$$\int_0^K g(y)\, dy = 0,$$

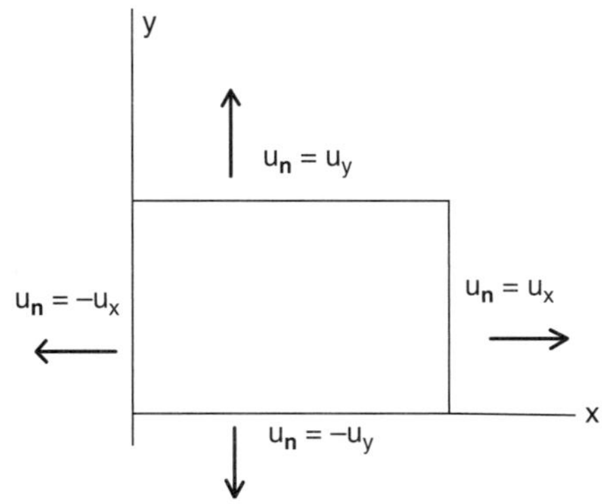

Figure 4.14: Normal derivatives on the sides of a rectangle $0 \leq x \leq L, 0 \leq y \leq k$.

and we assume that g satisfies this condition. As we work toward the solution, it will be seen why this condition is required.

To separate the variables, let $u(x,y) = X(x)Y(y)$ to obtain

$$X'' - \lambda X = 0, Y'' + \lambda Y = 0,$$

in which λ is the separation constant. Now,

$$u_y(x,0) = X(x)Y'(0) = 0,$$

so $Y'(0) = 0$. Similarly, $u_y(x,K) = 0$ implies that $Y'(K) = 0$. And $u_x(0,y) = 0$ gives us $X'(0) = 0$. The problems for X and Y are

$$X'' - \lambda X = 0; X'(0) = 0$$

and

$$Y'' + \lambda Y = 0; Y'(0) = Y'(K) = 0.$$

This is problem 1.18, so the problem for X has eigenvalues and eigenfunctions

$$\lambda_n = \frac{n^2\pi^2}{K^2}, Y_n(y) = \cos\left(\frac{n\pi y}{K}\right)$$

for $n = 0, 1, 2, \cdots$.

Now that we know the eigenvalues, the problem for X is

$$X'' - \frac{n^2\pi^2}{K^2}X = 0; X'(0) = 0,$$

where n can be any nonnegative integer. If $n = 0$, then

$$X(x) = cx + d,$$

and $X'(0) = 0$ implies that $c = 0$. Therefore $X_0(x)$ is a constant function. If $n = 1, 2, \cdots$, the differential equation for x has general solution

$$X(x) = ce^{n\pi x/K} + de^{-n\pi x/K}.$$

Then

$$X'(0) = \frac{n\pi}{K}(c - d) = 0,$$

so $c = d$ and $X_n(x)$ is a multiple of $\cosh(n\pi x/K)$.

Now we have functions

$$u_0(x, y) = \text{ constant}$$

and, for $n = 1, 2, \cdots$,

$$u_n(x, y) = \alpha_n \cos\left(\frac{n\pi y}{K}\right) \cosh\left(\frac{n\pi x}{K}\right).$$

These are harmonic and satisfy the homogeneous boundary conditions on the the top, lower and left sides. To satisfy the nonhomogeneous condition on the right side $x = L$, attempt a superposition

$$u(x, y) = \alpha_0 + \sum_{n=1}^{\infty} a_n \cos\left(\frac{n\pi y}{K}\right) \cosh\left(\frac{n\pi x}{K}\right). \qquad (4.15)$$

For the coefficients, we need

$$u_x(L, y) = \sum_{n=1}^{\infty} \alpha_n \left(\frac{n\pi}{K}\right) \sinh\left(\frac{n\pi L}{K}\right) \cos\left(\frac{n\pi y}{K}\right) = g(y).$$

Here we see why vanishing of $\int_0^K g(y)\,dy$ is a necessary condition for this problem to have a solution. This series is a Fourier cosine expansion of $g(y)$ on $[0, K]$. The constant term in this expansion is

$$\frac{2}{K}\int_0^K g(\xi)\,d\xi.$$

But we see in this solution that this expansion has constant term zero. The only way these statements can be consistent is for this integral to be zero, and this is the necessary condition for existence.

For the other coefficients, we need

$$\alpha_n \frac{n\pi}{K} \sinh(n\pi L/K) = \frac{2}{K}\int_0^K g(\xi) \cos\left(\frac{n\pi\xi}{K}\right) d\xi.$$

Then

$$\alpha_n = \frac{2}{n\pi \sinh(n\pi L/K)} \int_0^K g(\xi) \cos\left(\frac{n\pi\xi}{K}\right) d\xi.$$

With these coefficients, equation 4.15 is the solution of this Neumann problem. Here, α_0 is an arbitrary constant, not a Fourier cosine coefficient. An arbitrary constant is always attached to a solution of a Neumann problem because Laplace's equation and the boundary conditions all involve derivatives, so any added constant vanishes in the differentiations. Neumann problems do not have unique solutions.

Problems for Section 4.5.2

1. Solve

$$\begin{aligned}\nabla^2 u &= 0 \text{ for } 0 < x < 1, 0 < y < 1,\\ u_x(0,y) &= u_x(1,y) = 0 \text{ for } 0 < y < 1,\\ u_y(x,0) &= 4\cos(\pi x), u_y(x,1) = 0 \text{ for } 0 < x < 1.\end{aligned}$$

2. Solve

$$\begin{aligned}\nabla^2 u &= 0 \text{ for } 0 < x < 1, 0 < y < \pi,\\ u_x(0,y) &= y - \frac{\pi}{2}, u_x(1,y) = \cos(y) \text{ for } 0 < y < \pi,\\ u_y(x,0) &= u_y(x,\pi) = 0 \text{ for } 0 < x < 1.\end{aligned}$$

3. Solve

$$\begin{aligned}\nabla^2 u &= 0 \text{ for } 0 < x < \pi, 0 < y < \pi,\\ u_x(0,y) &= u_x(\pi,y) = 0 \text{ for } 0 < y < \pi,\\ u_y(x,0) &= \cos(3x), u_y(x,\pi) = 6x - 3\pi \text{ for } 0 < x < \pi.\end{aligned}$$

4. Use separation of variables to solve the mixed boundary value problem:

$$\begin{aligned}\nabla^2 u &= 0 \text{ for } 0 < x < \pi, 0 < y < \pi,\\ u(x,0) &= f(x), u(x,\pi) = 0 \text{ for } 0 < x < \pi,\\ u_x(0,y) &= u_x(\pi,y) = 0 \text{ for } 0 < y < \pi.\end{aligned}$$

   Does this problem have a unique solution?

5. Attempt a separation of variables to solve

$$\begin{aligned}\nabla^2 u &= 0 \text{ for } 0 < x < 1, 0 < y < 1,\\ u(x,0) &= u(x,1) = 0 \text{ for } 0 < x < 1,\\ u_x(0,y) &= 3y^2 - 2y, u_y(1,y) = 0 \text{ for } 0 < x < 1.\end{aligned}$$

4.5.3 Neumann Problem for a Disk

We will solve the Neumann problem for a disk about the origin:

$$\nabla^2 u(r,\theta) = 0 \text{ for } 0 \le r < \rho, -\pi \le \theta \le \theta,$$
$$\frac{\partial u}{\partial r}(\rho,\theta) = f(\theta) \text{ for } -\pi \le \theta \le \pi.$$

The boundary condition specifies $\partial u/\partial r$ because this is the normal derivative to the circle bounding the disk. At any point on this circle, the radial line from the origin through this point is normal to the circle.

A necessary condition for a solution to exist is that

$$\int_{-\pi}^{\pi} f(\theta)\, d\theta = 0,$$

and we will assume that $f(\theta)$ satisfies this condition. We also assume that $f(\theta)$ is periodic of period 2π, as we expect in polar coordinates.

As with the Dirichlet problem for a disk, attempt a solution

$$u(r,\theta) = \frac{1}{2}a_0 + \sum_{n=1}^{\infty} \left(a_n r^n \cos(n\theta) + b_n r^n \sin(n\theta)\right).$$

We must choose the coefficients to satisfy the boundary condition, so

$$\frac{\partial u}{\partial r}(\rho,\theta) = \sum_{n=1}^{\infty} \left(a_n n\rho^{n-1}\cos(n\theta) + b_n n\rho^{n-1}\sin(n\theta)\right) = f(\theta).$$

This is a Fourier expansion of $f(\theta)$ on $[-\pi,\pi]$, so the coefficients (taking into account the factors $n\rho^{n-1}$) are

$$a_n = \frac{1}{n\pi\rho^{n-1}} \int_{-\pi}^{\pi} f(\xi)\cos(n\xi)\, d\xi$$

and

$$b_n = \frac{1}{n\pi\rho^{n-1}} \int_{-\pi}^{\pi} f(\xi)\sin(n\xi)\, d\xi$$

for $n = 1, 2, \cdots$. The constant term in this expansion is zero because we are assuming that the integral of $f(\theta)$ over the interval is zero.

Upon inserting these coefficients, the solution is

$$u(r,\theta) = \frac{1}{2}a_0 + \frac{\rho}{\pi}\sum_{n=1}^{\infty} \frac{1}{n}\left(\frac{r}{\rho}\right)^n \int_{-\pi}^{\pi} [\cos(n\xi)\cos(n\theta) + \sin(n\xi)\sin(\theta)] f(\xi)\, d\xi. \quad (4.16)$$

a_0 is the arbitrary constant that can be added to any solution of a Neumann problem. It is denoted $a_0/2$ for convenience in an upcoming computation.

Reminiscent of the Dirichlet problem for a disk, we will manipulate equation 4.16 to obtain an integral solution of this Neumann problem. Using a trigonometric identity, write

$$u(r,\theta) = \frac{1}{2}a_0 + \frac{\rho}{\pi}\int_{-\pi}^{\pi}\left(\sum_{n=1}^{\infty}\frac{1}{n}\left(\frac{r}{\rho}\right)^n \cos(n(\theta-\xi))\right) f(\xi)\,d\xi. \qquad (4.17)$$

Let $z = Re^{i\zeta}$, so

$$z^n = R^n(\cos(n\zeta) + i\sin(n\zeta))$$

and

$$\sum_{n=1}^{\infty} R^n\cos(n\zeta) = \sum_{n=1}^{\infty}\operatorname{Re}(z^n) = \operatorname{Re}\left[\sum_{n=1}^{\infty} z^n\right].$$

If $0 \le R < 1$, then $|z| < 1$ and

$$\sum_{n=1}^{\infty} z^n = \frac{z}{1-z}.$$

Compute

$$\frac{1}{2} + \sum_{n=1}^{\infty} z^n = \frac{1}{2} + \frac{z}{1-z} = \frac{1+z}{2(1-z)}.$$

To extract the real part of this expression, first multiply numerator and denominator on the right by $1-\overline{z}$:

$$\frac{1}{2} + \sum_{n=1}^{\infty} z^n = \frac{1}{2}\frac{1-\overline{z}+z-z\overline{z}}{1-z-\overline{z}+z\overline{z}}.$$

But

$$z + \overline{z} = 2R\cos(\zeta),$$
$$z - \overline{z} = -2iR\sin(\zeta),$$

and

$$z\overline{z} = |z|^2 = R^2.$$

Therefore

$$\frac{1}{2} + \sum_{n=1}^{\infty} z^n = \frac{1}{2}\left(\frac{1-2iR\sin(\zeta)-R^2}{1-2R\cos(\zeta)+R^2}\right).$$

Because the denominator on the right is real, it is easy to identify the real part of both sides of this equation, obtaining

$$\frac{1}{2} + \sum_{n=1}^{\infty} R^n\cos(n\zeta) = \frac{1}{2}\left(\frac{1-R^2}{1+R^2-2R\cos(\zeta)}\right).$$

Then

$$\sum_{n=1}^{\infty} R^n \cos(n\zeta) = \frac{1}{2}\left(\frac{1-R^2}{1+R^2-2R\cos(\zeta)}\right) - \frac{1}{2}$$
$$= \frac{R\cos(\zeta) - R^2}{1+R^2-2R\cos(\zeta)}.$$

Upon dividing by R, we obtain

$$\sum_{n=1}^{\infty} R^{n-1}\cos(n\zeta) = \frac{\cos(\zeta) - R}{1+R^2-2R\cos(\zeta)}.$$

Then

$$\int_0^R \sum_{n=1}^{\infty} t^{n-1}\cos(n\zeta)\,dt = \sum_{n=1}^{\infty}\int_0^R t^{n-1}\cos(n\zeta)\,dt$$
$$= \sum_{n=1}^{\infty}\frac{1}{n}R^n\cos(n\zeta)$$
$$= \int_0^R \frac{\cos(\zeta)-t}{1+t^2-2t\cos(\zeta)}\,dt$$
$$= -\frac{1}{2}\ln(1+R^2-2R\cos(\zeta)).$$

Now put $R = r/\rho$ and $\zeta = \theta - \xi$:

$$\sum_{n=1}^{\infty}\frac{1}{n}\left(\frac{r}{\rho}\right)^n \cos(n(\theta-\xi)) = -\frac{1}{2}\ln\left(1+\frac{r^2}{\rho^2} - 2\frac{r}{\rho}\cos(\theta-\zeta)\right).$$

Finally we have an integral formula for the solution of the Neumann problem for a disk of radius ρ about the origin:

$$u(r,\theta) = \frac{1}{2}a_0 - \frac{\rho}{2\pi}\int_{-\pi}^{\pi}\ln\left(1+\frac{r^2}{\rho^2} - 2\frac{r}{\rho}\cos(\theta-\xi)\right) f(\xi)\,d\xi, \qquad (4.18)$$

in which a_0 is an arbitrary constant.

Problems for Section 4.5.3

1. Write two solutions of the following Neumann problem, one using equation 4.16 and one using 4.18:

$$\nabla^2 u(r,\theta) = 0 \text{ for } 0 \le r < \rho, -\pi \le \theta \le \pi,$$
$$\frac{\partial u}{\partial r}(\rho,\theta) = \sin(3\theta) \text{ for } -\pi \le \theta \le \pi.$$

2. Write two solutions of the following Neumann problem, one using equation 4.16 and one using 4.18:

$$\nabla^2 u(r,\theta) = 0 \text{ for } 0 \le r < \rho, -\pi \le \theta \le \pi,$$
$$\frac{\partial u}{\partial r}(\rho,\theta) = \cos(3\theta) \text{ for } -\pi \le \theta \le \pi.$$

3. By converting the problem to polar coordinates, solve:

$$\nabla^2 u(x,y) = 0 \text{ for } x^2+y^2 < 9,$$
$$\frac{\partial u}{\partial n} = 4xy \text{ if } x^2+y^2 = 9.$$

The solution should be in terms of rectangular coordinates.

4. Solve the Neumann problem:

$$\nabla^2 u(x,y) = 0 \text{ for } x^2+y^2 < 1,$$
$$\frac{\partial u}{\partial n} = x \text{ for } x^2+y^2 = 1.$$

The solution should be in terms of rectangular coordinates.

5. Solve the Neumann problem:

$$\nabla^2 u(x,y) = 0 \text{ for } x^2+y^2 < 1,$$
$$\frac{\partial u}{\partial n} = xy^2 \text{ if } x^2+y^2 = 1.$$

4.6 Poisson's Equation

Simeon Denis Poisson (1781–1840) was a French mathematician and physicist. The equation

$$u_{xx} + u_{yy} = F(x,y)$$

is called *Poisson's equation* after him. Boundary conditions consist of values the solution is to have on the boundary of the domain of interest. We may therefore think of this as a generalization of a Dirichlet problem.

In the absence of boundary data, if $u(x,y)$ is a solution of Poisson's equation and $h(x,y)$ is harmonic on D, then $u+h$ also satisfies Poisson's equation.

In some cases, it is possible to write a series or integral solution of Poisson's equation on a domain, with values given on the boundary of the domain. In particular, consider the following problem on a rectangle R, in which the solution of Poisson's equation is to vanish on the boundary of R:

$$u_{xx} + u_{yy} = F(x,y) \text{ for } 0 < x < a, 0 < y < b,$$
$$u(x,0) = u(x,b) = 0 \text{ for } 0 < x < a,$$
$$u(0,y) = u(a,y) = 0 \text{ for } 0 < y < b.$$

Assume that, for any y in $[0, b]$, $F(x, y)$, thought of as a function of x, has a Fourier sine expansion on $[0, a]$:

$$F(x,y) = \sum_{n=1}^{\infty} f_n(y) \sin\left(\frac{n\pi x}{a}\right).$$

We know these sine coefficients:

$$f_n(y) = \frac{2}{a}\int_0^a F(\xi, y) \sin\left(\frac{n\pi\xi}{a}\right) d\xi.$$

Now write

$$X_n(x) = \sin\left(\frac{n\pi x}{a}\right).$$

The strategy we will carry out is to find functions $u_n(x, y)$ satisfying

$$(u_n)_{xx} + (u_n)_{yy} = f_n(y)X_n(x).$$

If we can do this, then

$$u(x,y) = \sum_{n=1}^{\infty} u_n(x,y)$$

will satisfy $\nabla^2 u(x, y) = F(x, y)$.

To determine the functions $u_n(x, y)$, we will look for functions $Y_n(y)$ such that

$$u_n(x,y) = X_n(x)Y_n(y).$$

We know that $X_n(x)$ satisfies.

$$X_n'' + \frac{n^2\pi^2}{a^2}X_n = 0.$$

Then

$$\begin{aligned}(u_n)_{xx} + (u_n)_{yy} &= Y_n''X_n + Y_nX_n'' \\ &= Y_n''X_n - \frac{n^2\pi^2}{a^2}Y_nX_n.\end{aligned}$$

Then

$$\left(Y_n''(y) - \frac{n^2\pi^2}{a^2}Y_n(y)\right)X_n(x) = f_n(y)X_n(x),$$

so

$$Y_n''(y) - \frac{n^2\pi^2}{a^2}Y_n(y) = f_n(y). \tag{4.19}$$

Now $f_n(y)$ is a known function, in the sense that it is determined as a Fourier sine expansion of the given function $F(x, y)$, in terms of x, on $[0, a]$. In this expansion, y is thought of as fixed, so the coefficients are functions of y.

Since $u_n(x, 0) = u_n(x, b) = 0$, we must have

$$Y_n(0) = Y_n(b) = 0.$$

These boundary conditions, in conjunction with the differential equation 4.19, are enough to completely determine $Y_n(y)$. When this is done, the solution of the problem is

$$u(x,y) = \sum_{n=1}^{\infty} X_n(x)Y_n(y).$$

We will illustrate this method of solution for the rectangle $R : 0 \le x \le \pi, 0 \le y \le \pi$, with $F(x,y) = x^2y$. First compute

$$\begin{aligned} f_n(y) &= \frac{2}{\pi}\int_0^{\pi} \xi^2 y \sin(n\xi)\, d\xi \\ &= \frac{2y}{n^3\pi}[2\cos(n\pi) - 2 - n^2\pi^2\cos(n\pi)] = k_n y, \end{aligned}$$

where

$$k_n = \frac{2}{n^3\pi}[2((-1)^n - 1) - n^2\pi^2(-1)^n].$$

Next, solve

$$Y_n'' - n^2Y_n = f_n(y) = k_n y; Y_n(0) = Y_n(b) = 0$$

to obtain

$$Y_n(y) = \frac{4k_n}{n^2\sinh(4n)}\sinh(ny) - \frac{k_n y}{n^2}.$$

Then

$$u(x,y) = \sum_{n=1}^{\infty} Y_n(y)\sin(nx).$$

It is possible to write general integral solutions for Poisson's equation for a bounded domain D in two and three dimensions. In R^2, the solution is

$$u(x,y) = \iint_D F(\xi,\eta)\ln\left(\frac{1}{\rho((x,y),(\xi,\eta))}\right) d\xi\, d\eta,$$

in which (x,y) is an arbitrary point of D and $\rho((x,y),(\xi,\eta))$ is the distance between these two points:

$$\rho((x,y),(\xi,\eta)) = \sqrt{(x-\xi)^2 + (y-\eta)^2}.$$

In three dimensions the solution is a triple integral:

$$u(x,y,z) = \iiint_D F(\xi,\eta,\zeta)\frac{1}{\rho((x,y,z),(\xi,\eta,\zeta))}\, d\xi\, d\eta\, d\zeta,$$

where $\rho((x,y,z),(\xi,\eta,\zeta))$ is the distance in R^3 between these two points.

Derivations of these integral theorems are similar to those of the representation theorems for harmonic functions. As we might expect, the integrals can be evaluated in closed form only for special choices of D and F.

Problems for Section 4.6

For each of the following, write a solution of the Poisson problem for the given b and $F(x,y)$, with $a = \pi$.

1. $b = 3, F(x, y) = \sin(5x)$.

2. $b = 4, F(x, y) = \sqrt{2}\sin(12x) - 14\sin(3x)$.

3. $b = 1, F(x, y) = x$.

4. $b = 2\pi, F(x, y) = \sqrt{3}y$.

5. $b = 12, F(x, y) = xy$.

6. $b = 7, F(x, y) = x - y$.

4.7 Existence Theorem for a Dirichlet Problem

In this section we will prove an existence theorem for continuous solutions of Dirichlet problems. First, however, it is interesting to see how a Dirichlet problem can fail to have a continuous solution. We will sketch a physically motivated example in 3−space, from French mathematician Henri Lebesgue (1875–1941).

Begin by defining the surface to be used in the example. Let S be a sphere of radius 1 about the origin $(0, 0, 0)$. Suppose S is made of a thin elastic material. Take a long thin, blunt spike (or bodkin), and stand on the outside surface of the sphere at $(0, 1, 0)$. Push the spike into the sphere, toward the origin, being careful to stretch but not tear the sphere. The material of the sphere is deformed inward around the spike and forms a thin tunnel approaching the origin, as suggested by Figure 4.15. The shaded band drawn on the surface will be used soon.

Let Ω be the domain bounded by the deformed surface. The deformed surface itself is $\partial\Omega$.

Define a function f over $\partial\Omega$. Let $f(x, y, z) = 0$ for (x, y, z) on the spike part of the surface, up to the shaded band. At this band, proceeding from the spike outward over the shaded band, make $f(x, y, z)$ grow at a rapid rate until it reaches a number T on the outer edge of this band. Beyond the band, $f(x, y, z)$ has the constant value T over the rest of the spherical part of $\partial\Omega$. T itself can be chosen arbitrarily large. We can also make the spike thin enough, and the shaded band of sufficiently small width, that the surface area of the band and the spike are arbitrarily small.

We will argue that, for this function and surface, the Dirichlet problem

$$\nabla^2 u(x, y, z) = 0 \text{ on } \Omega,$$
$$u(x, y, z) = f(x, y, z) \text{ for } (x, y, z) \in \partial\Omega$$

can have no continuous solution. For suppose that $u(x, y, z)$ is a solution. Think of u as the steady-state temperature distribution throughout Ω. The temperature on the surface $\partial\Omega$ is a constant T except over the spike and the shaded band, which can be constructed to have as small a surface area as we want. We

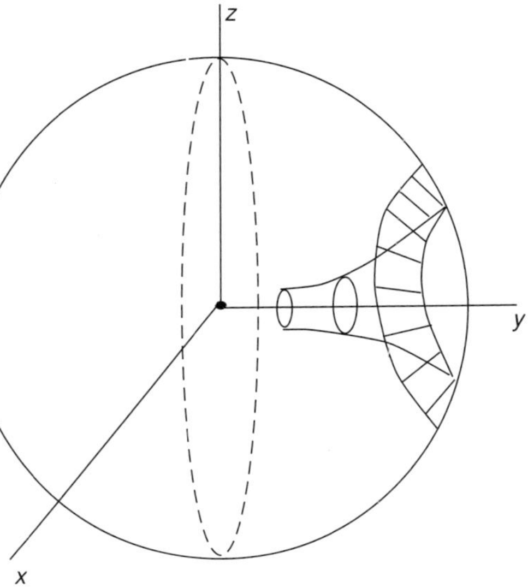

Figure 4.15: Lebesgue's example of a Dirichlet problem with no solution.

would therefore expect the steady-state temperature over Ω to be slightly less than this constant value T, say $T - \epsilon$ for some positive ϵ:

$$u(x, y, z) = T - \epsilon \text{ for } (x, y, z) \in \Omega.$$

But now u cannot be continuous on $\overline{\Omega}$, because $u(x, y, z) = 0$ at points on the spike, and $u(x, y, z) = T - \epsilon$ at points inside the sphere arbitrarily close to this spike. This Dirichlet problem has no continuous solution.

This example suggests that an existence theorem for continuous solutions of the Dirichlet problem must place some conditions on the domain, or perhaps its boundary, as well as on the boundary function prescribed on $\partial\Omega$.

We will now develop an existence theorem. The approach is called the *methode de balayage*, and was developed by French mathematician Henri Poincaré, Henri (1854–1912), who had a profound influence on the development of topology, differential equations, and dynamical systems. Because the proof was later refined by Perron, it is also known as the Poincaré-Perron method. We will develop the proof in R^2 because some details are easier to visualize there, but the idea extends to R^n.

The key is to construct a set of (not necessarily harmonic) functions called superharmonic functions, having the correct values on the boundary of the domain. From these a function that is harmonic on the domain, and takes on the correct boundary values, is manufactured, provided that at each point of the boundary a certain condition is satisfied.

As preparation for these constructions, we need two results named for Harnack. The first states that uniform convergence of a sequence of harmonic functions on the boundary of a domain ensures that the sequence converges uniformly on the closure of the entire domain, and, further, that the limit of the sequence is also harmonic.

Theorem 4.9 (Harnack's First Theorem) Let Ω be a bounded domain in the plane. Suppose $u_n(x,y)$ is a sequence of functions that are harmonic on Ω and continuous on $\overline{\Omega}$. Suppose that the sequence $u_n(x,y)$ converges uniformly on the boundary $\partial\Omega$. Then $u_n(x,y)$ converges uniformly on $\overline{\Omega}$ to a function that is harmonic on Ω

Proof Let $\epsilon > 0$. Because of the uniform convergence on $\partial\Omega$, there is a positive number N such that

$$|u_n(x,y) - u_m(x,y)| < \epsilon,$$

if $m, n \geq N$ and $(x,y) \in \partial\Omega$. Now apply the maximum principle to each $u_n(x,y)$ to conclude that

$$|u_n(x,y) - u_m(x,y)| < \epsilon,$$

if $n, m \geq N$, for all $(x,y) \in \overline{\Omega}$. This makes the convergence uniform on all of $\overline{\Omega}$.

Now suppose $u_n(x,y) \to U(x,y)$. All that remains is to show that U is harmonic on Ω.

To do this, let $P \in \Omega$ and let C be a circle of radius R centered at P, with R small enough that C and the disk it bounds are contained in Ω. Center polar coordinates at P to represent $u_n(x,y)$ at each (x,y) enclosed by C as a Poisson integral of the form

$$u_n(x,y) = \frac{1}{2\pi}\int_{-\pi}^{\pi} \frac{R^2 - \rho^2}{R^2 + \rho^2 - 2R\rho\cos(\theta - \xi)} u_n(Re^{i\xi})\, d\xi.$$

Because the convergence is uniform, we can take the limit as $n \to \infty$ inside the integral to write

$$U(x,y) = \frac{1}{2\pi}\int_{-\pi}^{\pi} \frac{R^2 - \rho^2}{R^2 + \rho^2 - 2R\rho\cos(\theta - \xi)} U(Re^{i\xi})\, d\xi,$$

and this proves that U is harmonic on this disk about P. Because P is any point of Ω, then U is harmonic on Ω.

Armed with this result, and a fact about compact sets, we can prove a stronger version of Harnack's first theorem. The fact we need about compact sets is this. If F is a compact set, and $D_1, D_2, \cdots$ are open disks that cover F (in the sense that every point of F is in one or more of the disks), then it is possible to cover F with a finite number $D_{i_1}, D_{i_2}, \cdots, D_{i_n}$ of these disks.

Theorem 4.10 (Harnack's Second Theorem) Ler $u_n(x,y)$ be a sequence of nonnegative functions that are harmonic on a bounded domain Ω of the plane. Suppose this sequence converges at some point $(a,b) \in \Omega$. Then $u_n(x,y)$

converges on Ω to a harmonic function, and this convergence is uniform on each compact subset of Ω.

Proof We will first show that $u_n(x, y)$ converges uniformly on every disk within Ω and centered at (a, b). Let D_1 be such a disk, having radius R. Suppose ϵ is positive such that the disk D_2 of radius $R + \epsilon$ about (a, b) is also contained in Ω. By Harnack's inequlaity,

$$\frac{R+\epsilon-\rho}{R+\epsilon+\rho}u_n(a,b) \leq u_n(\rho e^{i\theta}) \leq \frac{R+\epsilon+\rho}{R+\epsilon-\rho}u_n(a,b)$$

for any point in D_2, using polar coordinates centered at (a, b). This shows that convergence of the sequence of functions $u_n(x, y)$ at (a, b) implies convergence on $\overline{D_2}$. By Harnack's first theorem, the limit function is also harmonic on D_2.

We now want to prove the convergence of u_n at an arbitrary point P of Ω. The strategy for doing this is similar to that used to prove the maximum theorem. Connect (a, b) to P by a polygonal path L in Ω. For some positive number δ, each point of L is at least δ distance from $\partial\Omega$. Starting from the first circle, about (a, b), u_n converges uniformly in this circle, hence in the next circle and so on, until in a finite number of steps the last circle (about P) is reached. This proves convergence of u_n at P. Further, again by Harnack's first theorem, the limit function is harmonic about P. Finally, it must be shown that the convergence of u_n is uniform on each compact subset of Ω. Let F be such a subset. By the compactness of F, it is possible to cover F with a finite number of open disks $O_1, \cdots, O_m$, with the closure $\overline{O_j}$ of each of these disks entirely within Ω. But u_n converges at the center of each of these disks, hence, by what has already been proved, this sequence converges uniformly on each $\overline{O_j}$ and, therefore, also uniformly on F.

This completes a proof of Harnack's second theorem.

We are ready to begin the construction of the class of functions we will use to prove an existence theorem. For the remainder of this section, let Ω be a bounded domain in the plane, and let f be continuous on $\partial\Omega$. We will use the fact that a function is harmonic on an open set exactly when it is harmonic on every open disk within this set. This will enable us to exploit the fact that we know an integral solution of a Dirichlet problem on a disk.

Now suppose that K is any open disk and $K \subset \Omega$. Suppose v is continuous on $\overline{\Omega}$. Because the Dirichlet problem for a disk has a solution (the Poisson integral), there is a function $f_{K,v}$ which is harmonic on K, and

$$f_{K,v}(Q) = v(Q) \text{ for } Q \in \partial K.$$

Extend $f_{K,v}$ to all of $\overline{\Omega}$ by setting

$$f_{K,v}(P) = v(P) \text{ for } P \in \overline{\Omega} \text{ but } P \text{ not in } K.$$

This definition is illustrated in Figure 4.16.

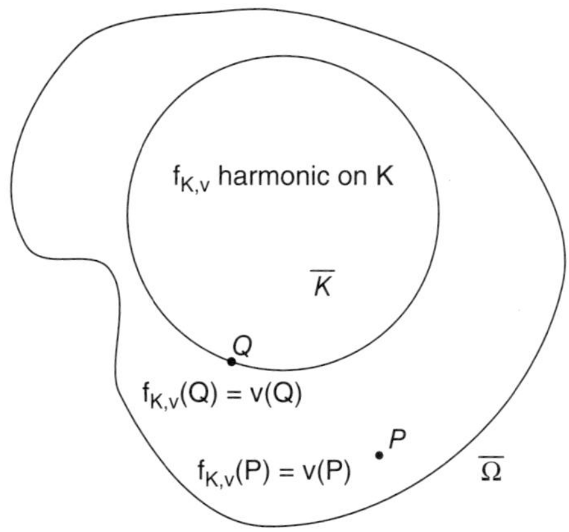

Figure 4.16: Extending $f_{(K,v)}$ first to ∂K, then to all of $\overline{\Omega}$ outside K.

In this way we associate with each continuous function on $\overline{\Omega}$, and each open disk K within Ω, a unique continuous function that is harmonic on K and agrees with v at points that are within the closure of Ω, but not in K. Think of $f_{K,v}$ as a function that is harmonic on K, and blends into v as the variable passes out of K across its boundary.

Introduce an ordering by setting

$$v \leq w \text{ if } v(P) \leq w(P) \text{ for all } P \in \Omega.$$

If $v \leq w$ on Ω, then $f_{K,v} \leq f_{K,w}$ for each open disk K within Ω. This fact can be seen using the Poisson integral formula. For suppose that K has radius R. If $P = \rho e^{i\theta}$ is in K, using polar coordinates with center at the center of K, then

$$f_{K,v}(P) = \frac{1}{2\pi}\int_{-\pi}^{\pi} \frac{R^2 - \rho^2}{R^2 + \rho^2 - 2R\rho\cos(\theta - \xi)} v(\rho e^{i\xi})\, d\xi.$$

Similarly,

$$f_{K,w}(P) = \frac{1}{2\pi}\int_{-\pi}^{\pi} \frac{R^2 - \rho^2}{R^2 + \rho^2 - 2R\rho\cos(\theta - \xi)} w(\rho e^{i\xi})\, d\xi.$$

Because $v(\xi) \leq w(\xi)$, the integrand in the expression for $f_{K,v}(P)$ is less than or equal to the integrand in the expression for $f_{K,w}(P)$. This proves that $f_{K,v} \leq f_{K,w}$ on K. Indeed,

$$f_{K,v}(P) = v(P) \leq w(P) = f_{K,w}(P)$$

for P on the boundary circle of K, so in fact $f_{K,v} \leq f_{K,w}$ on $\overline{K}$, not just on the open disk K.

If v is continuous on $\overline{\Omega}$, we say that v is *superharmonic* (on Ω) if $f_{K,v} \leq v$ for every open disk K within Ω. If this inequality is reversed, so that $v \leq f_{K,v}$, for each open disk K in Ω, then v is called *subharmonic*.

We will develop a sequence of ten statements about, or properties of, subharmonic and superharmonic functions. These will lead to an existence theorem for the Dirichlet problem on Ω.

1. Every harmonic function on Ω is both subharmonic and superharmonic on Ω.

This follows immediately from the observation that v is harmonic on Ω exactly when $f_{K,v} = v$ for each open disk K within Ω.

2. A finite sum of superharmonic (subharmonic) functions is superharmonic (subharmonic).

This is a consequence of the easily verified fact that

$$f_{K,v_1+v_2} = f_{K,v_1} + f_{K,v_2}.$$

3. If v is superharmonic and u is harmonic on Ω, then $v + u$ and $v - u$ are superharmonic on Ω.

To verify this, first note by property 1 that u is superharmonic, so by property 2, $v + u$ is superharmonic. Similarly, $-u$ is harmonic, so the sum $v - u$ is superharmonic.

4. If v is superharmonic and w is subharmonic, then $v - w$ is superharmonic.

To see why this is true, let K be an open disk within Ω. Because v is superharmonic, $f_{K,v} \leq v$ on K and because u is subharmonic, $w \leq f_{K,w}$ on K. Then $-f_{K,w} \leq -w$ on K, so

$$f_{K,v-w} = f_{K,v} - f_{K,w} \leq v - w$$

on K.

5. If v is superharmonic on $\overline{\Omega}$, then $v(\mathbf{x})$ assumes its minimum and maximum values for $\mathbf{x} \in \overline{\Omega}$ at a point of $\partial\Omega$.

This result is reminiscent of the maximum principle for harmonic functions. To prove it, begin with the fact that the continuous function v assumes a minimum value on the compact set $\overline{\Omega}$. Assume that v is not a constant function. It is enough to show that this minimum value cannot be achieved at a point of the open set Ω, because in this event the nonconstant v would have to achieve its minimum on the boundary, not at any point of the interior. Suppose instead that v has its minimum value at some point $Q \in \Omega$. For some $P \in \Omega$, $v(Q) < v(P)$. Draw an open disk K within Ω and centered at Q, but not containing P. Then

$$v(Q) < v(P) = f_{K,v}(P),$$

and this contradicts the fact that v is superharmonic. This proves property 5.

Up to this point, the ideas of superharmonic and subharmonic functions have not been related in any way to the given function f. This is done through the concepts of upper and lower functions. We say that v is an *upper function for* f on Ω if v is superharmonic and

$$f \leq v \text{ on } \partial\Omega.$$

Thus an upper function for f is a superharmonic function v that dominates f on the boundary of Ω, in the sense that, at each point of $\partial\Omega$, $v(P)$ is at least as large as $f(P)$. (Recall that $f(\mathbf{x})$ is defined only for $\mathbf{x} \in \partial\Omega$).

There must exist upper functions for f, because any constant function $v(P) = c$ that bounds f on $\overline{\Omega}$ is an upper function for f.

In similar fashion, a *lower function for* f on $\overline{\Omega}$ is a subharmonic function u such that

$$u \leq f \text{ on } \partial\Omega.$$

Next, define a function u on $\overline{\Omega}$ by setting, for $P \in \overline{\Omega}$,

$$u(P) = \text{ glb}\{v(P)|v \text{ is an upper function for } f \text{ on } \overline{\Omega}\}.$$

Here is what this means. If P is in the closure of Ω, then look at the set of all numbers $v(P)$, as v varies over all upper functions for f on $\overline{\Omega}$. This produces a set of numbers. This set of numbers is bounded below (because there are numbers, and $f(P)$ is one of them, that are less than or equal to every such number $v(P)$). Choose the greatest lower bound of these numbers. This is a lower bound that is at least as large as any other lower bound of this set of $v-$ values of P.

This defines a function u on $\overline{\Omega}$. We will show that u is harmonic on Ω. Looking ahead, it will turn out that this will not be enough. In order to show that $u = f$ on the boundary of Ω (so u is a solution of the Dirichlet problem) we will need a condition about the boundary of Ω. This condition will make itself known when we come to this stage of the proof.

The proof that u is harmonic on Ω is nontrivial, and is carried out using properties 6–9.

6. If v is an upper function and w a lower function for f on Ω, then

$$w(P) \leq u(P) \text{ for } P \in \overline{\Omega}.$$

Once one works through all the terminology that has been developed, statement 6 follows directly from statements 4 and 5, as follows. Because v is superharmonic and w is subharmonic, then by (4), $v - w$ is superharmonic, and by property 5, $v - w$ assumes its minimum value on $\partial\Omega$. But $w(P) \leq f(P) \leq v(P)$, for P on the boundary of Ω. This means that $v(P) - w(P) \geq 0$ on this boundary, so $v(P) - w(P) \geq 0$ on $\overline{\Omega}$. This proves statement 6.

7. If v is an upper function for f, and D_1 is an open disk contained in Ω, then $f_{D_1,v}$ is an upper function for f on Ω.

Denote $w = f_{D_1,v}$. We want to show that w is superharmonic and dominates f on the boundary of Ω.

First, if $P \in \partial\Omega$, then P is in $\overline{\Omega}$, but not in D_1, so

$$f(P) \leq v(P) = f_{D_1,v}(P) = w(P),$$

so w dominates f on the boundary of Ω.

It remains to be shown that w is superharmonic. Let K be any open disk within Ω. We must show that $f_{K,w}(P) \leq w(P)$ on K. The proof of this lies in considering possibilities for the respective positions of K and D_1 within Ω. There are four possibilities:

Case 1. $K \subset D_1$.

Case 2. $K \subset \overline{\Omega} - D_1$.

Case 3. $D_1 \subset K$.

Case 4. If none of cases 1 through 3 apply, then the boundaries of K and D_1 must have points in common.

Look at the details for each case. For case 1, suppose $P \in K$ and K has radius R. Using polar coordinates from the center of K, write $P = \rho e^{i\theta}$. Because $f_{K,w} =$ w on the boundary of K, then

$$f_{K,w}(P) = \frac{1}{2\pi}\int_{-\pi}^{\pi} \frac{R^2 - \rho^2}{R^2 + \rho^2 - 2R\rho\cos(\theta - \xi)} w(\rho e^{i\xi})\, d\xi.$$

But because $D_1 \subset K$, this integral is also the Poisson integral representation of $W(P)$. Therefore $f_{K,w} = w$ on K in case 1.

Details for cases 2 and 3 are similar. In case 4, suppose K and D_1 have points in common, but neither is a subset of the other. Because v is superharmonic, $w \leq v$ on $\partial\Omega$. This means that

$$f_{K,w} \leq f_{K,v} \text{ on } K.$$

But then, again because v is superharmonic,

$$f_{K,v} \leq v.$$

Because $w = v$ outside K,

$$f_{K,w} \leq w \text{ outside } D_1.$$

(This applies to case 3 as well). But w and $f_{K,w}$ are harmonic on $D_1 \cap K$, the set of points common to D_1 and K, so

$$f_{K,w} \leq w \text{ on } \partial(D_1 \cap K),$$

hence also on $D_1 \cap K$.

Next, we claim that the minimum of any finite set of upper functions is also an upper function.

8. Let $v_1, \cdots, v_n$ be upper functions for f on Ω. Let

$$v(P) = \min\{v_1(P), \cdots, v_n(P)\} \text{ for } P \in \overline{\Omega}.$$

Then v is an upper function for f.

To prove statement 8, first suppose that$P \in \partial\Omega$. Then $f(P) \leq v_j(P)$ for each j, so $f(P) \leq v(P)$. Therefore v dominates f on the boundary of Ω.

To show that v is superharmonic, suppose K is a disk within Ω. If $P \in K$, then $v(P) = v_j(P)$ for some j, so

$$v(P) = v_j(P) \geq f_{K,v_j}(P) = f_{K,v}(P).$$

This proves statement 8. Now we are ready to produce a candidate for a solution of the Dirichlet problem on Ω.

9. For $P \in \Omega$, define $u(P)$ to be the greatest lower bound of the numbers $v(P)$ over all upper functions v for f on Ω:

$$u(P) = \text{glb}\{v(P) | v \text{ is an upper function for } f \text{ on } \Omega\}.$$

We claim that u is harmonic on Ω. To prove this, it is enough to prove that u is harmonic on every open disk in Ω. Let K be such an open, having center P_0. Let ϵ be any positive number. Because of the way $u(P_0)$ is defined as a greatest lower bound, there is an upper function v_1 that is harmonic on K and satisfying

$$v_1(P_0) \leq u(P_0) + \epsilon.$$

Now take w_1 to be an upper function satisfying

$$w_1(P_0) \leq u(P_0) + \frac{\epsilon}{2}.$$

Define

$$v_2 = f_{K,\min(v_1,w_1)}.$$

By statements 7 and 8, v_2 is an upper function for f on Ω and is harmonic on K.

Continuing in this way, we obtain a decreasing sequence of upper functions $v_1, v_2, \cdots$ which are harmonic on K and bounded below (because every upper function is at least as large as every lower function). By Harnack's second theorem, the uniform limit v of the sequence $v_1, v_2, \cdots$ on K is harmonic on K.

To show that u is harmonic on Ω, we will show that $v = u$ on K, keeping in mind that K is any disk within Ω. Suppose that the inequality $v = u$ on K is false. Then there must be an upper function w and a point $Q \in K$ such that

$$w(Q) < v(Q).$$

Let the distance between Q and P_0 be r and consider the disk K_Q of radius r centered at P_0. Q is on the boundary of K_Q. Define

$$w_n = f_{K_Q, \min(w, v_n)}.$$

Each w_n is an upper function for f on Ω. We know that the sequence v_n converges uniformly to v on $\overline{K_Q}$. Therefore w_n also converges uniformly on $\overline{K_Q}$. By choosing n sufficiently large, we can make $w_n(P_0)$ as close as we like to $f_{K_Q, \min(w, v_n)}(P_0)$, and

$$f_{K_Q, \min(w, v_0)}(P_0) \leq v(P_0) = u(P_0).$$

But this contradicts the definition of $u(P_0)$ as the greatest lower bound of the upper functions for f on Ω, evaluated at P_0. This completes the proof that u is harmonic on Ω.

If all we wanted was a function that is harmonic on Ω, we would not need all of the machinery we have just developed. The issue now is to produce a harmonic function that agrees with f on the boundary of Ω. We will show that the harmonic function produced in statement 9 satisfies this requirement, if a certain condition is satisfied at each point of $\partial\Omega$.

To this end, we introduce the notion of a barrier function at a boundary point Q of Ω. This barrier function will be superharmonic and positive at all points of $\overline{\Omega}$ except Q, where the function will have the value 0.

More carefully, if $Q \in \partial\Omega$, a function ω_Q is called a *barrier function for* Ω *at* Q if:

(1) ω_Q is continuous on $\overline{\Omega}$,

(2) ω_Q is superharmonic,

(3) $\omega_Q(Q) = 0$, and

(4) $\omega_Q(P) > 0$ for all $P \in \overline{\Omega}$ with $P \neq Q$.

We will now prove that u and f agree at each boundary point Q at which a barrier function ω_Q exists.

10. Let $Q \in \partial\Omega$. Suppose that a barrier function ω_Q for Ω exists. Then $\omega_Q(Q) = f(Q)$.

To prove this, begin by letting ϵ be any positive number. Because f is continuous on $\partial\Omega$, there is some open disk D_Q about Q such that

$$f(Q) = \epsilon \leq f(P) \leq f(Q) + \epsilon$$

for P in the boundary of D_Q and also in the boundary of Ω. By assumption, there is a barrier function ω_Q. Define a new function F_U on $\overline{\Omega}$ by

$$F_U(P) = f(Q) + \epsilon + C\omega_Q(P).$$

Here C is a positive constant we will specify shortly.

We claim that F_U is superharmonic on Ω. The reason for this is that any constant function is superharmonic, and ω_Q is superharmonic, so (with C positive) $C\omega_Q$ is superharmonic, and a sum of superharmonic functions is superharmonic. This holds for any positive number C.

Next, notice that, by choosing C sufficiently large, $f(P) \leq F_U(P)$ for all $P \in \partial\Omega$. Therefore F_U is an upper function for f on Ω.

Now define

$$F_L = f(Q) - \epsilon - C\omega_Q(P).$$

By an argument like that for F_U, verify that for sufficiently large C, F_L is a lower function for f on $\partial\Omega$.

For the remainder of this proof, suppose that C has been chosen sufficiently large that F_U is an upper function and F_L is a lower function for f on Ω. Then, because $\omega_Q = Q$, we have

$$F_L(Q) = f(Q) - \epsilon \text{ and } F_U(Q) = f(Q) + \epsilon.$$

Then

$$f(Q) - \epsilon = F_L(Q) \leq u(Q) \leq F_U(Q) + \epsilon.$$

Because ϵ can be chosen arbitrarily small, this implies that $u(Q) = f(Q)$, completing the proof.

We will summarize this discussion in an existence theorem.

Theorem 4.11 (Existence of a Continuous Solution of a Dirichlet Problem) Let Ω be a bounded domain in the plane. Let f be a continuous real-valued function on $\partial\Omega$. Suppose, for each $Q \in \partial\Omega$, there is a barrier function ω_Q. Then there is a continuous solution u such that:

$$\nabla^2 u(x, y) = 0 \text{ on } \Omega,$$
$$u(x, y) = f(x, y) \text{ for } (x, y) \in \partial\Omega.$$

The definitions and arguments used in proving this theorem pass over almost verbatim to the Dirichlet problem in R^n.

As an application of the theorem, we will prove the existence of a continuous solution of a Dirichlet problem in the plane, for a bounded domain Ω, assuming that $\partial\Omega$ is a simple closed curve, and that the given boundary function is continuous.

All we need to do is show that these conditions are sufficient to ensure the existence of a barrier function at each point of $\partial\Omega$. Suppose first that Ω is entirely within the right quarter plane $x > 0, y > 0$. Let $Q = (x_0, y_0) \in \partial\Omega$. Let δ be the diameter of Ω. This diameter is a measure of the width of a set of points in the plane, and is defined to be the greatest lower bound of the set of all distances between pairs of points of Ω. In the case that Ω is a bounded domain, this is a positive number. Using δ, define

$$\varphi(x, y) = \ln\left(\frac{\sqrt{x^2 + y^2}}{2\delta}\right)$$

and

$$\psi(x, y) = \arctan\left(\frac{y}{x}\right).$$

Heere is a candidate for a barrier function:

$$\omega_Q(x, y) = \frac{\varphi(x - x_0, y - y_0)}{(\varphi(x - x_0, y - y_0))^2 + (\psi(x - x_0, y - y_0))^2}.$$

By the definition of δ,

$$(x - x_0)^2 + (y - y_0)^2 \leq \delta^2$$

for each $(x, y) \in \overline{\Omega}$. Therefore, $\varphi(x, y) < 0$ so $\omega_Q(x, y) > 0$ for $(x, y) \in \overline{\Omega}$ and $(x, y) \neq (x_0, y_0)$.

Now $\omega_Q(x, y)$ is not defined at (x_0, y_0). However, it is easy to verify that

$$\omega_Q(x, y) \to 0 \text{ as } (x, y) \to (x_0, y_0).$$

We can therefore define

$$\omega_Q(x_0, y_0) = 0.$$

It is now routine to verify that ω_Q is a barrier function at Q for each boundary point Q of Ω. By the existence theorem 4.11, the Dirichlet problem for a bounded domain in the right quarter plane has a solution if the function specified on the boundary is continuous.

If Ω does not lie entirely in the right quarter plane, translate Ω to this quadrant and use the fact that translates of harmonic functions are harmonic. The translated problem will have a solution, and this can be translated back to a solution of the problem for the original domain.

Problems for Section 4.7

1. Fill in the details of a proof that v is harmonic on Ω if and only if $f_{K,v}$ is harmonic on every open disk in Ω.

2. Prove the assertion made in statement 2 that

$$f_{K,v_1+v_2} = f_{K,v_1} + f_{K,v_2}.$$

3. Complete the proof of statement 7 for cases 2 and 3.

4. (Substantial Project) The proof of the existence theorem 4.11 focused on superharmonic functions and upper functions. Develop a proof using subharmonic functions and lower functions, with the solution $u(P)$ at P defined as the least upper bound of the solutions $v(P)$ taken over all lower functions of f on Ω. A derivation can follow the treatment using upper functions, but reversing inequalities where appropriate to base the solution on lower functions.

Chapter 5

Fourier Integral Methods of Solution

To solve problems on bounded intervals, we often use Fourier series. For problems on a line or half line, where the space variable is unbounded, we may turn to Fourier integrals.

Such problems can have both theoretical and modeling implications. To a radio astronomer, radio waves in space may be modeled as traveling along an infinite line. And to a sonar operator in a submarine, the ocean may look like an infinite medium.

We will begin with an informal development of the Fourier integral and sine and cosine integrals, then look at their uses in solving initial-boundary value problems.

5.1 The Fourier Integral of a Function

Suppose $f(x)$ is piecewise smooth on every interval $[-L, L]$, and that f is *absolutely integrable*, which means that $\int_{-\infty}^{\infty} |f(x)|\, dx$ converges.

Instead of writing a Fourier series representation of $f(x)$ on an interval $[-L, L]$, we want to attempt a representation that is in some sense valid for all x.

Choose any $L > 0$ and start with the Fourier expansion of $f(x)$ on $[-L, L]$, ignoring $f(x)$ for $|x| > L$. This expansion is

$$\frac{1}{2L}\int_{-L}^{L} f(\xi)\,d\xi$$
$$+\sum_{n=1}^{\infty}\left[\frac{1}{L}\int_{-L}^{L} f(\xi)\cos\left(\frac{n\pi\xi}{L}\right)\cos\left(\frac{n\pi x}{L}\right)d\xi\right.$$
$$\left.+\frac{1}{L}\int_{-L}^{L} f(\xi)\sin\left(\frac{n\pi\xi}{L}\right)\sin\left(\frac{n\pi x}{L}\right)d\xi\right].$$

We would like to let $L \to \infty$ in this expression to attempt a representation of $f(x)$ over the entire real line. To determine this limit (at least intuitively), let $\omega_n = n\pi/L$ to write

$$\Delta\omega = \omega_n - \omega_{n-1} = \frac{\pi}{L}.$$

Upon substitution, the Fourier series of $f(x)$ on $[-L, L]$ becomes:

$$\frac{1}{2\pi}\left(\int_{-L}^{L} f(\xi)\,d\xi\right)\Delta\omega$$
$$+\frac{1}{\pi}\sum_{n=1}^{\infty}\left[\left(\int_{-L}^{L} f(\xi)\cos(\omega_n\xi)\,d\xi\right)\cos(\omega_n x)\right.$$
$$\left.+\left(\int_{-L}^{L} f(\xi)\sin(\omega_n\xi)\,d\xi\right)\sin(\omega_n x)\right]\Delta\omega. \tag{5.1}$$

Now let $L \to \infty$. Then $\Delta\omega \to 0$. Because $\int_{-L}^{L} f(\xi)\,d\xi$ converges to a finite number as $L \to \infty$, in this limit,

$$\frac{1}{2\pi}\left(\int_{-L}^{L} f(\xi)\,d\xi\right)\Delta\omega \to 0.$$

With a stretch of the imagination, the other two terms in equation 5.1 resemble a Riemann sum for a definite integral. As $L \to \infty$, these two terms approach the limit

$$\frac{1}{\pi}\int_{0}^{\infty}\left[\left(\int_{-\infty}^{\infty} f(\xi)\cos(\omega\xi)\,d\xi\right)\cos(\omega x)\right.$$
$$\left.+\left(\int_{-\infty}^{\infty} f(\xi)\sin(\omega\xi)\,d\xi\right)\sin(\omega x)\right]d\omega. \tag{5.2}$$

This is called the Fourier integral representation of $f(x)$ on the real line. It is possible to prove that, under the assumptions made about f, it converges to

$$\frac{1}{2}(f(x-) + f(x+))$$

for every real x. If f is continuous at x, this integral representation converges to $f(x)$.

We will summarize this discussion in a form that may clarify the analogy between the Fourier integral on the line and the Fourier series of a function on an interval. If f is absolutely integrable and piecewise smooth on each interval $[-L, L]$, then the *Fourier integral representation* of $f(x)$ is

$$\frac{1}{2}(f(x-) + f(x+)) = \int_0^\infty (A_\omega \cos(\omega x) + B_\omega \sin(\omega)) \, d\omega, \tag{5.3}$$

in which the *Fourier integral coefficients* of $f(x)$ are

$$A_\omega = \frac{1}{\pi} \int_{-\infty}^{\infty} f(\xi) \cos(\omega\xi) \, d\xi \text{ and } B_\omega = \frac{1}{\pi} \int_{-\infty}^{\infty} f(\xi) \sin(\omega\xi) \, d\xi. \tag{5.4}$$

By combining terms and using a trigonometric identity in the expression 5.2, as in the derivation of equation 1.19 for Dirichlet's formula, this Fourier integral representation can also be written

$$\frac{1}{2}(f(x-) + f(x+)) = \frac{1}{\pi} \int_0^\infty \int_{-\infty}^{\infty} f(\xi) \cos(\omega(\xi - x)) \, d\xi \, d\omega. \tag{5.5}$$

Example 5.1 Let

$$f(x) = \begin{cases} 0 & \text{for } |x| > 1, \\ 1 & \text{for } |x| \le 1. \end{cases}$$

f is continuous, with a continuous derivative, for all x except $x = \pm 1$, where the function has jump discontinuities. Further, f is absolutely integrable, because

$$\int_{-\infty}^{\infty} |f(x)| \, ds = \int_{-1}^{1} dx = 2.$$

To write the Fourier integral of $f(x)$, compute the coefficients:

$$\begin{aligned} A_\omega &= \frac{1}{\pi} \int_{-\infty}^{\infty} f(\xi) \cos(\omega\xi) \, d\xi \\ &= \frac{1}{\pi} \int_{-1}^{1} \cos(\omega\xi) \, d\xi \\ &= \frac{2\sin(\omega)}{\pi\omega}, \end{aligned}$$

and $B_\omega = 0$ without any integration because $f(\xi)\sin(\omega\xi)$ is an odd function of ξ on $[-1, 1]$.

The Fourier integral of $f(x)$ is

$$\frac{2}{\pi} \int_0^\infty \frac{\sin(\omega)}{\omega} \cos(\omega x) d\omega.$$

This converges to

$$\begin{cases} 1 & \text{for } -1 < x < 1, \\ 1/2 & \text{for } x = \pm 1, \\ 0 & \text{for } x < -1 \text{ and for } x > 1. \end{cases}$$

We can get some feeling for the convergence of the Fourier integral of this function by approximating some values. At $x = 1$ the Fourier integral is

$$\frac{2}{\pi}\int_0^\infty \frac{\sin(\omega)}{\omega}\cos(\omega)\,d\omega,$$

and this converges to

$$\frac{1}{2}(f(1-) + f(1+)) = \frac{1}{2}(1+0) = \frac{1}{2}.$$

To check this convergence, approximate the Fourier integral by using a numerical routine to approximate the integral

$$\frac{1}{\pi}\int_0^k \frac{\sin(\omega)}{\omega}\cos(\omega)\,d\omega$$

for large values of k. For $k = 50$, this integral is approximately 0.4972718105; for $k = 100$ we obtain 0.4992316037; for $k = 200$, we get 0.5004197049; and for $k = 400$, we get 0.5001778591. These approximations appear to approach 1/2 as k is chosen larger.

5.1.1 Fourier Cosine and Sine Integrals

If $g(x)$ is piecewise smooth and absolutely integrable on $x \geq 0$, we can represent it as a both a Fourier cosine integral and as a Fourier sine integral, analogous to Fourier cosine and sine series on an interval $[0, L]$.

For a cosine integral, extend $g(x)$ to an even function $g_e(x)$ defined over the entire real line, by setting

$$g_e(x) = \begin{cases} g(x) & \text{if } x \geq 0, \\ g(-x) & \text{if } x < 0. \end{cases}$$

Now expand $g_e(x)$ in a Fourier integral on the entire line. The sine coefficients B_ω are all zero because $g_e(x)\sin(\omega x)$ is an odd function, so we are left with just the cosine terms. For these, the coefficients can be written as integrals over $[0, \infty)$.

In summary, the *Fourier cosine integral* of $g(x)$ *on* $x \geq 0$ is

$$\int_0^\infty a_\omega \cos(\omega x)\,d\omega,$$

in which the *Fourier cosine integral coefficients* of $g(x)$ are

$$a_\omega = \frac{2}{\pi}\int_0^\infty g(\xi)\cos(\omega\xi)\,d\xi.$$

This cosine integral representation converges to

$$\begin{cases} \frac{1}{2}(g(x-)+g(x+)) & \text{if } x > 0 \ , \\ g(0+) & \text{if } x = 0. \end{cases}$$

For the Fourier sine integral representation of $g(x)$ on $x \geq 0$, extend $g(x)$ to an odd function $g_o(x)$ on the real line by setting

$$g_o(x) = \begin{cases} g(x) & \text{if } x \geq 0, \\ -g(-x) & \text{if } x < 0. \end{cases}$$

Now the cosine coefficients are zero because $g_o(x)\cos(\omega x)$ is odd, and we obtain the *Fourier sine integral* of $g(x)$ *on* $x \geq 0$:

$$\int_0^\infty b_\omega \sin(\omega x)\, d\omega,$$

in which the *Fourier sine integral coefficients* of $g(x)$ are

$$b_\omega = \frac{2}{\pi}\int_0^\infty g(\xi)\sin(\omega\xi)\, d\xi.$$

This sine integral converges to

$$\begin{cases} \frac{1}{2}(g(x-)+g(x+)) & \text{if } x > 0, \\ 0 & \text{if } x = 0. \end{cases}$$

Example 5.2 (**Laplace's Integrals**) Let $k > 0$ and let $g(x) = e^{-kx}$ for $x \geq 0$.

First, $g(x)$ is continuous with a continuous derivative for $x > 0$, and

$$\int_0^\infty e^{-kx}\, dx = \frac{1}{k}$$

so g is absolutely integrable on $x \geq 0$.

For the cosine integral expansion of $g(x)$, compute the coefficients

$$a_\omega = \frac{2}{\pi}\int_0^\infty e^{-k\xi}\cos(\omega\xi)\, d\xi = \frac{2}{\pi}\frac{k}{k^2+\omega^2}.$$

Making use of the convergence theorem for cosine integrals, and the continuity of e^{-kx}, this cosine integral converges to $g(x)$ for $x \geq 0$ and we can write

$$e^{-kx} = \frac{2k}{\pi}\int_0^\infty \frac{1}{k^2+\omega^2}\cos(\omega x)\, d\omega \text{ for } x \geq 0.$$

For the sine integral expansion, compute the coefficients

$$b_\omega = \frac{2}{\pi}\int_0^\infty e^{-k\xi}\sin(\omega\xi)\, d\xi = \frac{2}{\pi}\frac{\omega}{k^2+\omega^2}.$$

Because e^{-kx} is continuous on $x \geq 0$, the convergence theorem for Fourier sine integrals tells us that

$$\frac{2}{\pi}\int_0^\infty \frac{\omega}{k^2+\omega^2}\sin(\omega x)\,d\omega = \begin{cases} e^{-kx} & \text{if } x > 0, \\ 0 & \text{if } x = 0. \end{cases}$$

These integral representations of e^{-kx} for $x \geq 0$ are called *Laplace's integrals*, because the coefficient a_ω of the cosine integral is $2/\pi$ times the Laplace transform of $\sin(\omega x)$, and the coefficient b_ω of the sine integral is $2/\pi$ times the Laplace transform of $\cos(\omega x)$.

The remainder of this chapter is devoted to Fourier integral solutions of heat, wave, Dirichlet and Neumann problems. We will also solve Burgers' equation, which is a diffusion equation with a convection term.

Problems for Section 5.1

In each of problems 1–6, write the Fourier integral representation of the function, and determine what this integral converges to.

1. $f(x) = \begin{cases} x & \text{for } -\pi \leq x \leq \pi, \\ 0 & \text{for } |x| > \pi. \end{cases}$

2. $f(x) = \begin{cases} \cos(x) & \text{for } -\alpha \leq x \leq \alpha, \\ 0 & \text{for } |x| > \alpha, \end{cases}$ with α a positive number.

3. $f(x) = e^{-|x|}$.

4. $f(x) = xe^{-|x|}$.

5. $f(x) = \begin{cases} |x| & \text{for } -\alpha \leq x \leq \alpha, \\ 0 & \text{for } |x| > \alpha, \end{cases}$ with α a positive number.

6. $f(x) = \begin{cases} k & \text{for } 0 < x \leq \alpha, \\ -k & \text{for } -\alpha \leq x \leq 0, \\ 0 & \text{for } |x| > \alpha. \end{cases}$ Here k and α are positive numbers.

7. Obtain the integral in equation 5.5 from the Fourier integral representation 5.2.

8. Show that the Fourier integral representation of $f(x)$ can be written

$$\frac{1}{2\pi}\int_{-\infty}^\infty \int_{-\infty}^\infty e^{i\omega(x-\xi)} f(\xi)\,d\xi\,d\omega.$$

In each of problems 9–14, write the Fourier sine and the Fourier cosine integral representation of the function. Determine what each integral converges to.

9. $g(x) = \begin{cases} \sinh(x) & \text{for } 0 \le x \le k, \\ 0 & \text{for } x > k, \end{cases}$ with k a positive number.

10. $g(x) = \begin{cases} \cos(\pi x) & \text{for } 0 \le x \le 1, \\ 0 & \text{for } x > 0. \end{cases}$

11. $g(x) = e^{-x}\cos(x)$.

12. $g(x) = xe^{-3x}$.

13. $g(x) = \begin{cases} k & \text{for } 0 \le x \le \alpha, \\ 0 & \text{for } x > \alpha, \end{cases}$ with k and α positive numbers.

14. $f(x) = \begin{cases} x^2 & \text{for } 0 \le x \le \alpha, \\ 0 & \text{for } x > \alpha, \end{cases}$ with α a positive constant.

15. Use the Laplace integrals to compute the Fourier cosine integral of

$$g(x) = \frac{1}{1+x^2}$$

and the Fourier sine integral of

$$h(x) = \frac{x}{1+x^2}.$$

16. Assume that

$$\lim_{x\to\infty} g(x) = \lim_{x\to\infty} g'(x) = 0.$$

Suppose also that $g''(x)$ exists for $x \ge 0$ and $g'(0) = 0$. Show that the integrand in the Fourier cosine integral representation of $g''(x)$ is equal to $-\omega^2$ times the integrand in the Fourier cosine integral representation of $g(x)$.

17. Suppose

$$h(x) = \frac{2}{\pi}\int_0^\infty a(\omega)\cos(\omega x)\,d\omega,$$

where

$$a(\omega) = \int_0^\infty h(t)\cos(\omega t)\,dt.$$

(a) Show that

$$x^2 h(x) = \frac{1}{\pi}\int_0^\infty (-a''(\omega))\cos(\omega x)\,d\omega.$$

(b) Show that

$$xh(x) = \frac{1}{\pi}\int_0^\infty (-a'(\omega))\sin(\omega x)\,d\omega.$$

5.2 The Heat Equation on the Real Line

We will solve the problem

$$u_t = ku_{xx} \text{ for } -\infty < x < \infty, t > 0,$$
$$u(x,0) = f(x) \text{ for } -\infty < x < \infty. \tag{5.6}$$

This problem models a temperature distribution in an unbounded medium, having one space dimension. There is no bound on the space variable, hence no boundary condition, but there is an initial condition.

The heat equation looks the same with or without boundary conditions, so put a separation of variables. Put $u(x,t) = X(x)T(t)$ to obtain

$$X'' + \lambda X = 0, T' + \lambda kT = 0.$$

For a problem on a bounded interval, the boundary conditions provided conditions on $X(x)$ at 0 and L. These were used to find the eigenvalues, then the eigenfunctions. Here there are no boundary conditions. Try cases on λ.

If $\lambda = 0$, then

$$X(x) = cx + d.$$

There is no information given by boundary conditions to tell us anything about c and d. However, it is useful in many contexts to impose the additional condition that the solution must be bounded. Because x can be any real number, this requires that $c = 0$, leaving $X(x) =$ constant in this case.

If $\lambda < 0$, set $\lambda = -\omega^2$ with $\omega > 0$. Then

$$X(x) = ce^{\omega x} + de^{-\omega x}.$$

For this to be bounded as $x \to \infty$, we need $c = 0$. And for $X(x)$ to be bounded at $x \to -\infty$, we must choose $d = 0$. This case yields only the trivial solution for X.

If $\lambda > 0$, say $\lambda = \omega^2$ with $\omega > 0$, then

$$X(x) = c\cos(\omega x) + d\sin(\omega x).$$

This is bounded for any numbers c and d. We have no reason to restrict ω, other than in this case $\omega > 0$, so every positive real number is an eigenvalue. Corresponding eigenfunctions are sums of constants times $\cos(\omega x)$ and $\sin(\omega x)$.

We can combine the first and third cases by writing the eigenvalues as $\lambda = \omega^2$, with $\omega \geq 0$, and the eigenfunctions as

$$X_\omega(x) = A_\omega \cos(\omega x) + B_\omega \sin(\omega x) \text{ for } \omega \geq 0.$$

The problem for T is

$$T' + \omega^2 kT = 0,$$

with solutions that are constant multiples of

$$T_\omega(t) = e^{-\omega^2 kt}.$$

For each nonnegative real number ω, the functions

$$u_\omega(x,t) = (A_\omega \cos(\omega x) + B_\omega \sin(\omega x))e^{-\omega^2 kt}$$

satisfy the heat equation.

When we reached this point for a problem on a bounded interval, we had what are called *discrete eigenvalues*, $\lambda_n = n^2\pi^2/L^2$, one for each positive integer n. We summed the corresponding solutions $u_n(x,t)$ over n and used the initial condition to find the coefficients as a Fourier sine or cosine expansion of the initial temperature function.

We cannot "sum" over all nonnegative reals ω, so replace the sum $\sum_{n=1}^{\infty}$ with an integral $\int_0^\infty \cdots d\omega$ to write

$$u(x,t) = \int_0^\infty (A_\omega \cos(\omega x) + B_\omega \sin(\omega x))e^{-\omega^2 kt}\, d\omega. \tag{5.7}$$

The initial condition requires that the coefficients must be chosen so that

$$u(x,0) = f(x) = \int_0^\infty (A_\omega \cos(\omega x) + B_\omega \sin(\omega x))\, d\omega.$$

This tells us to choose the coefficients as the Fourier integral coefficients of $f(x)$:

$$A_\omega = \frac{1}{\pi}\int_{-\infty}^{\infty} f(\xi)\cos(\omega\xi)\, d\xi \text{ and } B_\omega = \frac{1}{\pi}\int_{-\infty}^{\infty} f(\xi)\sin(\omega\xi)\, d\xi.$$

With these coefficients, equation 5.7 is the bounded solution of the problem.

Example 5.3 Suppose the initial temperature in problem 5.6 is

$$f(x) = e^{-|x|}.$$

The solution is given by equation 5.7 and all we have to do is compute the Fourier integral coefficients. These are

$$A_\omega = \frac{1}{\pi}\int_{-\infty}^{\infty} e^{-|\xi|}\cos(\omega\xi)\, d\xi = \frac{2}{\pi}\frac{1}{1+\omega^2}$$

and

$$B_\omega = \frac{1}{\pi}\int_{-\infty}^{\infty} e^{-|\xi|}\sin(\omega\xi)\, d\xi = 0.$$

The solution is

$$u(x,t) = \frac{2}{\pi}\int_0^\infty \frac{1}{1+\omega^2}\cos(\omega x)e^{-\omega^2 kt}\, d\omega.$$

One can question the value of such a solution, because it is unclear how to evaluate $u(x,t)$ at any point x and time t. However, numerical approximations

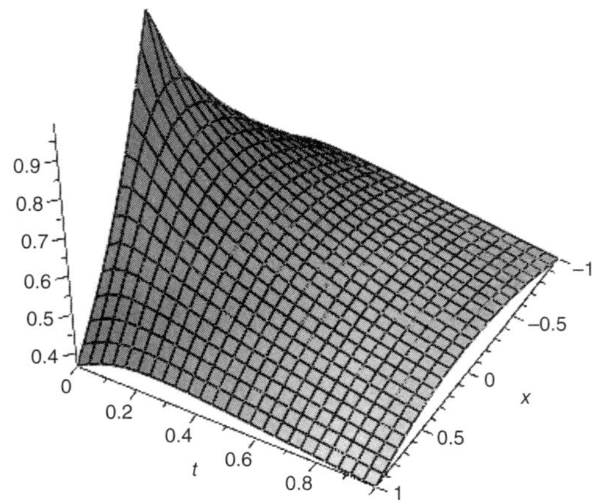

Figure 5.1: Surface plot of the temperature function in example 5.3 for $-1 \leq x \leq 1, 0 \leq t \leq 1$.

can yield interesting and revealing results. Figure 5.1 shows an approximate surface plot for $-1 \leq x \leq 1$ and $0 \leq t \leq 1$.

5.2.1 A Reformulation of the Integral Solution

We will look at another way to write the solution 5.7. Insert the formulas for the Fourier integral coefficients of the initial temperature function into equation 5.7 and use the trigonometric identity

$$\cos(\omega\xi)\cos(\omega x) + \sin(\omega\xi)\sin(\omega x) = \cos(\omega(\xi - x))$$

to write

$$\begin{aligned} u(x,t) &= \int_0^\infty (A_\omega \cos(\omega x) + B_\omega \sin(\omega x))e^{-\omega^2 kt}\, d\omega \\ &= \frac{1}{\pi}\int_0^\infty \left[\left(\int_{-\infty}^\infty f(\xi)\cos(\omega\xi)\, d\xi\right)\cos(\omega x)\right. \\ &+ \left.\left(\int_{-\infty}^\infty f(\xi)\sin(\omega\xi)\, d\xi\right)\sin(\omega x)\right] e^{-\omega^2 kt}\, d\omega \\ &= \frac{1}{\pi}\int_0^\infty \int_{-\infty}^\infty f(\xi)\left[\cos(\omega\xi)\cos(\omega x) + \sin(\omega\xi)\sin(\omega x)\right] e^{-\omega^2 kt}\, d\xi\, d\omega \\ &= \frac{1}{\pi}\int_0^\infty \int_{-\infty}^\infty f(\xi)\cos(\omega(\xi - x))e^{-\omega^2 kt}\, d\xi\, d\omega. \end{aligned}$$

In the last line, interchange the order of integration, and use the fact that the integrand is an even function in ω to write

$$\int_0^\infty \cdots\, d\omega = \frac{1}{2}\int_{-\infty}^\infty \cdots\, d\omega.$$

We obtain

$$u(x,t) = \frac{1}{2\pi}\int_{-\infty}^{\infty}\left(\int_{-\infty}^{\infty}\cos(\omega(\xi - x))e^{-\omega^2 kt}\,d\omega\right) f(\xi)\,d\xi.$$

We now claim that the integral with respect to ω, which is enclosed in large parentheses, can be evaluated in closed form. The key is the integral

$$\int_{-\infty}^{\infty} e^{-\zeta^2}\cos\left(\frac{\alpha\zeta}{\beta}\right)d\zeta = \sqrt{\pi}e^{-\alpha^2/4\beta^2}. \tag{5.8}$$

This integral can be found in tables, and the integration can also be carried out using residues and complex integration. An evaluation based on a differential equation is outlined in problem 11 for this section. To make use of the integral 5.8, let

$$\xi = \omega\sqrt{kt}, \alpha = x - \xi, \text{ and } \beta = \sqrt{kt}.$$

Then

$$d\zeta = \sqrt{kt}\,d\omega \text{ and } \frac{\alpha\zeta}{\beta} = \omega(x-\xi),$$

so

$$\begin{aligned}\int_{-\infty}^{\infty} e^{-\omega^2 kt}\cos(\omega(\xi - x))\,d\omega &= \int_{-\infty}^{\infty} e^{-\zeta^2}\cos\left(\frac{\alpha\zeta}{\beta}\frac{1}{\sqrt{kt}}\right)d\zeta \\ &= \frac{1}{\sqrt{kt}}\sqrt{\pi}e^{-(x-\xi)^2/4kt}.\end{aligned}$$

Now the solution is

$$u(x,t) = \frac{1}{2\pi}\sqrt{\frac{\pi}{kt}}\int_{-\infty}^{\infty} e^{-(x-\xi)^2/4kt} f(\xi)\,d\xi,$$

or

$$u(x,t) = \frac{1}{2\sqrt{\pi kt}}\int_{-\infty}^{\infty} e^{-(x-\xi)^2/4kt} f(\xi)\,d\xi. \tag{5.9}$$

It is routine to check that this function satisfies the heat equation by differentiating under the integral sign. This is allowed because the integral converges uniformly in x and t. To show that $u(x,0) = f(x)$, make the change of variables

$$\zeta = \frac{\xi - x}{\sqrt{4kt}}$$

to obtain

$$u(x,t) = \frac{1}{\sqrt{\pi}}\int_{-\infty}^{\infty} f(x + \sqrt{4kt}\zeta)e^{-\zeta^2}\,d\zeta.$$

Because f is assumed to be a bounded function, this integral converges uniformly in x and t, so the integral and the limit can be interchanged:

$$\lim_{t\to 0+} u(x,t) = \frac{1}{\sqrt{\pi}}\int_{-\infty}^{\infty}\left(\lim_{t\to 0+} f(x + \sqrt{4kt}\zeta)\right)e^{-\zeta^2}\,d\zeta.$$

But

$$\lim_{t\to 0+} f(x+\sqrt{4kt}\zeta) = f(x),$$

if f is continuous. Further, $f(x)$ passes through the integration with respect to ζ, so we finally obtain

$$\lim_{t\to 0+} u(x,t) = u(x,0) = \frac{1}{\sqrt{\pi}}\int_{-\infty}^{\infty} e^{-\zeta^2}\, d\zeta = f(x)$$

because

$$\int_{-\infty}^{\infty} e^{-\zeta^2}\, d\zeta = \sqrt{\pi}.$$

This integral is widely used in statistics and can be found in tables, or evaluated by various techniques.

Example 5.4 Suppose

$$f(x) = \begin{cases} 3 & \text{for } 0 \le x \le 2, \\ 0 & \text{for } x < 0 \text{ and for } x > 2, \end{cases}$$

and $k = 1/4$. The solution is

$$u(x,t) = \frac{1}{\sqrt{\pi t}}\int_0^2 3e^{-(x-\xi)^2/t}\, d\xi.$$

Figure 5.2 shows graphs of $u(x,1/4)$, $u(x,1)$ and $u(x,3)$. Figure 5.3 is a surface plot of $u(x,t)$ for $-5 \le x \le 5$ and $0 \le t \le 4$.

5.2.2 The Heat Equation on a Half Line

For a problem on a half line, consider

$$\begin{aligned} u_t &= ku_{xx} \text{ for } x > 0, t > 0, \\ u(0,t) &= 0 \text{ for } t > 0, \\ u(x,0) &= f(x) \text{ for } x > 0. \end{aligned} \tag{5.10}$$

This problem is posed on a half line $x \ge 0$, which has a boundary point. For this reason a boundary condition is given at $x = 0$ for all times.

Setting $u(x,t) = X(x)T(t)$, we obtain

$$X'' + \lambda X = 0, T' + \lambda kT = 0.$$

However, unlike the problem on the entire line, we now have the condition

$$u(0,t) = X(0)T(t) = 0,$$

which implies that $X(0) = 0$.

Look at cases on λ.

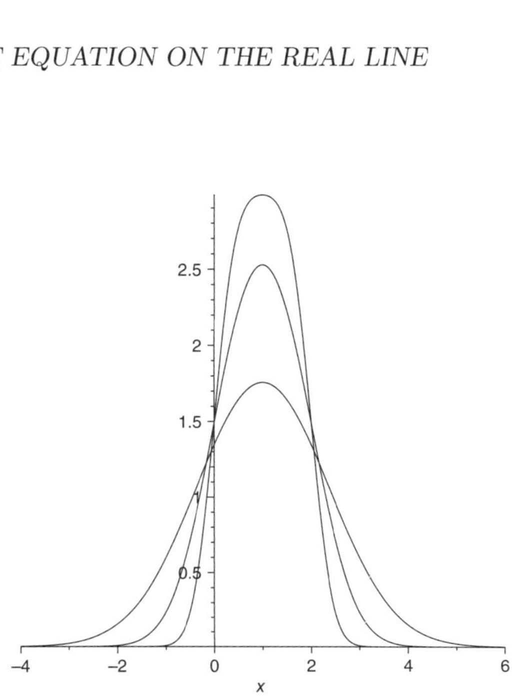

Figure 5.2: Graphs of $u(x, 1/4), u(x, 1)$ and $u(x, 3)$ in example 5.4.

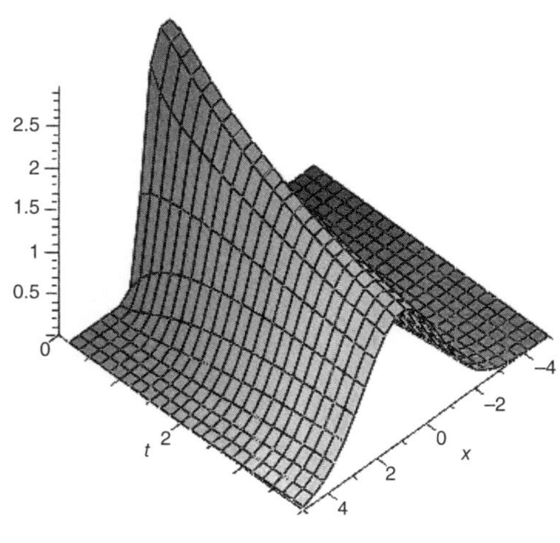

Figure 5.3: Surface plot of $u(x, t)$ in example 5.4.

If $\lambda = 0$, then $X(x) = cx + d$. Now $X(0) = d = 0$, so $X(x) = cx$, with c any constant. As with the problem on the line, we will look for bounded solutions. Because cx is bounded on the half line only if $c = 0$, this case has only the trivial solution.

If $\lambda < 0$, say $\lambda = \omega^2$ with $\omega > 0$, then

$$X(x) = ce^{\omega x} + de^{-\omega x}.$$

For a bounded solution, let $c = 0$. But then $X(0) = d = 0$, so this case has only the trivial solution.

The only remaining case is that $\lambda > 0$, say $\lambda = \omega^2$ with $\omega > 0$. Now

$$X(x) = c\cos(\omega x) + d\sin(\omega x).$$

This solution is bounded, for any constants c and d. Because $X(0) = c = 0$, we are left with

$$X(x) = d\sin(\omega x).$$

With $\lambda = \omega^2$, the equation for T is

$$T' + \omega^2 kT = 0,$$

with solutions that are constant multiples of

$$T(t) = e^{-\omega^2 kt}.$$

Now, for each $\omega > 0$, we have functions

$$u_\omega(x,t) = b_\omega \sin(\omega x)e^{-\omega^2 kt}$$

satisfying the heat equation and the boundary condition. To satisfy the initial condition, set

$$u(x,t) = \int_0^\infty b_\omega \sin(\omega x)e^{-\omega^2 kt}\,d\omega. \tag{5.11}$$

Then

$$u(x,0) = f(x) = \int_0^\infty b_\omega \sin(\omega)\,d\omega.$$

The coefficients b_ω are therefore the Fourier sine integral coefficients of the initial temperature function on $x \geq 0$:

$$b_\omega = \frac{2}{\pi}\int_0^\infty f(\xi)\sin(\omega\xi)\,d\xi.$$

Example 5.5 We will solve the problem 5.10 on the half line $x \geq 0$, with initial temperature function

$$f(x) = x^2e^{-4x} \text{ for } x \geq 0$$

and boundary condition

$$u(0,t) = 0 \text{ for } t \geq 0.$$

The solution is given by 5.11, with

$$\begin{aligned} b_\omega &= \frac{2}{\pi}\int_0^\infty \xi^2 e^{-4\xi}\sin(\omega\xi)\,d\xi \\ &= \frac{4\omega}{\pi}\frac{48-\omega^2}{16+\omega^2}. \end{aligned}$$

The solution is

$$u(x,t) = \frac{4}{\pi}\int_0^\infty \frac{\omega(48-\omega^2)}{16+\omega^2}\sin(\omega x)e^{-\omega^2 kt}\,d\omega.$$

There is a version of the integral solution of problem 5.6 on the entire line that can be adapted to an integral solution of the problem 5.10 on the half line. Begin with solution 5.11 by Fourier sine integral and insert the sine coefficients to obtain

$$u(x,t) = \frac{2}{\pi}\int_0^\infty \int_0^\infty f(\xi)\sin(\omega\xi)\sin(\omega x)e^{-\omega^2 kt}\,d\xi\,d\omega.$$

Use the identity

$$\sin(\omega\xi)\sin(\omega x) = \frac{1}{2}(\cos(\omega(x-\xi)) - \cos(\omega(x+\xi)))$$

and interchange the order of integration to obtain

$$\begin{aligned} u(x,t) = &\frac{1}{\pi}\int_0^\infty \left(\int_0^\infty e^{-\omega^2 kt}\cos(\omega(x-\xi))\,d\omega\right) f(\xi)\,d\xi \\ &- \frac{1}{\pi}\int_0^\infty \left(\int_0^\infty e^{-\omega^2 kt}\cos(\omega(x+\xi))\,d\omega\right) f(\xi)\,d\xi. \end{aligned}$$

Now use the fact that

$$\frac{1}{\pi}\int_0^\infty e^{-\omega^2 kt}\cos(\alpha\omega)\,d\omega = \frac{1}{2\sqrt{\pi kt}}e^{-\alpha^2/4kt},$$

as can be obtained from the integral 5.8 by a change of variables. This enables us to write the solution for the problem on the half line as

$$u(x,t) = \frac{1}{2\sqrt{\pi kt}}\int_0^\infty \left[e^{-(x-\xi)^2/4kt} - e^{-(x+\xi)^2/4kt}\right] f(\xi)\,d\xi. \tag{5.12}$$

In the 19th century, physicist Lord Kelvin (1824–1907) used this expression to derive his estimate of the age of the Earth. This story is told in the next section.

Problems for Section 5.2

1. Suppose the initial temperature function $f(x)$ for problem 5.6 is an odd function. Show that the solution is an odd function in the space variable.

2. Suppose, for some positive number a, the initial temperature function in problem 5.6 is zero for $|x| > a$ and positive for $|x| < a$. Show that $u(x,t) > 0$ for all $t > 0$. Contrast this behavior of the solution with solutions of the wave equation on the line, when the initial position function vanishes outside an interval $|x| \leq a$.

In each of problems 3–7, solve problem 5.6 for the given initial temperature function. Write one solution using the Fourier integral and use the integral solution 5.9 for a different version of the solution. Try graphing some curves $u(x,t)$ for specific values of t, as well as surfaces $u(x,t)$.

3. $f(x) = e^{-4|x|}$.

4. $f(x) = \begin{cases} \sin(x) & \text{for } |x| \leq \pi, \\ 0 & \text{for } |x| > \pi. \end{cases}$

5. $f(x) = \begin{cases} x & \text{for } 0 \leq x \leq 4, \\ 0 & \text{for } x < 0 \text{ and for } x > 4. \end{cases}$

6. $f(x) = \begin{cases} e^{-x} & \text{for } -1 \leq x \leq 1, \\ 0 & \text{for } |x| > 1. \end{cases}$

7. $f(x) = \begin{cases} 1 & \text{for } 0 \leq x \leq 1, \\ -1 & \text{for } -1 \leq x < 0, \\ 0 & \text{for } |x| > 1. \end{cases}$

8. Let
$$f(x) = \begin{cases} 1 & \text{for } -1 \leq x \leq 1, \\ 0 & \text{for } |x| > 1. \end{cases}$$
Solve problem 5.6 with this initial temperature function. Next, expand the integrand of this solution in a Maclaurin series and integrate term by term to obtain a series solution for $u(x,t)$.

9. Try to use separation of variables to solve the problem
$$u_t = ku_{xx} \text{ for } -\infty < x < \infty, t > 0,$$
$$u_x(x,0) = \begin{cases} \cos(\pi x) & \text{for } -1 \leq x \leq 1, \\ 0 & \text{for } |x| > 1. \end{cases}$$

10. Let
$$v(x,t) = xt^{-3/2}e^{-x^2/4t}.$$
Show that $v(x,t)$ is a solution of problem 5.6 for $k = 1$ and $f(x) = 0$ for all x (noting that $\lim_{t\to 0+} v(x,t) = 0$). However, $u(x,t) = 0$ for all x and $t \geq 0$ is also a solution. Does this say that problem 5.6 does not have a unique solution?

11. Derive the integral 5.8 as follows. Let
$$F(x) = \int_0^\infty e^{-\zeta^2}\cos(x\zeta)\,d\zeta.$$
Show that
$$F'(x) = -\frac{x}{2}F(x).$$
Solve this differential equation to obtain $F(x)$ to within a constant. Determine this constant by using the standard integral
$$\int_0^\infty e^{-\zeta^2}\,d\zeta = \frac{1}{2}\sqrt{\pi}.$$

In each of problems 12–16, write solutions of the problem on the half line for the given initial temperature $f(x)$, one using equation 5.11, then equation 5.12. Generate some graphs of $u(x,t)$ for selected values of t, as well as graphs of surfaces $u(x,t)$ for different ranges of values of x and t.

12. $f(x) = e^{-\alpha x}$, with α a positive constant.

13. $f(x) = xe^{-\alpha x}$, with α a positive constant.

14. $f(x) = \begin{cases} 1 & \text{for } 0 \leq x \leq h, \\ 0 & \text{for } x > h. \end{cases}$ Here h is a positive constant.

15. $f(x) = \begin{cases} x & \text{for } 0 \leq x \leq 2, \\ 0 & \text{for } x > 2. \end{cases}$

16. $f(x) = \begin{cases} 1 & \text{for } 0 \leq x \leq h, \\ -1 & \text{for } h < x \leq 2h, \\ 0 & \text{for } x > 2h. \end{cases}$

    h is a positive constant.

17. Use separation of variables to solve
$$u_t = ku_{xx} \text{ for } x > 0, t > 0,$$
$$u_x(x,0) = \begin{cases} A & \text{for } 0 \leq x \leq h, \\ 0 & \text{for } x > h. \end{cases}$$
Here h and A are positive constants.

18. For each positive integer n, let problem P_n consist of the heat equation $u_t = u_{xx}$, for all x and $t > 0$, together with the conditions $u_x(0,t) = 0$ and

$$u(0,t) = \varphi_n(t) = \frac{2}{n}\sin(2n^2t).$$

Show that P_n has the solution

$$u_n(x,t) = \frac{1}{n}\left(e^{nx}\sin(2n^2t + nx) + e^{-nx}\sin(2n^2t - nx)\right).$$

However, show that $u_n(x,t) \to 0$ as $n \to \infty$, while it is possible to choose x and t so that, for arbitrarily large n, $u_n(x,t)$ can be made as large in magnitude as we like. Conclude that solutions of the problem

$$u_t = u_{xx} \text{ for } -\infty < x < \infty, t > 0,$$
$$u(0,t) = \varphi(t), u_x(0,t) = 0 \text{ for } t > 0,$$

do not depend continuously on the data.

5.3 The Debate over the Age of the Earth

The solution of the heat equation in a semi-infinite medium played an important role in one of the most important scientific controversies ever to take place. The entire story is told in Joe D. Burchfield's book *Lord Kelvin and the Age of the Earth* (Macmillan, 1975). It is also the subject of Chapter 6, "Lord Kelvin versus Geologists and Biologists," in Hal Hellman's book *Great Feuds in Science* (Wiley, 1988).

In the 18th century it was widely believed that the Earth and everything on it had been created at about the same time. This put the age of the Earth at the same age as the human race, and this was estimated to be several thousand years. Many of these estimates were based on biblical studies, making lists of families in the Bible and estimating the age of each generation. One chronology, constructed in the third century A.D. by Julius Africanus (c. 160–c. 240), was based on the cosmic week, with each cosmic day spanning one thousand years. Africanus arrived at a creation date of about 5500 B.C. for the Earth, which would make it about 7500 years old today.

In the 19th century, scientists began to suspect that some geological processes, such as the formation of mountains and canyons, must have taken a very long time. Estimates of the age of the Earth were extended to tens of thousands of years. At the extreme, some geologists held that the Earth's history extended infinitely far back, as there did not appear to be any evidence of a beginning.

Then, in the period 1830–1833, Sir Charles Lyell's (1797 - 1875) three-volume work *Principles of Geology* was published. In it Lyell argued that processes such as erosion and sedimentation in rivers had to occur over extended periods of time. Unlike his predecessors and colleagues, Lyell thought in terms of many millions of years, perhaps hundreds of millions, a bold step beyond anything previously imagined. This view became known as the *uniformatarian theory* because it disputed the belief that early violent geological activity had caused the present-day Earth to form over a relatively short time and maintained that the Earth reached its present state over an extended period of gradual, nearly uniform changes.

In 1859, Charles Darwin (1809–1882) published his classic *On the Origin of Species*. Darwin was very much a supporter of Lyell's view because evolution required longer periods of time than the several thousand years the 18th century scientists had been willing to allow.

In the 1850's, Lord Kelvin (Sir William Thomson, 1824–1907) entered the arena with a completely different perspective, that of a physicist. Kelvin was instrumental in the development of thermodynamics, and was sensitive to arguments involving generation and conservation of heat energy. He began by considering the Sun as the primary source of energy for our planet. Because nothing was known at this time of nuclear processes, it was at first believed that the Sun must be a gigantic chemical reaction. But this notion was soon discarded for want of an adequate source of fuel. Kelvin turned to the idea that the Sun generates heat energy converted from gravitational potential energy as the Sun contracts. This led him to an initial estimate that the Sun, and therefore the Earth, could not be more than 100 million years old.

Soon Kelvin expanded this argument. He had observed from data taken from mines extending deep into the Earth that the temperature of our planet increases with depth. He reasoned that this must be caused by a continual loss of heat energy from the interior of the planet by conduction into the upper crust. However, it was also observed that this upper crust does not increase in temperature on a year-to-year basis. Therefore Kelvin reasoned that there must be a continual loss of heat energy from the Earth. Further, Kelvin, and many geologists, believed that the Earth was a solid ball that had solidified at a uniform temperature. This meant that, if one could make an estimate of this initial temperature, and assuming a continual loss of heat energy, it should be possible by determining the rate of heat loss and current temperatures in the crust, to estimate the age of the Earth with some accuracy.

Unlike geologists, biologists and other natural philosophers of this time, who were not well versed in mathematics, Kelvin was well prepared to carry out the details. He began by assuming that the Earth is an approximately homogeneous sphere, say of radius R. If $u(\mathbf{x}, t)$ is the temperature distribution, giving the temperature at point $\mathbf{x}$ and time t, with the origin as the center of the sphere, then u satisfies the three-dimensional heat equation

$$\frac{\partial u}{\partial t} = k\nabla^2 u \text{ for } |\mathbf{x}| < R \text{ and } t > 0.$$

Kelvin had worked with this equation while engaged in studies of heat conduction leading to his discoveries in thermodynamics. But now he became convinced that very near the Earth's surface, the temperature had remained (nearly) constant and was on average approximately its initial value. This led him to neglect the curvature of the Earth and model heat conduction through the Earth as being along the half line from its center outward. Now let $u(x,t)$ be the temperature at distance x from the center at time t. This could be thought of as a one-dimensional problem on a half line and the model for $u(x,t)$ is:

$$\begin{aligned} u_t &= ku_{xx} \text{ for } x > 0, t > 0, \\ u(x,0) &= c \text{ for } x > 0, \\ u(0,t) &= 0 \text{ for } t > 0. \end{aligned}$$

Kelvin solved this problem, obtaining

$$u(x,t) = \frac{c}{2\sqrt{\pi kt}} \int_0^\infty \left[e^{-(x-\xi)^2/4kt} - e^{-(x+\xi)^2/4kt} \right] d\xi.$$

Kelvin had no way of measuring the Earth's temperature at large depths. However, by taking measurements as he descended into mine shafts, he estimated the rate of increase of temperature as one descended from the surface toward the center. This suggested that he compute the rate of change of temperature with depth. From the solution, he computed

$$\frac{\partial u}{\partial x} = \frac{c}{2\sqrt{\pi kt}} \int_0^\infty \left[-\frac{x-\xi}{2kt} e^{-(x-\xi)^2/4kt} + \frac{x+\xi}{2kt} e^{-(x+\xi)^2/4kt} \right] d\xi.$$

Then

$$\begin{aligned} \frac{\partial u}{\partial x}(0,t) &= \frac{c}{2\sqrt{\pi kt}} \int_0^\infty \frac{\xi}{kt} e^{-\xi^2/4kt}\, d\xi \\ &= \frac{c}{2\sqrt{\pi kt}} \left[-e^{-\xi^2/4kt} \right]_0^\infty \\ &= \frac{c}{\sqrt{\pi kt}}. \end{aligned}$$

Solve this for the time to obtain

$$t = \frac{1}{\pi k} \left(\frac{c}{u_x(0,t)} \right)^2.$$

Kelvin used measurements made in mine shafts to estimate $u_x(0,t)$. He also took numerous measurements of the temperature of rocks and molten lava to estimate c and k. Allowing for errors, these estimates led him to take the age of the Earth to be between 100 and 400 million years.

These numbers are very different from the numbers we accept today. For one thing, Kelvin had no way of taking into account nuclear processes generating energy from the sun, or radioactive minerals generating heat within the Earth. More significantly, his concept of the heat conduction within the Earth was flawed and he did not take into account convection in the partly fluid interior of the planet. The geology of his time simply could not give him an accurate picture of interior diffusion.

Many geologists disagreed with Kelvin's conclusions, but they did not have the mathematical expertise to dispute his analysis. Indeed, Kelvin's mathematics, coupled with his considerable and deserved reputation, made him a formidable opponent in the debate.

Despite his primary involvement with physics and thermodynamics, Kelvin maintained in interest in geology and the age of the Earth until his death early in the 20th century. Perhaps as a result of his influence and this debate, many geologists became more attuned to the careful collection and mathematical analysis of data based on geological samples. To its benefit, the subject became more quantitative in its practice.

The story has an amusing footnote. In 1904, Ernest Rutherford's (1871–1937) studies of radium led to his discovery of heat generated by radioactive processes. One day Rutherford was scheduled to lecture on his work, and he expected Kelvin to be present. Somewhat in awe of the grand old patriarch of physics, Rutherford was worried about Kelvin's reaction to this new factor accounting for heat energy within the Earth, which Kelvin himself had not known to include in his estimates. Somewhat to Rutherford's relief, Kelvin appeared to doze off during the first few minutes of the talk. But then Rutherford began to approach the crucial point and Kelvin suddenly sprang to attention. Rutherford mentioned Kelvin's estimate of the age of the Earth, but then diplomatically emphasized that no one could have foreseen this newly found source of heat energy. Kelvin settled back and seemed satisfied, and all was well.

5.4 Burgers' Equation

We will look at a diffusion equation that includes a convection term. The second-order, nonlinear partial differential equation

$$u_t + uu_x = ku_{xx}$$

for $-\infty < x < \infty$, is called *Burgers' equation*. The uu_x term is a convection, or mixing, term in which it is assumed that the effect of the convection at x at any time t is measurable as a product of the fluid or energy density $u(x,t)$ at x and that time t, multiplied by the rate of change of this density with respect to the space variable.

We seek solutions of this equation subject to an initial condition

$$u(x, 0) = f(x).$$

Begin with an observation. If we substitute $u = w_x$ into Burgers' equation, we get

$$w_{xt} + w_x w_{xx} = kw_{xxx}.$$

We can integrate this with respect to x to obtain

$$w_t + \frac{1}{2}w_x^2 = kw_{xx}.$$

Now define v by

$$v = e^{-w/2k}.$$

Then

$$w = -2k\ln(v),$$

and a routine differentiation yields

$$w_t + \frac{1}{2}w_x^2 - kw_{xx} = 0 = -\frac{2k}{v}(v_t - kv_{xx}).$$

This means that $v_t = kv_{xx}$, so v satisfies the standard one-dimensional heat equation.

Thus far, we have transformed

$$u \Rightarrow w \Rightarrow v.$$

Putting these two steps together, we have the composite transformation

$$u = -\frac{2kv_x}{v}. \tag{5.13}$$

This is the *Cole-Hopf transformation.* Using this, we have

$$u(x,0) = -\frac{2kv_x(x,0)}{v(x,0)} = f(x).$$

Integrate this to obtain

$$\int_0^x f(\xi)\,d\xi = -2k\ln(v(x,0)),$$

hence

$$v(x,0) = e^{(-1/2k)\int_0^x f(\xi)\,d\xi} = g(x). \tag{5.14}$$

We can therefore solve Burgers' equation on the real line, subject to $u(x,0) = f(x)$, by using the Cole-Hopf transform to convert this problem to an initial value problem for v:

$$\begin{aligned}
u_t + uu_x &= ku_{xx} \text{ for } -\infty < x < \infty, t > 0,\\
u(x,0) &= f(x) \text{ for } -\infty < x < \infty\\
&\Longrightarrow\\
v_t &= kv_{xx} \text{ for } -\infty < x < \infty, t > 0,\\
v(x,0) &= g(x) \text{ for } -\infty < x < \infty,
\end{aligned}$$

where $g(x)$ is given by equation 5.14.

We know the solution of this problem for v:

$$v(x,t) = \frac{1}{2\sqrt{\pi kt}}\int_{-\infty}^{\infty} g(\xi)e^{-(x-\xi)^2/4kt}\,d\xi.$$

Using this, compute

$$v_x(x,t) = -\frac{1}{2\sqrt{\pi kt}} \int_{-\infty}^{\infty} g(\xi) \left(\frac{x-\xi}{2kt} \right) e^{-(x-\xi)^2/4kt} \, d\xi.$$

Now we have the solution $u(x,t)$ of Burgers' problem from equation 5.13:

$$\begin{aligned} u(x,t) &= -\frac{2kv_x(x,t)}{v(x,t} \\ &= \frac{\int_{-\infty}^{\infty} g(\xi)((x-\xi)/t)e^{-(x-\xi)^2/4kt} \, d\xi}{\int_{-\infty}^{\infty} g(\xi)e^{-(x-\xi)^2/4kt} \, d\xi}. \end{aligned} \tag{5.15}$$

This is the solution we sought, as a quotient of integrals.

As an example, suppose $k = 1/2$ and $f(x) = 1$. Now

$$\frac{1}{2k} \int_0^x f(\xi) \, d\xi = x$$

so

$$g(x) = e^{-x}.$$

The solution to this problem for Burgers' equation is

$$u(x,t) = \frac{\int_{-\infty}^{\infty} e^{-\xi} \left(\frac{x-\xi}{t} \right) e^{-(x-\xi)^2/2t} \, d\xi}{\int_{-\infty}^{\infty} e^{-\xi} e^{-(x-\xi)^2/2t} \, d\xi}.$$

5.4.1 Traveling-Wave Solutions of Burgers' Equation

It is perhaps surprising that a diffusion equation can have traveling-wave solutions. By this, we mean solutions of the form $u(x,t) = h(x - ct$ for some function h of a single variable and some number c. Such solutions appear as a wave moving to the right along the $x-$axis.

We will show how to explicitly determine such a solution, under certain conditions. Look for a function h of one variable, having two continuous derivatives, and also for a number, c, that makes $u(x,t) = h(x - ct)$ a solution of Burgers' equation. We will require that $h(x)$ have a limit L_+ at ∞ and a limit L_- at $-\infty$, and that $0 < L_+ < L_-$. These conditions imply that $h'(x) \to 0$ as $x \to \pm\infty$.

To begin the search, substitute $h(x - ct)$ into Burgers' equation to get

$$-ch' + hh' = kh''.$$

Here h' denotes $h'(\xi)$ with $\xi = x - ct$.

Integrate this equation to obtain

$$-ch + \frac{1}{2}h^2 = kh' + a,$$

with a a constant of integration. Solve for h':

$$h' = \frac{1}{k}\left(\frac{1}{2}h^2 - ch - a\right). \tag{5.16}$$

Let $x \to \infty$ in this equation to get

$$\frac{1}{k}\left(\frac{1}{2}L_+^2 - cL_+ - a\right) = 0.$$

And let $x \to -\infty$ in equation 5.16 to get

$$\frac{1}{k}\left(\frac{1}{2}L_-^2 - cL_- - a\right) = 0.$$

Then

$$a = \frac{1}{2}L_+^2 - cL_+ = \frac{1}{2}L_-^2 - cL_-,$$

so

$$c(L_+ - L_-) = \frac{1}{2}L_+^2 - \frac{1}{2}L_-^2 = \frac{1}{2}(L_+ - L_-)(L_+ + L_-).$$

This gives us

$$c = \frac{1}{2}(L_+ + L_-).$$

Then

$$a = \frac{1}{2}L_+^2 - cL_+ = \frac{1}{2}L_+^2 - \frac{1}{2}(L_+ + L_-)L_+ = -\frac{1}{2}L_+L_-.$$

Then

$$\begin{aligned} h' &= \frac{1}{k}\left(\frac{1}{2}h^2 - ch - a\right) \\ &= \frac{1}{k}\left(\frac{1}{2}h^2 - \frac{L_+ + L_-}{2}h + \frac{1}{2}L_+L_-\right) \\ &= \frac{1}{2k}(h - L_+)(h - L_-). \end{aligned}$$

Denoting the independent variable as x, we now have

$$-2k\frac{dh}{dx} = (h - L_+)(L_- - h).$$

This is a separable ordinary differential equation for h. Use a partial fractions decomposition to write

$$\frac{-2}{L_- - L_+}\left(\frac{1}{h - L_+} + \frac{1}{L_- - h}\right) dh = \frac{1}{k}\, dx.$$

Integrate to obtain

$$\frac{2}{L_- - L_+}\ln\left(\frac{L_- - h}{h - L_+}\right) = \frac{x}{k} + \alpha.$$

We can choose the constant of integration α to be zero if we set $h(0) = c$, so

$$\frac{L_- - h(0)}{h(0) - L_+} = 1.$$

Now

$$\ln\left(\frac{L_- - h}{h - L_+}\right) = \frac{(L_- - L_+)x}{2k}.$$

Finally, we have

$$h(x) = \frac{L_- + L_+ e^{(L_- - L_+)x/2k}}{1 + e^{(L_- - L_+)x/2k}}.$$

The traveling wave solution we have been seeking is

$$\begin{aligned} u(x,t) &= h(x - ct) \\ &= \frac{L_- + L_+ e^{(L_- - L_+)(x-ct)/2k}}{1 + e^{(L_- - L_+)(x-ct)/2k}}. \end{aligned}$$

As an example, if we specify that $L_+ = 1, L_- = 3$ and $k = 1/2$, then $c = 2$ and a traveling wave solution of

$$u_t + uu_x = \frac{1}{2}u_{xx}$$

is

$$u(x,t) = \frac{3 + 2e^{2(x-ct)}}{1 + e^{2(x-ct)}}.$$

Figure 5.4 shows graphs of $u(x, 0)$, $u(x, 1)$, $u(x, 3)$ and $u(x, 5)$, showing the wave moving to the right as time increases. Figure 5.5 is a surface plot of $u(x, t)$ for $-8 \le x \le 8$ and $0 \le t \le 5$.

Problems for Section 5.4

In each of problems 1–3, find an expression for the solution of Burgers' equation with $u(x, 0) = f(x)$ for all real x.

1. $f(x) = \begin{cases} 1 - x & \text{for } 0 \le x \le 1, \\ 1 + x & \text{for } -1 \le x \le 0, \\ 0 & \text{for } x < -1 \text{ and for } x > 1. \end{cases}$

2. $f(x) = \begin{cases} 1 & \text{for } -1 \le x \le 1, \\ 0 & \text{for } |x| > 1. \end{cases}$

3. $f(x) = \begin{cases} -2 & \text{for } -3 \le x \le 0, \\ 5 & \text{for } 0 < x \le 2, \\ 0 & \text{for } x < -3 \text{ and for } x > 2. \end{cases}$

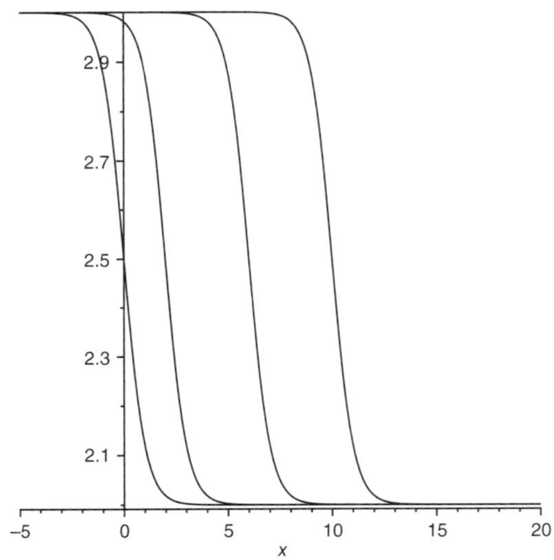

Figure 5.4: Traveling-wave solutions of Burgers' equation moving to the right.

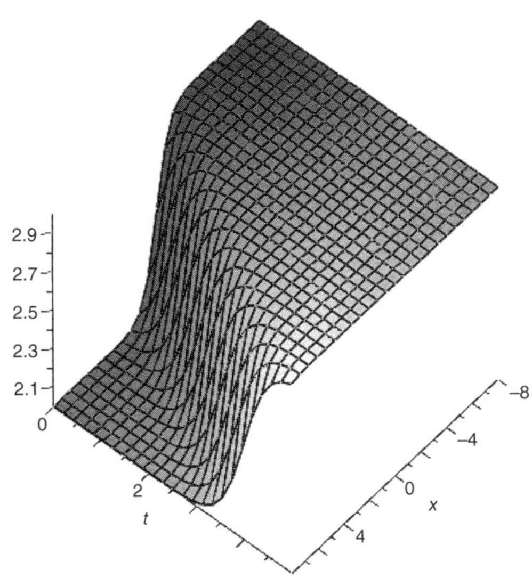

Figure 5.5: Surface plot of $u(x, t)$ for Burgers' equation.

4. Show that the solution of Burgers' equation on the line, with $u(x,0) = f(x)$, can be written as

$$u(x,t) = \frac{\int_{-\infty}^{\infty}((x-\xi)/t)e^{-H(\xi,x,t)/2k}\,d\xi}{\int_{-\infty}^{\infty}e^{-H(\xi,x,t)/2k}\,d\xi},$$

where

$$H(\xi,x,t) = \frac{(x-\xi)^2}{2t} + \int_0^x f(w)\,dw.$$

5. Consider Burgers' equation for $0 < x < 1$, subject to the conditions $u(0,t) = u(1,t) = 0$, $u(x,0) = f(x)$. The quantity

$$E(t) = \int_0^1 (u(x,t))^2\,dx$$

is called the *energy* of the system at time t. Show that the energy function is bounded for all $t \geq 0$. *Hint*: Compute $E'(t)$ as an integral, substitute for u_t in this integral from the partial differential equation and integrate by parts, making use of the initial conditions.

6. Show that $u(x,t) = c - a\tanh(a(x-ct)/2)$ satisfies Burgers' equation on the line for any positive numbers a and c. Graph this solution with $a = 1$ and $c = 1.5$ for a succession of times, $t = 0.02, 0.3, 0.7, 1.2, 2.5, 3$, and 5.

In problems 7–9, determine a traveling-wave solution of Burgers' equation on the real line for the given constants. Graph $u(x,0)$ and $u(x,t)$ for several values of the time.

7. $L_+ = 5, L_- = 10, k = 3$.

8. $L_+ = 4, L_- = 20, k = 5$.

9. $L_+ = 1, L_- = 2, k = 8$.

10. Under the conditions assumed in deriving a traveling wave solution $u(x,t) = h(x-ct)$, attempt to derive traveling wave solutions $u(x,t) = w(x+ct)$, which move to the left on the line as t increases.

5.5 The Cauchy Problem for the Wave Equation

Previously, we solved the Cauchy problem for the wave equation

$$\begin{aligned} u_{tt} &= c^2 u_{xx} \text{ for } -\infty < x < \infty, t > 0, \\ u(x,0) &= \varphi(x), u_t(x,0) = \psi(x) \text{ for } -\infty < x < \infty, \end{aligned} \tag{5.17}$$

using characteristics and d'Alembert's solution. This problem can also be solved using the Fourier integral. We will illustrate the idea for the case that the string is released from rest ($\psi(x) = 0$).

The problem is

$$u_{tt} = c^2 u_{xx} \text{ for } -\infty < x < \infty, t > 0,$$
$$u(x,0) = \varphi(x), u_t(x,0) = 0 \text{ for } -\infty < x < \infty. \tag{5.18}$$

Let $u(x,t) = X(x)T(t)$ to obtain

$$X'' + \lambda X = 0, T'' + \lambda c^2 T = 0.$$

In the absence of boundary conditions, there are no further conditions on X. The initial condition $u_t(x,0) = X(x)T'(0) = 0$ requires that $T'(0) = 0$. Therefore we have

$$X'' + \lambda X = 0$$

and

$$T'' + \lambda c^2 T = 0; T'(0) = 0.$$

The problem for X is the same as that encountered for the heat equation on the line. As usual, we seek a bounded solution. The eigenvalues and eigenfunctions are

$$\lambda = \omega^2 \text{ with } \omega \geq 0$$

and

$$X_\omega(x) = A_\omega \cos(\omega x) + B_\omega \sin(\omega x).$$

The differential equation for T has solutions

$$T(t) = k\cos(\omega ct) + d\sin(\omega ct).$$

Because $T'(0) = \omega d = 0$, then $d = 0$ for $\omega > 0$, and we must choose T as constant multiples of

$$T_\omega(t) = \cos(\omega ct).$$

For $\omega \geq 0$, we now have bounded functions

$$u_\omega(x,t) = (A_\omega \cos(\omega x) + B_\omega \sin(\omega x))\cos(\omega ct),$$

which satisfy the wave equation and the zero initial velocity condition. To satisfy the initial displacement condition, write a superposition over $\omega \geq 0$:

$$u(x,t) = \int_0^\infty u_\omega(x,t)\,d\omega$$
$$= \int_0^\infty [A_\omega \cos(\omega x) + B_\omega \sin(\omega x)]\cos(\omega ct)\,d\omega. \tag{5.19}$$

Now we have to choose the coefficients to satisfy

$$u(x,0) = \varphi(x) = \int_0^\infty [A_\omega \cos(\omega x) + B_\omega \sin(\omega x)]\,d\omega$$

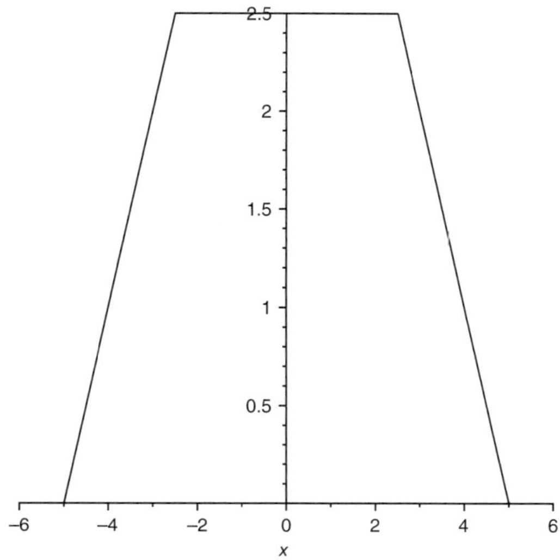

Figure 5.6: Initial displacement function in example 5.6, with $K = 5$.

for all real x. This is a Fourier integral representation of $\varphi(x)$ on the real line, so choose

$$A_\omega = \frac{1}{\pi}\int_{-\infty}^{\infty} \varphi(\xi)\cos(\omega\xi)\,d\omega \text{ and } B_\omega = \frac{1}{\pi}\int_{-\infty}^{\infty} \varphi(\xi)\sin(\omega\xi)\,d\xi.$$

Example 5.6 Solve the problem 5.18 with $c = 5$ and initial displacement function

$$\varphi(x) = \begin{cases} x + K & \text{for } -K \leq x \leq -K/2, \\ K/2 & \text{for } -K/2 \leq x \leq K/2, \\ -x + K & \text{for } K/2 \leq x \leq K \\ 0 & \text{for } x < -K \text{ and for } x > K. \end{cases}$$

K is a positive constant. Figure 5.6 shows this initial displacement function for the case that $K = 5$.

The solution is given by equation 5.19, and all we have to do is insert the Fourier integral coefficients of $\varphi(x)$. These are

$$\begin{aligned} A_\omega &= \frac{1}{\pi}\int_{-\infty}^{\infty} \varphi(\xi)\cos(\omega\xi)\,d\xi \\ &= \frac{1}{\pi}\int_{-K}^{K} \varphi(\xi)\cos(\omega\xi)\,d\xi \\ &= \frac{2}{\pi\omega^2}\left(\cos\left(\frac{K\omega}{2}\right) - \cos(K\omega)\right) \end{aligned}$$

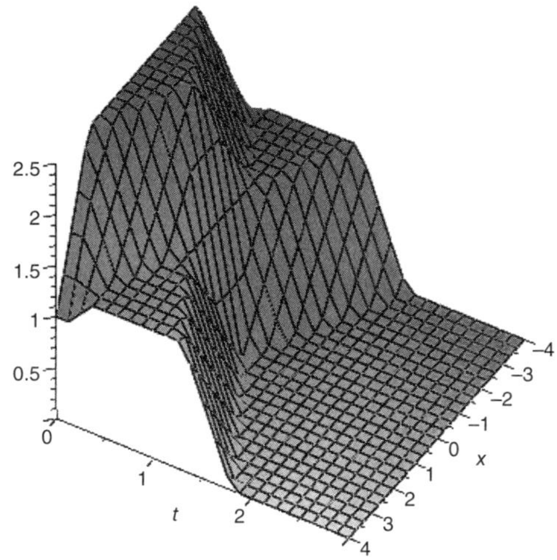

Figure 5.7: $u(x,t)$ in example 5.6, with K = 5.

and $B_\omega = 0$ because $\varphi(x)\sin(\omega x)$ is an odd function on the real line.

The solution is

$$\begin{aligned} u(x,t) &= \int_0^\infty A_\omega \cos(\omega x)\sin(5\omega t)\,d\omega \\ &= \int_0^\infty \frac{2}{\pi\omega^2}(\cos(K\omega/2) - \cos(K\omega))\cos(\omega x)\cos(5\omega t)\,d\omega. \end{aligned}$$

In the case that $K = 5$, this solution is

$$u(x,t) = \int_0^\infty \frac{2}{\pi\omega^2}(\cos(5\omega/2) - \cos(5\omega))\cos(\omega x)\cos(5\omega t)\,d\omega.$$

Part of an approximate graph of the surface $u(x,t)$ is shown in Figure 5.7.

For a wave problem on a half line $x \geq 0$, modeling an infinite string with one end, the problem is similar to the problem on the real line, except now there is one boundary condition. A typical such problem, with fixed left end, has the form

$$\begin{aligned} u_{tt} &= c^2 u_{xx} \text{ for } x > 0, t > 0, \\ u(0,t) &= 0 \text{ for } t \geq 0, \\ u(x,0) &= \varphi(x), u_t(x,0) = \psi(x) \text{ for } x > 0. \end{aligned}$$

For the case that $\psi(x)$ is identically zero, a solution by separation of variables proceeds as in the problem for the real line, except now there is the boundary condition

$$u(0,t) = X(0)T(t) = 0,$$

so $X(0) = 0$. We obtain the functions

$$X_\omega(x) = \sin(\omega x), T_\omega(t) = \cos(\omega ct),$$

leading us to try the solution

$$u(x,t) = \int_0^\infty b_\omega \sin(\omega x)\cos(\omega ct)\, d\omega.$$

To satisfy the initial condition, choose the coefficients so that

$$u(x,0) = \varphi(x) = \int_0^\infty b_\omega \sin(\omega x)\, d\omega,$$

hence choose

$$b_\omega = \frac{2}{\pi}\int_0^\infty \phi(\xi)\sin(\xi\omega)\, d\xi.$$

Problems for Section 5.5

In each of problems 1–6, solve the wave problem 5.18 on the line, with the given initial position function.

1 $\varphi(x) = e^{-|x|}$

2. $\varphi(x) = \begin{cases} 1 - x^2 & \text{for } -1 \leq x \leq 1, \\ 0 & \text{for } |x| > 1. \end{cases}$

3. $\varphi(x) = \begin{cases} \sin(x) & \text{for } -\pi \leq x \leq \pi, \\ 0 & \text{for } |x| > \pi. \end{cases}$

4. $\varphi(x) = \begin{cases} 4 - |x| & \text{for } -4 \leq x \leq 4, \\ 0 & \text{for } |x| > 4. \end{cases}$

5. $\varphi(x) = \begin{cases} e^{-x} & \text{for } x \geq 0, \\ 0 & \text{for } x < 0. \end{cases}$

6. $\varphi(x) = \begin{cases} \sin^2(x) & \text{for } -4\pi \leq x \leq 4\pi, \\ 0 & \text{for } |x| > 4\pi. \end{cases}$

7. Write a (bounded) integral solution of the Cauchy problem for the wave equation on the real line for the case that the initial position function is identically zero and the initial velocity is $\psi(x)$.

In each of problems 8–11, solve the initial value problem posed in problem 7, for the given initial velocity function.

8. $\psi(x) = e^{-|x|}$.

9. $\psi(x) = \begin{cases} k & \text{for } -\alpha \leq x \leq \alpha, \\ 0 & \text{for } |x| > \alpha. \end{cases}$ Here k and α are positive constants.

10. $\psi(x) = \begin{cases} \cos(x) & \text{for } -\pi/2 \leq x \leq \pi/2, \\ 0 & \text{for } |x| > \pi/2. \end{cases}$

11. $\psi(x) = \begin{cases} x & \text{for } -5 \leq x \leq 5, \\ 0 & \text{for } |x| > 5. \end{cases}$

12. Solve the problem on the half line if $\varphi(x) = xe^{-3x}$ for $x \geq 0$ and $\psi(x) = 0$.

13. Solve the problem on the half line if $\varphi(x) = 0$ and

$$\psi(x) = \begin{cases} \sin(x) & \text{for } 0 \leq x \leq \pi, \\ 0 & \text{for } x > \pi. \end{cases}$$

14. Solve the problem on the half line if $\varphi(x) = 0$ for $x \geq 0$, and

$$\psi(x) = \begin{cases} K & \text{for } 0 \leq x \leq K, \\ -x + K & \text{for } K \leq x \leq 2K, \\ 0 & \text{for } x \geq K. \end{cases}$$

5.6 Laplace's Equation on Unbounded Domains

The Fourier integral can also be used to some Dirichlet and Neumann problems on unbounded domains.

5.6.1 Dirichlet Problem for the Upper Half Plane

We will solve the Dirichlet problem for the upper half-plane consisting of all (x, y) with $y \geq 0$:

$$\begin{aligned} \nabla^2 u(x, y) &= 0 \text{ for } -\infty < x < \infty, y > 0, \\ u(x, 0) &= f(x) \text{ for } -\infty < x < \infty. \end{aligned} \tag{5.20}$$

The boundary of the upper half-plane is the horizontal axis, hence the boundary condition.

Let $u(x, y) = X(x)Y(y)$ to obtain

$$X'' + \lambda X = 0, Y'' - \lambda Y = 0.$$

The problem for X, with no boundary conditions, is familiar from the heat and wave equations on the line, and we have

$$\lambda = \omega^2, X_\omega(x) = A_\omega \cos(\omega x) + B_\omega \sin(\omega x).$$

Then

$$Y'' - \omega^2 Y = 0$$

with the general solution

$$Y_\omega(y)e = ce^{\omega y} + de^{-\omega y}.$$

For a bounded solution, take $c = 0$, so $Y_\omega(y)$ is a multiple of $e^{-\omega y}$.

For $\omega \geq 0$, we now have functions

$$u_\omega(x, y) = (A_\omega \cos(\omega x) + B_\omega \sin(\omega x))e^{-\omega y}$$

which are harmonic on the upper half-plane. To satisfy the boundary condition, set

$$u(x, y) = \int_0^\infty (A_\omega \cos(\omega x) + B_\omega \sin(\omega x))e^{-\omega y}\, d\omega. \tag{5.21}$$

This is the solution if we choose the coefficients so that

$$u(x, 0) = f(x) = \int_0^\infty (A_\omega \cos(\omega x) + B_\omega \sin(\omega x))\, d\omega.$$

Thus choose A_ω and B_ω to be the Fourier integral coefficients of $f(x)$ on $(-\infty, \infty)$.

It is perhaps surprising that, if we insert these Fourier coefficients into the general form 5.21 of the solution, we obtain a relatively simple expression. By reversing the order of integration and using a trigonometric identity, as we have done before, write

$$\begin{aligned} &u(x, y) = \\ &\frac{1}{\pi}\int_0^\infty \int_{-\infty}^\infty [\cos(\omega\xi)\cos(\omega x) + \sin(\omega\xi)\sin(\omega x)]e^{-\omega y} f(\xi)\, d\xi\, d\omega \\ &= \frac{1}{\pi}\int_{-\infty}^\infty \left(\int_0^\infty \cos(\omega(\xi - x))e^{-\omega y}\, d\omega\right) f(\xi)\, d\xi. \end{aligned}$$

We can carry out the inner integration with respect to ω:

$$\begin{aligned} &\int_0^\infty e^{-\omega y}\cos(\omega(\xi - x))\, d\omega \\ &= \left[\frac{e^{-\omega y}}{y^2 + (\xi - x)^2}(-y\cos(\omega(\xi - x)) + (\xi - x)\sin(\omega(\xi - x)))\right]_0^\infty \\ &= \frac{y}{y^2 + (\xi - x)^2}. \end{aligned}$$

The solution is therefore

$$u(x, y) = \frac{y}{\pi}\int_{-\infty}^\infty \frac{f(\xi)}{y^2 + (\xi - x)^2}\, d\xi. \tag{5.22}$$

Example 5.7 Find a function $u(x,y)$ that is harmonic on the upper half-plane, and satisfies

$$u(x,0) = f(x) = 1 \text{ for all real } x.$$

By equation 5.22, the solution is

$$\begin{aligned}
u(x,y) &= \frac{y}{\pi}\int_{-\infty}^{\infty} \frac{1}{y^2+(\xi-x)^2}\,d\xi \\
&= \frac{y}{\pi}\lim_{K\to\infty}\int_{-K}^{K} \frac{1}{y^2+(\xi-x)^2}\,d\xi \\
&= \frac{y}{\pi}\lim_{K\to\infty}\frac{1}{y}\left[\arctan\left(\frac{K+x}{y}\right)+\arctan\left(\frac{K-x}{y}\right)\right] \\
&= \frac{y}{\pi}\frac{1}{y}\left[\frac{\pi}{2}+\frac{\pi}{2}\right] \\
&= 1.
\end{aligned}$$

Of course! The constant function $u(x,y) = 1$ is harmonic on the upper half plane and equals 1 on the x–axis. This problem does not require any of the analysis performed for its solution, but it does demonstrate that the formula works, at least in a simple case.

Example 5.8 Solve the Dirichlet problem for the upper half-plane if

$$u(x,0) = f(x) = \begin{cases} k & \text{for } -a \le x \le a, \\ 0 & \text{for } |x| > a. \end{cases}$$

Here a and k are positive constants.

The solution is

$$\begin{aligned}
u(x,y) &= \frac{y}{\pi}\int_{-\infty}^{\infty} \frac{f(\xi)}{y^2+(\xi-x)^2}\,d\xi \\
&= \frac{y}{\pi}\int_{-a}^{a} \frac{k}{y^2+(\xi-x)^2}\,d\xi \\
&= \frac{k}{\pi}\left[\arctan\left(\frac{a+x}{y}\right)+\arctan\left(\frac{a-x}{y}\right)\right].
\end{aligned}$$

5.6.2 Dirichlet Problem for the Right Quarter Plane

We will solve a special Dirichlet problem

$$\begin{aligned}
\nabla^2 u(x,y) &= 0 \text{ for } x>0, y>0, \\
u(x,0) &= f(x) \text{ for } x \ge 0, \\
u(0,y) &= 0 \text{ for } y > 0 \qquad (5.23)
\end{aligned}$$

for the right quarter plane. The boundary of this domain consists of the nonnegative $x-$axis and the nonnegative $y-$axis, accounting for the boundary conditions. The more general problem in which $u(0, y) = g(y)$ can be done by separation of variables, but we will pursue the given special case to demonstrate an idea.

Notice that the right quarter plane can be obtained by folding the upper half-plane across the vertical axis. This suggests that we may be able to use the solution 5.22 for the upper half-plane to obtain the solution for the right quarter plane. in which the data function on the positive $y-$axis is zero.

To do this, we will manufacture a Dirichlet problem for the upper half-plane in such a way that, when we restrict x to be nonnegative, we obtain the solution of the current problem.

Begin by defining

$$g(x) = \begin{cases} f(x) & \text{for } x \geq 0, \\ \text{any} & \text{for } x < 0, \end{cases}$$

where by *any* we mean we will fill in $g(x)$ for $x < 0$ later when it becomes clear what we need. Now the problem

$$\begin{aligned} \nabla^2 u(x, y) &= 0 \text{ for } -\infty < x < \infty, y > 0, \\ u(x, 0) &= g(x) \end{aligned}$$

for the upper half-plane has the solution

$$u_{\text{hp}}(x, y) = \frac{y}{\pi} \int_{-\infty}^{\infty} \frac{g(\xi)}{y^2 + (\xi - x)^2} \, d\xi.$$

Write this as

$$u_{\text{hp}}(x, y) = \frac{y}{\pi} \left[\int_{-\infty}^{0} \frac{g(\xi)}{y^2 + (\xi - x)^2} \, d\xi + \int_{0}^{\infty} \frac{g(\xi)}{y^2 + (\xi - x)^2} \, d\xi \right].$$

Put $w = -\xi$ in the left integral in $u_{\text{hp}}(x, y)$:

$$\int_{-\infty}^{0} \frac{g(\xi)}{y^2 + (\xi - x)^2} \, d\xi = \int_{\infty}^{0} \frac{g(-w)}{y^2 + (w + x)^2} (-1) \, dw.$$

Write ξ as the variable of integration, and reverse the limits of integration, so this last integral is

$$\int_{0}^{\infty} \frac{g(-\xi)}{y^2 + (\xi + x)^2} \, d\xi.$$

Then

$$\begin{aligned} u_{\text{hp}}(x, y) &= \frac{y}{\pi} \int_{0}^{\infty} \left(\frac{g(-\xi)}{y^2 + (\xi + x)^2} + \frac{g(\xi)}{y^2 + (\xi - x)^2} \right) d\xi \\ &= \frac{y}{\pi} \int_{0}^{\infty} \left(\frac{g(-\xi)}{y^2 + (\xi + x)^2} + \frac{f(\xi)}{y^2 + (\xi - x)^2} \right) d\xi. \end{aligned}$$

Now fill in the "any" in the definition of g. The function $u_{\text{hp}}(x,y)$ will vanish on the positive $y-$axis (on which $x=0$) if we make $g(-\xi)=-g(\xi)$ for $\xi<0$. This occurs if we set

$$g(x)=-f(-x) \text{ for } x<0.$$

This means we should make g an odd extension of f to the entire real line. When this is done,

$$u_{\text{hp}}(x,y)=\frac{y}{\pi}\int_0^\infty\left(\frac{1}{y^2+(\xi-x)^2}-\frac{1}{y^2+(\xi+x)^2}\right)f(\xi)\,d\xi \tag{5.24}$$

is the solution of this Dirichlet problem for the upper half plane. But u_{hp} is also harmonic on the right quarter plane, vanishes when $x=0$, and equals $f(x)$ is $x\geq 0$. Therefore equation 5.24 provides the solution of the Dirichlet problem for the right quarter plane in which the boundary function on the nonnegative vertical axis is zero.

Example 5.9 Solve the Dirichlet problem

$$\begin{aligned}\nabla^2(x,y)&=0 \text{ for } x>0,y>0,\\ u(0,y)&=0 \text{ for } y\geq 0,\\ u(x,0)&=xe^{-x} \text{ for } x\geq 0.\end{aligned}$$

We have a formula 5.24 for the solution:

$$u(x,y)=\frac{y}{\pi}\int_0^\infty\left(\frac{1}{y^2+(\xi-x)^2}-\frac{1}{y^2+(\xi+x)^2}\right)\xi e^{-\xi}\,d\xi.$$

This integral cannot be evaluated in closed form, but values at specific points (x,y) can be approximated by numerical integrations.

As another example, look at the Dirichlet problem 5.23 when $u(x,0)=f(x)=1$ for $x\geq 0$. Now the solution is

$$u(x,y)=\frac{y}{\pi}\int_0^\infty\frac{1}{y^2+(\xi-x)^2}\,d\xi-\frac{y}{\pi}\int_0^\infty\frac{1}{y^2+(\xi+x)^2}\,d\xi.$$

Both of these integrals can be evaluated. The first is

$$\begin{aligned}\int_0^\infty\frac{1}{y^2+(\xi-x)^2}\,d\xi&=\int_0^\infty\frac{1}{y^2\left[1+\left(\frac{\xi-x}{y}\right)^2\right]}\,d\xi\\ &=\frac{1}{y}\left(\frac{\pi}{2}-\arctan\left(-\frac{x}{y}\right)\right)\\ &=\frac{\pi}{2y}+\frac{1}{y}\arctan\left(\frac{x}{y}\right).\end{aligned}$$

Similarly,

$$\int_0^\infty \frac{1}{y^2+(\xi+x)^2}\,d\xi = \frac{\pi}{2y} - \frac{1}{y}\arctan\left(\frac{x}{y}\right).$$

The solution is y/π times the difference of these two integrals:

$$u(x,y) = \frac{2}{\pi}\arctan\left(\frac{x}{y}\right).$$

It is routine to check that $u(x,y)$ is harmonic on the right quarter plane, that $u(0,y)=0$, and, for $x>0$,

$$\lim_{y\to 0+}\frac{2}{\pi}\arctan\left(\frac{x}{y}\right) = \frac{2}{\pi}\frac{\pi}{2} = 1.$$

We will conclude this section with an example of a Neumann problem for an unbounded domain, specifically, the upper half-plane

5.6.3 A Neumann Problem for the Upper Half Plane.

We will solve the problem

$$\begin{aligned} \nabla^2 u(x,y) &= 0 \text{ for } -\infty < x < \infty, y > 0, \\ u_y(x,0) &= f(x) \text{ for } -\infty < x < \infty. \end{aligned} \tag{5.25}$$

The boundary of the upper half plane consists of all points on the x−axis, and $u_y(x,0$ is the normal derivative on this boundary.

Although $\partial\Omega$ is not a closed curve in this type of problem, it is still possible to show that a necessary condition for the problem to have a solution is that the integral of the normal derivative on the boundary must be zero. In this case, this is

$$\int_{-\infty}^\infty f(\xi)\,d\xi = 0.$$

We will assume that $f(x)$ satisfies this condition.

There is an elegant way of reducing this Neumann problem to one we have already solved. Let $v = u_y$. Observe that

$$\begin{aligned} \nabla^2 v &= v_{xx} + v_{yy} = (u_y)_{xx} + (u_y)_{yy} \\ &= (u_{xx})_y + (u_{yy})_y = (\nabla^2 u)_y = 0 \text{ for } -\infty < x < \infty, y > 0 \end{aligned}$$

and

$$v(x,0) = u_y(x,0) = f(x) \text{ for } -\infty < x < \infty.$$

This means that v is a solution of a Dirichlet problem for the upper half-plane, if u is a solution of the Neumann problem 5.25. We know the solution of this Dirichlet problem for v:

$$v(x,y) = \frac{y}{\pi}\int_{-\infty}^\infty \frac{f(\xi)}{y^2+(\xi-x)^2}\,d\xi.$$

Because $v = u_y$, recover u from this expression for v by integrating with respect to y:

$$\begin{aligned} u(x,y) &= \frac{1}{\pi}\int\left(\int_{-\infty}^{\infty}\frac{y}{y^2+(\xi-x)^2}f(\xi)\,d\xi\right)dy \\ &= \frac{1}{\pi}\int_{-\infty}^{\infty}\left(\int\frac{y}{y^2+(\xi-x)^2}\,dy\right)f(\xi)\,d\xi \\ &= \frac{1}{2\pi}\int_{-\infty}^{\infty}\ln(y^2+(\xi-x)^2)f(\xi)\,d\xi + c, \end{aligned} \tag{5.26}$$

in which c is an arbitrary constant.

Problems for Section 5.6

In each of problems 1–4, write a solution of the Dirichlet problem 5.20 for the upper half-plane for the given $f(x)$.

1. $f(x) = e^{-6|x|}$.

2. $f(x) = \begin{cases} K & \text{for } 0 \le x \le \alpha, \\ 0 & \text{for } |x| > \alpha, \end{cases}$ where K is any real number and $\alpha > 0$.

3. $f(x) = \begin{cases} \cos(x) & \text{for } -\pi/2 \le x \le \pi/2, \\ 0 & \text{for } |x| > \pi/2. \end{cases}$

4. $f(x) = \begin{cases} 1 & \text{for } -1 \le x \le \alpha, \\ 2 & \text{for } \alpha < x \le 2\alpha, \\ 0 & \text{for } x < -1 \text{ or } x > 2\alpha \end{cases}$, with α any positive number.

5. Write a solution of the Dirichlet problem for the right quarter plane for $f(x) = x\cos(x)$ for $x \ge 0$.

6. Write a solution of the Dirichlet problem for the right quarter plane for

$$f(x) = \begin{cases} 1 & \text{for } 0 \le x \le 3, \\ -1 & \text{for } 3 \le x \le 5, \\ 0 & \text{for } x > 5. \end{cases}$$

7. Solve the Dirichlet problem for the left quarter plane:

$$\begin{aligned} \nabla^2 u(x,y) &= 0 \text{ for } x < 0, y > 0, \\ u(0,y) &= 0 \text{ for } y > 0, \\ u(x,0) &= f(x) \text{ for } x < 0. \end{aligned}$$

8. Solve the Dirichlet problem for the quarter plane $x < 0, y < 0$:

$$\begin{aligned}\nabla^2 u(x,y) &= 0 \text{ for } x < 0, y < 0,\\ u(0,y) &= 0 \text{ for } y < 0,\\ u(x,0) &= f(x) \text{ for } x < 0.\end{aligned}$$

9. Solve the Dirichlet problem:

$$\begin{aligned}\nabla^2 u(x,y) &= 0 \text{ for } x < 0, y > 0,\\ u(0,y) &= k \text{ for } y > 0,\\ u(x,0) &= f(x) \text{ for } x < 0,\end{aligned}$$

in which k is a constant.

10. Solve the problem

$$\begin{aligned}\nabla^2 u(x,y) &= 0 \text{ for } x < 0, y > 0,\\ u(0,y) &= 0 \text{ for } y > 0,\\ u(x,0) &= x \text{ for } -5 \le x \le 0,\\ u(x,0) &= 0 \text{ for } x < -5.\end{aligned}$$

11. Solve the Neumann problem for the lower half-plane:

$$\begin{aligned}\nabla^2 u(x,y) &= 0 \text{ for } -\infty < x < \infty, y < 0,\\ u_y(x,0) &= f(x) \text{ for } -\infty < x < \infty.\end{aligned}$$

12. Solve the Neumann problem for the right quarter plane:

$$\begin{aligned}\nabla^2 u(x,y) &= 0 \text{ for } x > 0, y > 0,\\ u_x(0,y) &= 0 \text{ for } y > 0,\\ u_y(x,0) &= f(x) \text{ for } x > 0.\end{aligned}$$

13. Solve the Neumann problem for the right half-plane:

$$\begin{aligned}\nabla^2 u(x,u) &= 0 \text{ for } x > 0, -\infty < y < \infty,\\ u_y(0,y) &= g(y) \text{ for } -\infty < y < \infty.\end{aligned}$$

Assume that g is bounded and continuous. Is there any other condition required of g?

14. Solve the Neumann problem for the left half-plane:

$$\begin{aligned}\nabla^2 u(x,y) &= 0 \text{ for } x < 0, -\infty < y < \infty,\\ u_x(0,y) &= g(y) \text{ for } -\infty < y < \infty.\end{aligned}$$

15. Solve the mixed boundary-value problem for the right quarter plane:

$$\begin{aligned} \nabla^2 u(x,y) &= 0 \text{ for } x>0, y>0, \\ u(0,y) &= 0 \text{ for } y \geq 0, \\ u_y(x,0) &= f(x) \text{ for } x>0. \end{aligned}$$

16. Show that, for any positive integer n, the problem

$$\begin{aligned} \nabla^2 u(x,y) &= 0 \text{ for all } x,y \\ u(x,0) &= 0 \text{ for all } x, \\ u_y(x,0) &= \frac{1}{n}\sinh(nx) \text{ for all } x \end{aligned}$$

has the solution

$$u_n(x,y) = \frac{1}{n^2}\sinh(nx)\sin(ny).$$

Show that, as $n \to \infty$, $u_n(x,0) \to 0$. However, $u_n(x,y)$ assumes arbitrarily large values for certain (x,y) with y arbitrarily small.

17. Consider the problem

$$\begin{aligned} \nabla^2 u(x,y) &= 0 \text{ for } -\infty < x < \infty, y>0, \\ u(,0) = 0, u_y(x,0) &= ne^{-\sqrt{n}}\sin(nx) \text{ for } y>0, \end{aligned}$$

in which n is a positive integer. Show that

$$u_n(x,y) = e^{-\sqrt{n}}\sin(nx)e^{ny}$$

is a solution for each positive integer n, and that $u_n(x,y_0)$ can be made arbitrarily large in magnitude for any $y_0 > 0$, by choosing n sufficiently large.

Chapter 6

Solutions Using Eigenfunction Expansions

6.1 A Theory of Eigenfunction Expansions

We have solved many initial-boundary value problems in which separation of variables has led to sines and/or cosines as eigenfunctions, in terms of which we had to write some given function (such as an initial temperature, position or velocity) in a Fourier expansion.

We will now solve initial-boundary value problems in which "new" functions, or special functions, have to be created as eigenfunctions. Further, we will have to expand some given function in a series of these new eigenfunctions. This section provides an environment in which we can deal with these issues.

Begin with a general form of the differential equation for the eigenfunctions. Separation of variables in many kinds of important problems leads to a differential equation of the form

$$(ry')' + (q + \lambda p)y = 0, \tag{6.1}$$

where r, q, and p are given, continuous functions on the relevant interval, say $[a, b]$ or (a, b), and $r(x) > 0$ and $p(x) > 0$ for $a < x < b$.

This is the *Sturm-Liouville differential equation*. Of course, any differential equation can be written in different ways, so we will refer to the form given in equation 6.1 as the *standard form* of the Sturm-Liouville equation.

A *Sturm-Liouville problem* consists of this differential equation, together with boundary conditions. We will consider two kinds of boundary conditions.

A *regular Sturm-Liouville problem* on $[a, b]$ consists of equation 6.1, with $r(a)$ and $r(b)$ both nonzero, and the boundary conditions

$$A_1y(a) + A_2y'(a) = 0, B_1y(b) + B_2y'(b) = 0, \tag{6.2}$$

where A_1, A_2, B_1, and B_2 are given numbers, with A_1 and A_2 not both zero, and B_1 and B_2 not both zero.

A *singular Sturm-Liouville problem* consists of equation 6.1, with boundary conditions that take one of the following forms:

1. $r(a) = 0$ replaces the boundary condition at a in 6.2,

or

2. $r(b) = 0$ replaces the boundary condition at b in 6.2,

or

3. $r(a) = r(b) = 0$ replaces both boundary conditions in 6.2.

We will also impose the condition that solutions must be bounded on $[a, b]$.

For either kind of Sturm-Liouville problem, a value of λ for which there is a nontrivial solution is called an *eigenvalue* of the problem, and a corresponding nontrivial solution is called an *eigenfunction* associated with that eigenvalue.

Example 6.1 The Sturm-Liouville problem

$$y'' + \lambda y = 0; y(0) = y(L) = 0$$

is regular on $[0, L]$, with $r(x) = p(x) = 1$ and $q(x) = 0$. The eigenvalues are $\lambda_n = n^2\pi^2/L^2$ and the eigenfunctions are $y_n(x) = \sin(n\pi x/L)$, for each positive integer n.

Example 6.2 The differential equation

$$[(1 - x^2)y']' + \lambda y = 0 \tag{6.3}$$

is called *Legendre's equation*. Here $r(x) = 1 - x^2$, $q(x) = 0$ and $p(x) = 1$. We usually consider this equation on $[-1, 1]$. Because $r(1) = r(-1) = 0$, this problem is singular, and there are no boundary conditions. We will find eigenvalues and eigenfunctions of this system when we develop Legendre polynomials.

Example 6.3 *Bessel's equation of order* ν, with ν any real number, is

$$(xy')' + \left(\lambda x - \frac{\nu}{x}\right) y = 0. \tag{6.4}$$

Now $r(x) = p(x) = x$ and $q(x) = -\nu/x$. This problem is usually solved on an interval $0 < x \leq b$, so a boundary condition $B_1 y(b) + B_2 y'(b) = 0$ is specified at the right end of the interval. We will find eigenvalues and eigenfunctions for this problem when we have Bessel functions.

It can be shown that a regular or singular Sturm-Liouville problem on $[a, b]$ has an infinite and unbounded sequence of eigenvalues, which can be ordered so that

$$\lambda_1 < \lambda_2 < \cdots < \lambda_n \cdots .$$

With this ordering,

$$\lim_{n\to\infty} \lambda_n = \infty.$$

We have seen this behavior in the eigenvalues $n^2\pi^2/L^2$ of some problems involving the wave and heat equations.

Now suppose, for each positive integer n, $y_n(x)$ is an eigenfunction associated with the eigenvalue λ_n of a Sturm-Liouville problem on $[a, b]$. We want to explore the possibility of expanding a given function $f(x)$ in a series of these eigenfunctions:

$$\sum_{n=1}^{\infty} c_n y_n(x)$$

on the relevant interval. We want some way of choosing the coefficients so that this series converges to $f(x)$ on $[a, b]$, if this is possible.

This is the problem we faced with Fourier series. What led us to the coefficients in that case was the fortunate circumstance that integrals of products of distinct eigenfunctions on $[-L, L]$ were zero. This was the orthogonality condition noted in Section 1.2.

Now a miracle occurs! There is a similar orthogonality relationship that holds for eigenfunctions of Sturm-Liouville problems in general.

Theorem 6.1 (Weighted Orthogonality of the Eigenfunctions) Let $y_n(x)$ and $y_m(x)$ be eigenfunctions of a Sturm-Liouville problem on $[a, b]$, corresponding to distinct eigenvalues λ_n and λ_m. Then

$$\int_a^b p(x)y_n(x)y_m(x)\,dx = 0.$$

This integral relationship is called the *weighted orthogonality* of the eigenfunctions, and $p(x)$ is called the *weight function*. In the case of Fourier series, $p(x) = 1$, but in general $p(x)$ is read as the coefficient of λ in the standard form (6.1) of the Sturm-Liouville differential equation.

We will prove the theorem and then show how this property is used for the coefficients for an eigenfunction expansion.

Proof By the definition of eigenvalues and eigenfunctions,

$$(ry_n')' + (q + \lambda_n p)y_n = 0$$

and

$$(ry_m')' + (q + \lambda_m p)y_m = 0.$$

Multiply the first equation by y_m, the second by y_n, and subtract:

$$(ry_n')'y_m - (ry_m')'y_n = (\lambda_m - \lambda_n)py_ny_m.$$

Then

$$\int_a^b [(ry_n')'y_m - (ry_m')'y_n]\,dx = (\lambda_m - \lambda_n)\int_a^b p(x)y_n(x)y_m(x)\,dx.$$

The left side of this equation can be integrated by parts to obtain

$$\begin{aligned} &r(b)[y_m(b)y_n'(b) - y_n(b)y_m'(b)] - r(a)[y_m(a)y_n'(a) - y_n(a)y_m'(a)] \\ &= (\lambda_m - \lambda_n)\int_a^b p(x)y_n(x)y_m(x)\,dx. \end{aligned} \tag{6.5}$$

To complete the proof, we will show that the left side of equation 6.5 is zero. Because $\lambda_n \neq \lambda_m$, this will imply that the integral on the right side is zero, which is what we want to prove.

Suppose first that the Sturm-Liouville problem is regular. Then, at a we have

$$\begin{aligned} A_1y_n(a) + A_2y_n'(a) &= 0, \\ A_1y_m(a) + A_2y_m'(a) &= 0. \end{aligned}$$

Think of these two equations as defining a 2×2 homogeneous system of linear equations in the unknowns $\tilde{x}$ and $\tilde{y}$:

$$\begin{aligned} y_n(a)\tilde{x} + y_n'(a)\tilde{y} &= 0, \\ y_m(a)\tilde{x} + y_m'(a)\tilde{y} &= 0. \end{aligned}$$

This system has a nontrivial solution, namely $\tilde{x} = A_1, \tilde{y} = A_2$, which are not both zero by assumption. Therefore the determinant of the system must be zero:

$$\begin{vmatrix} y_n(a) & y_n'(a) \\ y_m(a) & y_m'(a) \end{vmatrix} = 0.$$

This makes

$$y_n(a)y_m'(a) - y_m(a)y_n'(a) = 0.$$

Reasoning in the same way at b, we also have

$$\begin{vmatrix} y_n(b) & y_n'(b) \\ y_m(b) & y_m'(b) \end{vmatrix} = y_n(b)y_m'(b) - y_m(b)y_n'(b) = 0.$$

Together, these make the left side of equation 6.5 zero, hence the integral on the right side is also zero, as we wanted.

A similar argument applies if the Sturm-Liouville problem is singular. For example, if $r(a) = r(b) = 0$, then the left side of equation 6.5 is zero. If $r(a) = 0$ and $r(b) \neq 0$, then the boundary condition at b, together with vanishing of

$r(x)$ at a, again causes the left side of equation 6.5 to be zero, and similarly if $r(b) = 0$ but $r(a) \neq 0$.

By exploiting this weighted orthogonality, following the argument we used for Fourier series, we can derive a likely formula for c_n in the expansion of $f(x)$ in a series of these eigenfunctions. Suppose

$$f(x) = \sum_{n=1}^{\infty} c_n y_n(x).$$

Multiply both sides by $p(x)y_m(x)$ and integrate:

$$\int_a^b p(x)f(x)y_m(x)\,dx = \sum_{n=1}^{\infty} c_n \int_a^b p(x)y_n(x)y_m(x)\,dx..$$

Because of the weighted orthogonality of the eigenfunctions, each term in the sum on the right is zero except for the $n = m$ term, leaving

$$\int_a^b p(x)f(x)y_m(x)\,dx = c_m \int_a^b p(x)y_m^2(x)\,dx.$$

This suggests that we should choose

$$c_m = \frac{\int_a^b p(x)f(x)y_m(x)\,dx}{\int_a^b p(x)y_m^2(x)\,dx} \text{ for } m = 1, 2, \cdots. \tag{6.6}$$

These are the coefficients in the expansion of $f(x)$ in the eigenfunctions of the particular Sturm-Liouville problem.

As with Fourier series, finding a likely form for the coefficients does not itself prove convergence. The theorem for general eigenfunction expansions is like that for Fourier series. This should not be surprising, since it includes the Fourier convergence theorem as a special case.

Theorem 6.2 (Convergence of Eigenfunction Expansions) Let the eigenvalues of a Sturm-Liouville problem on $[a, b]$ be $\lambda_1, \lambda_2, \cdots$. Let $y_n(x)$ be an eigenfunction associated with λ_n. Let $f(x)$ be piecewise smooth on (a, b), and let c_n be specified by equation 6.6. Then, for $a < x < b$,

$$\sum_{n=1}^{\infty} c_n y_n(x) = \frac{1}{2}(f(x-) + f(x+).$$

In particular, if f is continuous at a point x in (a, b), then this series converges to $f(x)$.

Example 6.4 The Sturm-Liouville problem

$$(xy')' + \frac{\lambda}{x}y = 0; y(1) = y(3) = 0,$$

is regular on $[1, 3]$. We will expand $f(x) = 1$ in a series of the eigenfunctions on $[1, 3]$.

The differential equation is in standard Sturm-Liouville form, and from the coefficient of λ we read the weight function $p(x) = 1/x$. To solve the differential equation, however, it is easiest to write it in the form

$$x^2y'' + xy' + \lambda y = 0$$

which we recognize as a Cauchy-Euler equation, with general solution

$$y(x) = a\cos(\sqrt{\lambda}\ln(x)) + b\sin(\sqrt{\lambda}\ln(x)).$$

Now

$$y(1) = a = 0,$$

so

$$y(3) = b\sin(\sqrt{\lambda}\ln(3)) = 0.$$

To have a nontrivial solution, $\sqrt{\lambda}\ln(3)$ must be an integer multiple of π:

$$\sqrt{\lambda}\ln(3) = n\pi \text{ for } n = 1, 2, \cdots .$$

The eigenvalues are

$$\lambda_n = \left(\frac{n\pi}{\ln(3)}\right)^2.$$

Corresponding eigenfunctions are

$$y_n(x) = \sin\left(\frac{n\pi}{\ln(3)}\ln(x)\right).$$

From equation 6.6, the nth coefficient in the expansion of $f(x)$ is

$$c_n = \frac{\int_1^3 \frac{1}{x} f(x) y_n(x)\,dx}{\int_1^3 \frac{1}{x} y_n(x)^2\,dx}.$$

These integrations are routine and we obtain

$$c_n = \frac{2}{n\pi}(1 - (-1)^n) = \begin{cases} 4/n\pi & \text{if } n \text{ is odd,} \\ 0 & \text{if } n \text{ is even.} \end{cases}$$

Because all the even-indexed coefficients are zero, retain only the odd-indexed coefficients in the series by replacing n with $2n - 1$, obtaining the eigenfunction series

$$\sum_{n=1}^{\infty} \frac{4}{(2n-1)\pi} \sin\left(\frac{(2n-1)\pi\ln(x)}{\ln(3)}\right) = 1 \text{ for } 1 < x < 3.$$

This series converges to $f(x)$ on $(1, 3)$ because $f(x)$ is continuous. Figure 6.1 shows a graph of $f(x)$ and the $150th$ partial sum of this series.

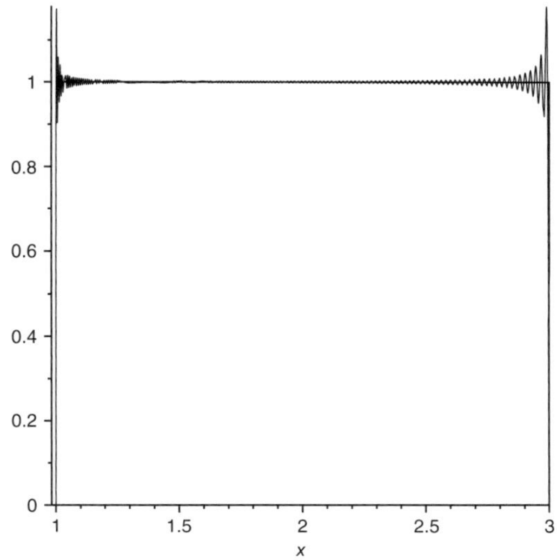

Figure 6.1: $f(x)$ and the 150th partial sum of the eigenfunction expansion in example 6.4.

In example 6.4, the partial sum of 150 terms appears to fit the function reasonably well on the interval of the expansion. In particular cases, it may take a partial sum of many more terms of an eigenfunction expansion to closely approximate the function. To illustrate this point, look at the regular Sturm-Liouville problem

$$y'' + 6y' + (5 + \lambda)y = 0; y(0) = y(2) = 0$$

on $[0, 2]$. This problem has eigenvalues and eigenfunctions

$$\lambda_n = 4 + \frac{n^2\pi^2}{4}, y_n(x) = e^{-3x} \sin\left(\frac{n\pi x}{2}\right).$$

To expand a given $g(x)$ in a series of these eigenfunctions on $[0, 2]$, first write the differential equation in standard Sturm-Liouville form. Multiply it by the integrating factor $e^{\int 6\,dx} = e^{6x}$ to obtain

$$(e^{6x}y')' + (5 + \lambda)e^{6x}y = 0,$$

from which we read the weight function $p(x) = e^{6x}$. The eigenfunction expansion of $g(x)$ on $[0, 2]$ has coefficients

$$c_n = \frac{\int_0^2 e^{6x} g(x) e^{-3x} \sin(n\pi x/2)\, dx}{\int_0^2 e^{6x} \left(e^{-3x} \sin(n\pi x/2)\right)^2\, dx},$$

or

$$c_n = \frac{\int_0^2 e^{3x} g(x) \sin(n\pi x/2)\, dx}{\int_0^2 \sin^2(n\pi x/2)\, dx}.$$

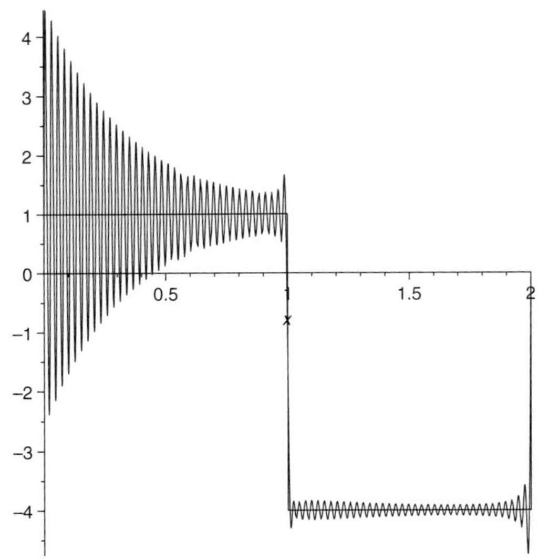

Figure 6.2: $g(x)$ and the 150th partial sum of the eigenfunction expansion.

Look at a particular $g(x)$, say

$$g(x) = \begin{cases} 1 & \text{for } 0 \leq x < 1, \\ -4 & \text{for } 1 \leq x \leq 2. \end{cases}$$

Then

$$c_n = \frac{1}{36 + n^2\pi^2} \left(2n\pi(1 + 4e^8(-1)^n) + 60e^3 \sin(n\pi/2) - 10e^3 n\pi \cos(n\pi/2)\right).$$

Figures 6.2–6.4 compare, respectively, $g(x)$ with the partial sum of 150, 500, and 2000 terms. The eigenfunction series appears to converge to the function, but this convergence is very slow.

Before putting these ideas to use in solving initial-boundary value problems, we will place them in a more abstract setting, which will reveal additional properties of the coefficients of eigenfunction expansions.

6.1.1 A Closer Look at Expansion Coefficients

We will be concerned with piecewise smooth functions on some interval $[a, b]$. Let $PS[a, b]$ be the set of all such functions. Before looking at eigenfunction expansions of these functions, first focus on $PS[a, b]$ itself.

The set $PS[a, b]$ is called the *space of piecewise continuous functions* on the interval because, if f and g are in this set, so is $\alpha f + \beta g$ for any constants α and β. $PS[a, b]$ has the structure of a real vector space.

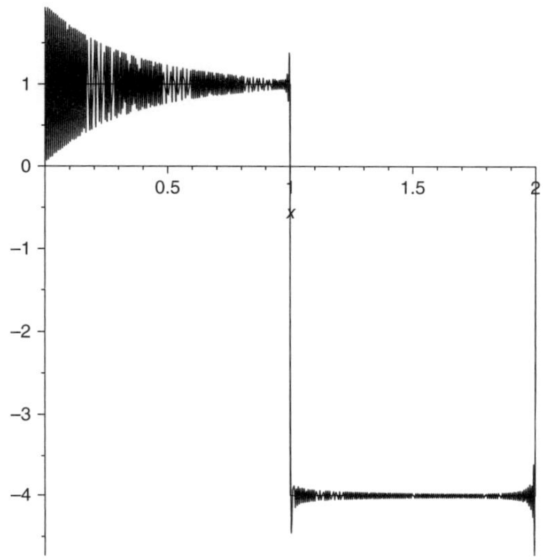

Figure 6.3: $g(x)$ and the 500th partial sum.

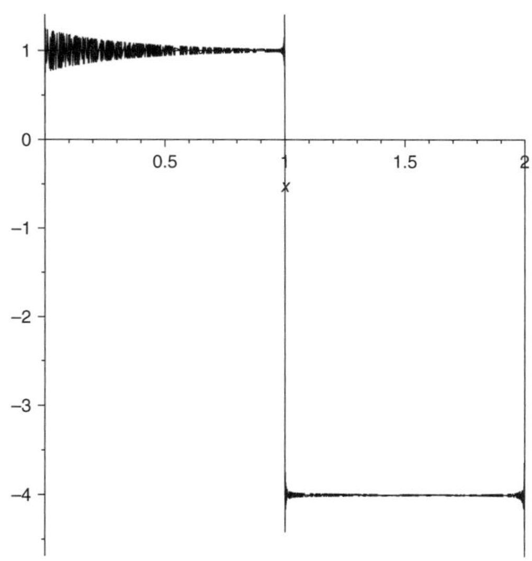

Figure 6.4: $g(x)$ and the 2000th partial sum.

We will draw analogies between $PS[a,b]$ and the space R^2 of vectors in the plane, enabling us to bring a sense of geometric intuition to $PS[a,b]$.

In the plane, we have a dot product of two vectors. For vectors $\mathbf{A} = a_1\mathbf{i}+a_2\mathbf{j}$ and $\mathbf{B} = b_1\mathbf{i} + b_2\mathbf{j}$,

$$\mathbf{A} \cdot \mathbf{B} = a_1 b_1 + a_2 b_2.$$

The norm of $\mathbf{A}$ is

$$\| \mathbf{A} \| = \sqrt{\mathbf{A} \cdot \mathbf{A}} = \sqrt{a_1^2 + a_2^2}.$$

This is the distance from the origin to the point (a_1, a_2), while

$$\begin{aligned} \| \mathbf{A} - \mathbf{B} \| &= \| (a_1 - b_1)\mathbf{i} + (a_2 - b_2)\mathbf{j} \| \\ &= \sqrt{(a_1 - b_1)^2 + (a_2 - b_2)^2} \end{aligned}$$

is the distance between the points (a_1, a_2) and (b_1, b_2).

Adapt these ideas to $PS[a,b]$. Begin with a Sturm-Liouville problem, say SL. Suppose problem SL has weight function $p(x)$. If f and g are any functions in $PS[a,b]$, define the *dot product* of f and g (with respect to problem SL) to be

$$f \cdot g = \int_a^b p(x) f(x) g(x)\, dx.$$

This operation behaves in many ways that are similar to the dot product of vectors in R^2. In particular,

$$\begin{aligned} f \cdot g &= g \cdot f, \\ (cf) \cdot g &= f \cdot cg = c(f \cdot g) \text{ for any constant } c, \\ (cf + dg \cdot h &= c(f \cdot h) + d(g \cdot h). \end{aligned}$$

It is also true that the dot product of f with itself is positive:

$$f \cdot f = \int_a^b p(x)(f(x))^2\, dx > 0,$$

unless the function is identically zero.

The square root of the dot product of a function with itself plays a special role in defining a notion of distance in $PS[a,b]$. Define the norm of f in $PS[a,b]$ to be

$$\| f \| = \sqrt{f \cdot f}.$$

Most often this is written

$$\| f \|^2 = f \cdot f.$$

We may think of $\| f \|$ as the distance in $PS[a,b]$ between f and the zero function. If g is also in $PS[a,b]$, then the *distance between f and g* is defined to be

$$\begin{aligned} \| f - g \| &= \sqrt{(f - g) \cdot (f - g)} \\ &= \int_a^b p(x)(f(x) - g(x))^2\, dx. \end{aligned}$$

This endows $PS[a, b]$ with a metric, or distance function, based on a given Sturm-Liouville problem. We use this metric to measure how "close" one function is to another in $PS[a, b]$.

In R^2 a vector of length 1 is called a unit vector. In $PS[a, b]$, we can divide a nonzero function f by its norm to obtain the *normalized function*

$$\frac{f}{\| f \|},$$

which has length, or norm, 1. Normalized functions in $PS[a, b]$ play the role of unit vectors in R^2.

Now suppose all of the eigenvalues of problem SL are

$$\lambda_1, \lambda_2, \cdots, \lambda_n, \cdots,$$

with corresponding eigenfunctions

$$\varphi_1, \varphi_2, \cdots, \varphi_n, \cdots.$$

These eigenfunctions are not unique, because any nonzero constant multiple of an eigenfunction associated with λ_n is also an eigenfunction for λ_n. However, for this discussion, imagine these eigenfunctions to be fixed, and normalize them by dividing each by its norm, obtaining eigenfunctions

$$\Phi_1, \Phi_2, \cdots, \Phi_n, \cdots,$$

each of length 1 in the function space.

Now let $f \in PS[a, b]$. We are interested in the following question. For a given positive integer N, how should we choose numbers $k_1, k_2, \cdots, k_N$ so that the sum

$$S_N(x) = \sum_{n=1}^{N} k_n \Phi_n(x)$$

is the closest function to f in $PS[a, b]$? That is, we want numbers $k_1, \cdots, k_N$ to minimize the distance between f and a linear combination $S_N(x)$ of $\Phi_1, \cdots, \Phi_N$.

If we write out this distance, the problem is one of choosing the constants $k_1, \cdots, k_N$ to minimize

$$\int_a^b p(x) \left(f(x) - S_N(x) \right)^2 dx.$$

When a choice of $k_1, \cdots, k_N$ minimizes this integral, we say that this $S_N(x)$ is the *best mean square approximation of* f (relative to this set of N normalized eigenfunctions).

In determining how to choose $k_1, \cdots, k_N$, keep in mind that

$$\int_a^b p(x)\Phi_n(x)\Phi_m(x)\, dx = \begin{cases} 0 & \text{if } n \neq m, \\ 1 & \text{if } n = m. \end{cases}$$

In the eigenfunction expansion $\sum_{n=1}^{\infty} c_n \Phi_n(x)$ of $f(x)$, we therefore have

$$c_n = \int_a^b p(x) f(x) \Phi_n(x)\, dx.$$

Now, returning to the question of choosing $k_1, \cdots, k_N$, compute

$$\begin{aligned}
&\int_a^b p(x)\left(f(x) - S_N(x)\right)^2 dx \\
&= \int_a^b p(x) \left[f(x) - \sum_{n=1}^{N} k_n \Phi_n(x)\right]^2 dx \\
&= \int_a^b p(x) \left[(f(x))^2 - 2\sum_{n=1}^{N} k_n f(x)\Phi(x) + \sum_{n=1}^{N}\sum_{m=1}^{N} k_n k_m \Phi_n(x)\Phi_m(x)\right] dx \\
&= \int_a^b p(x)(f(x))^2\, dx - 2\sum_{n=1}^{N} k_n \int_a^b p(x) f(x)\Phi(x)\, dx \\
&+ \sum_{n=1}^{N}\sum_{m=1}^{N} k_n k_m \int_a^b p(x)\Phi_n(x)\Phi_m(x)\, dx \\
&= \| f \|^2 - 2\sum_{n=1}^{N} k_n c_n + \sum_{n=1}^{N} k_n^2 \\
&= \| f \|^2 - 2\sum_{n=1}^{N} k_n c_n + \sum_{n=1}^{N} k_n^2 + \sum_{n=1}^{N} c_n^2 - \sum_{n=1}^{N} c_n^2 \\
&= \| f \|^2 + \sum_{n=1}^{N} (k_n - c_n)^2 - \sum_{n=1}^{N} c_n^2.
\end{aligned}$$

From the last line, it is immediate that the integral of interest (first line) is a minimum if

$$\sum_{n=1}^{N} (k_n - c_n)^2 = 0$$

hence each $k_n = c_n$. This means that $\sum_{n=1}^{N} k_n \Phi_n(x)$ is the best mean square approximation to $f(x)$ exactly when the k_n's are the first N coefficients $c_1, \cdots, c_N$ in the eigenfunction expansion of f with respect to these eigenfunctions.

We have proved the following.

Theorem 6.3 (Best Mean-Square Approximation) Among all choices of real numbers $k_1, \cdots, k_N$, the one that minimizes the distance between $f(x)$ and $\sum_{n=1}^{N} k_n \Phi_n(x)$ is

$$k_1 = c_1, k_2 = c_2, \cdots, k_N = c_N.$$

This result holds for Fourier series as well.

The calculation done in establishing the theorem also shows that

$$\sum_{n=1}^{N} c_n^2 \leq \| f \|^2 .$$

Because this is true for every positive integer N, we conclude that the nonnegative series $\sum_{n=1}^{\infty} c_n^2$ converges, and also that

$$\sum_{n=1}^{\infty} c_n^2 \leq \| f \|^2 . \tag{6.7}$$

This is known as *Bessel's inequality*. Further, because the general term of any convergent series has limit zero as $n \to \infty$,

$$\lim_{n\to\infty} c_n^2 = 0,$$

hence also

$$\lim_{n\to\infty} c_n = 0.$$

The sequence (as opposed to series) of eigenfunction expansion coefficients tends to zero as n tends to infinity. (Be careful here. This does not mean that these coefficients are decreasing, just that they have limit zero).

We are now ready to develop two Sturm-Liouville problems leading to special functions, enabling us to solve some initial-boundary value problems that were previously beyond our reach.

Problems for Section 6.1

In each of problems 1–6, find all of the eigenvalues and, for each eigenvalue, an eigenfunction, of the Sturm-Liouville problem. Determine the coefficients to expand the given function in a series of these eigenfunctions on the interval, and determine the sum of this series at points within the interval. Graph some partial sums of this expansion and compare them with a graph of the function.

1. $y'' - 2y' + (3+\lambda)y = 0; y(0) = y(1) = 0, f(x) = -3.$
2. $y'' + \lambda y = 0; y(0) = y'(2) = 0, f(x) = 2x.$
3. $y'' + \lambda y = 0; y'(0) = y(1) = 0, f(x) = \sin(2x).$
4. $y'' + \lambda y = 0; y'(0) = y'(3) = 0, f(x) = e^{-x}.$
5. $y'' - 3y' + (2+3\lambda)y = 0; y(0) = y(2) = 0, f(x) = x.$
6. $y'' + y' + (1+4\lambda)y = 0; y(0) = y(1) = 0, f(x) = x.$
7. Suppose $p(x) = 1$. Let f and g be in $PS[a,b]$. Show that

$$\int_a^b \int_a^b (f(x)g(y) - g(x)f(y))^2 \, dx \, dy = 2\left(\| f \|^2 \| g \|^2 - (f \cdot g)^2\right).$$

Use this to verify the Schwarz inequality:

$$|f \cdot g| \leq \| f \| \| g \| .$$

6.2 Bessel Functions

We will develop some facts about Bessel's equation and Bessel functions that we will use to solve a variety of problems.

The differential equation

$$x^2y'' + xy' + (\lambda x^2 - \nu^2)y = 0 \tag{6.8}$$

is one form of *Bessel's equation of order* ν, where ν can be any number. The designation *order* here refers solely to ν, and not to the differential equation, which is of second order.

The Sturm-Liouville form of Bessel's equation is

$$(xy')' + \left(\lambda x - \frac{\nu^2}{x}\right)y = 0, \tag{6.9}$$

which is also known as Bessel's equation.

One solution can be obtained using the method of Frobenius. Let

$$y = \sum_{n=1}^{\infty} a_n x^{n+r}$$

with the $a_n's$ and r to be determined, and $a_0 \neq 0$. Substitute this series into equation 6.8 and collect the terms involving x^r, x^{r+1}, and so on to obtain

$$[r(r-1) + r - \nu^2]a_0x^r + [r(r+1) + r + 1 - \nu^2]a_1x^{r+1}$$
$$+ \sum_{n=2}^{\infty} \left[[(n+r)(n+r-1) + (n+r) - \nu^2]a_n + a_{n-2}\right] x^{n+r} = 0.$$

Because the right side of this equation is zero for all x in some interval about zero, the coefficient of each power of x on the left side must be zero. From the coefficient of x^r,

$$r^2 - \nu^2 = 0,$$

so $r = \pm\nu$. Put $r = \nu$ into the coefficient of x^{r+1} to obtain

$$(2\nu + 1)a_1 = 0.$$

Therefore $a_1 = 0$. From the coefficient of x^{n+r}, we now have

$$[(n+\nu)(n+\nu-1) + (n+\nu) - \nu^2]a_n + a_{n-2} = 0$$

for $n = 2, 3, \cdots$. Therefore

$$a_n = -\frac{1}{n(n+2\nu)}a_{n-2}.$$

This is a recurrence relation from which we can recover each of the coefficients as a multiple of either a_0 or A_1. First, because $a_1 = 0$,

$$a_3 = a_5 = \cdots = a_{2n+1} = \cdots = 0.$$

All the odd-indexed coefficients are zero. For the even-indexed coefficients, replace n with $2n$ and work back from a_{2n} to a_{2n-2} to a_{2n-4} and so on, eventually writing a_{2n} in terms of a_0:

$$\begin{aligned} a_{2n} &= -\frac{1}{2n(2n+2\nu)}a_{2n-2} \\ &= -\frac{1}{2^2 n(n+\nu)}a_{2n-2} \\ &= -\frac{1}{2^2(n+\nu)}\frac{-1}{2(n-1)[2(n-1)+2\nu]}a_{2n-4} \\ &= \cdots = \frac{(-1)^n}{2^{2n}n!(1+\nu)(2+\nu)\cdots(n+\nu)}a_0. \end{aligned}$$

This gives us the solution

$$y(x) = a_0 \sum_{n=0}^{\infty} \frac{(-1)^n}{2^{2n}n!(1+\nu)(2+\nu)\cdots(n+\nu)}x^{2n+\nu}.$$

This is a nontrivial solution of Bessel's equation of order ν for any $a_0 \neq 0$ and $\nu \geq 0$. Any such solution is called a *Bessel function of order* ν.

Because Bessel functions are widely used, it is useful to agree on a choice of a_0 to define a standard solution. This is done using the *gamma function*, which is defined by

$$\Gamma(x) = \int_0^{\infty} t^{x-1}e^{-t}\,dt.$$

This integral converges for $x > 0$. What it brings to this discussion is its *factorial property*, which is that, for $x > 0$,

$$\Gamma(x+1) = x\Gamma(x). \tag{6.10}$$

This is easily proved by integration by parts.

By a straightforward integration,

$$\Gamma(1) = \int_0^{\infty} e^{-t}\,dt = 1.$$

Using the factorial property, compute

$$\begin{aligned} \Gamma(2) &= \Gamma(1+1) = 1\cdot\Gamma(1) = 1, \\ \Gamma(3) &= \Gamma(2+1) = 2\Gamma(2) = 2, \\ \Gamma(4) &= \Gamma(3+1) = 3\Gamma(3) = 3\cdot 2 = 3!, \\ \Gamma(5) &= \Gamma(4+1) = 4\Gamma(4) = 4(3!) = 4!, \end{aligned}$$

and so on. In general, if n is a positive integer, then

$$\Gamma(n+1) = n!.$$

This is the reason for the term *factorial property.*

If x is not an integer, then $\Gamma(x+1) = x\Gamma(x)$ is not an integer. However, using the factorial property, we can write

$$\begin{aligned}\Gamma(n+\nu+1) &= (n+\nu)\Gamma(n+\nu) \\ &= (n+\nu)(n+\nu-1)\Gamma(n+\nu-1) \\ &= \cdots = (1+\nu)(2+\nu)\cdots(n-1+\nu)(n+\nu)\Gamma(\nu+1).\end{aligned}$$

Then

$$(1+\nu)(2+\nu)\cdots(n+\nu) = \frac{\Gamma(n+\nu+1)}{\Gamma(\nu+1)}.$$

Put this into the solution of Bessel's equation to write it as

$$y(x) = a_0 \sum_{n=0}^{\infty} \frac{(-1)^n \Gamma(\nu+1)}{2^{2n} n! \Gamma(n+\nu+1)} x^{2n+\nu}.$$

It is customary to choose

$$a_0 = \frac{1}{2^{\nu}\Gamma(\nu+1)}.$$

obtaining the particular solution

$$J_\nu(x) = \sum_{n=0}^{\infty} \frac{(-1)^n}{2^{2n+\nu} n! \Gamma(n+\nu+1)} x^{2n+\nu}.$$

This is *Bessel function of the first kind of order* ν. The series defining $J_\nu(x)$ converges for all x.

Because Bessel's equation is of second order, there is a second, linearly independent solution. This may take several forms, depending on the following cases.

Case 1. If 2ν is not an integer, then $J_{-\nu}(x)$ is also a solution of Bessel's equation and is not a constant multiple of $J_\nu(x)$, so the general solution is

$$y = \alpha J_\nu(x) + \beta J_{-\nu}(x).$$

Case 2. If 2ν is an odd positive integer, say $2\nu = 2n+1$, then $\nu = n + 1/2$ and both $J_{n+1/2}(x)$ and $J_{-n-1/2}$ are independent solutions. Now the general solution is

$$y = \alpha J_{n+1/2}(x) + \beta J_{-n-1/2}(x).$$

Case 3. If $\nu > 0$, but neither case 1 or case 2 holds, then 2ν is an integer, but not of the form $2n+1$. Now $J_{-\nu}(x) = (-1)^\nu J_\nu(x)$, so $J_{-\nu}(x)$ and $J_\nu(x)$ are not linearly independent.

In this case we must invoke the method of Frobenius again and seek a second, linearly independent solution of the form

$$Y_\nu(x) = J_\nu(x)\ln(x) + \sum_{n=1}^{\infty} c_n^* x^n.$$

This is a *Bessel function of the second kind of order* ν.

For example,

$$Y_0(x) = J_0(x)\ln(x) + \sum_{n=0}^{\infty} \frac{(-1)^{n+1}}{2^{2n}(n!)^2}\phi(n)x^{2n},$$

where

$$\phi(n) = \sum_{j=1}^{n} \frac{1}{j} = 1 + \frac{1}{2} + \cdots + \frac{1}{n}.$$

Because of the logarithm term, $Y_\nu(x)$ is not defined on any interval containing zero. Further, it is unbounded on any interval in which the variable can approach 0. Because of this, we will not include Bessel functions of the second kind in solving some problems.

Figure 6.5 shows $J_0(x)$ and Figure 6.6 is a graph of $J_1(x)$. Figure 6.7 shows $J_2(x), J_3(x)$, and $J_4(x)$. Notice that these consecutively numbered Bessel functions appear to "interlace" as x increases. This is a property we will discuss later. Note also that these graphs cross the axis, suggesting the existence of zeros of Bessel functions. This will be important later in expanding functions in series of Bessel functions.

Figure 6.8 is a graph of $Y_0(x)$ (upper graph) and $Y_1(x)$.

6.2.1 Variations on Bessel's Equation

We will see in applications that transformed versions of Bessel's equation often occur. In particular, consider the differential equation

$$y'' - \left(\frac{2a-1}{x}\right)y' + \left(b^2c^2x^{2c-2} + \frac{a^2-\nu^2c^2}{x^2}\right)y = 0. \tag{6.11}$$

In the case that $\nu = n$ is a positive integer, a routine but lengthy calculation shows that the general solution is

$$y(x) = c_1x^aJ_n(bx^c) + c_2x^aY_n(bx^c).$$

If ν is not an integer, then

$$y(x) = c_1x^aJ_\nu(bx^c) + c_2x^aJ_{-\nu}(bx^c).$$

Example 6.5 We will solve

$$y'' - \left(\frac{2\sqrt{7}-1}{x}\right)y' + \left(9x^4 - \frac{11}{x^2}\right)y = 0.$$

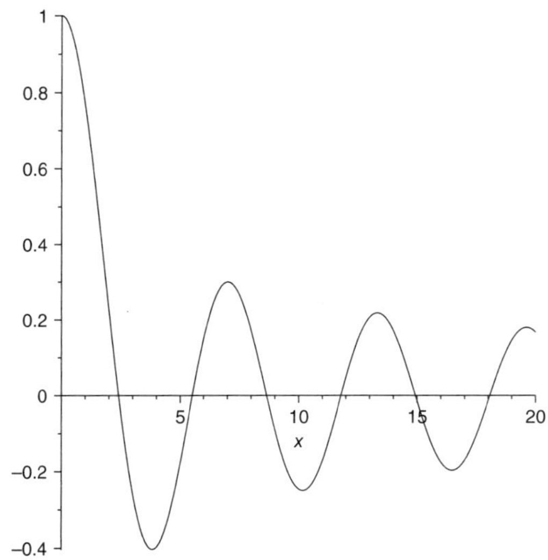

Figure 6.5: Graph of $J_0(x)$.

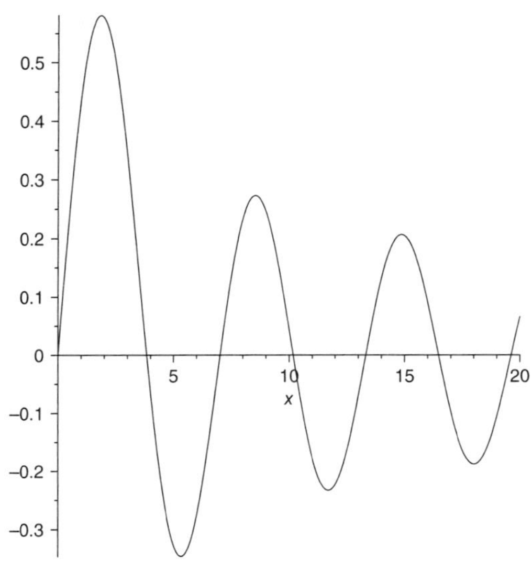

Figure 6.6: $J_1(x)$

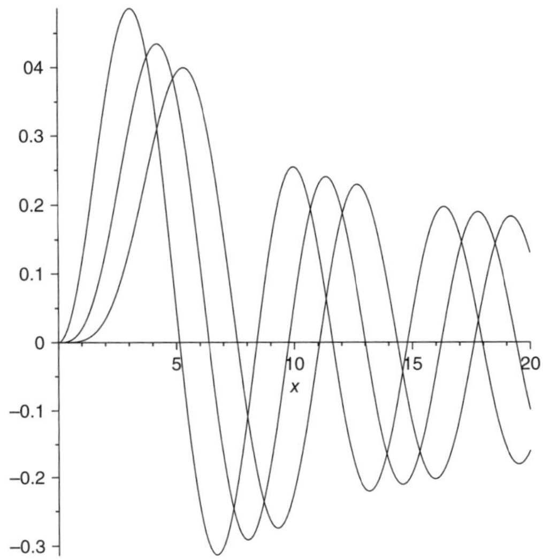

Figure 6.7: $J_2(x), J_3(x)$ and $J_4(x)$.

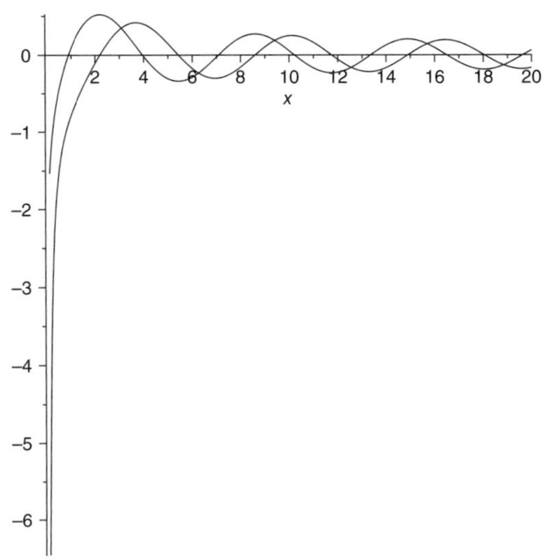

Figure 6.8: $Y_0(x)$ and $Y_1(x)$.

This equation fits the pattern of equation 6.11 if

$$a = \sqrt{7} - 1, b^2c^2 = 9, 2c - 2 = 4, \text{ and } a^2 - \nu^2c^2 = -11.$$

Immediately, $c = 3$. Then $b^2c^2 = 9b^2 = 9$ so $b = 1$. Finally, we need

$$a^2 - \nu^2c^2 = 7 - 9\nu^2 = -11$$

so $\nu = \sqrt{2}$. The general solution is

$$y = c_1 x^{\sqrt{7}} J_{\sqrt{2}}(x^3) + c_2 x^{\sqrt{7}} J_{-\sqrt{2}}(x^3).$$

6.2.2 Recurrence Relations

There are many recurrence relations for Bessel functions. Here are three that we will use.

Theorem 6.4 (Recurrence Relations)

$$\frac{d}{dx}(x^\nu J_\nu(x)) = x^\nu J_{\nu-1}(x). \tag{6.12}$$

$$\frac{d}{dx}(x^{-\nu} J_\nu(x)) = -x^{-\nu} J_{\nu+1}(x). \tag{6.13}$$

And, for $x > 0$,

$$\frac{2\nu}{x} J_\nu(x) = J_{\nu+1}(x) + J_{\nu-1}(x). \tag{6.14}$$

The first two relations can be proved by manipulating the series for $J_\nu(x)$. For the first,

$$\begin{aligned}
\frac{d}{dx}(x^\nu J_\nu(x)) &= \frac{d}{dx}\left[x^\nu \sum_{n=0}^{\infty} \frac{(-1)^n}{2^{2n+\nu} n! \Gamma(n+\nu+1)} x^{2n+\nu}\right] \\
&= \frac{d}{dx}\left[\sum_{n=0}^{\infty} \frac{(-1)^n}{2^{2n+\nu} n! \Gamma(n+\nu+1)} x^{2n+2\nu}\right] \\
&= \sum_{n=0}^{\infty} \frac{(-1)^n 2(n+\nu)}{2^{2n+\nu} n! (n+\nu)\Gamma(n+\nu)} x^{2n+2\nu-1} \\
&= x^\nu \sum_{n=0}^{\infty} \frac{(-1)^n}{2^{2n+\nu-1} n! \Gamma(n+\nu)} x^{2n+\nu-1} = x^\nu J_{\nu-1}(x).
\end{aligned}$$

Equation 6.13 is proved similarly, For equation 6.14, carry out the differentiations in equations 6.12 and 6.13

$$x^\nu J'_\nu(x) + \nu x^{\nu-1} J_\nu(x) = x^\nu J_{\nu-1}(x)$$

and

$$x^{-\nu}J_\nu'(x) - \nu x^{-\nu-1}J_\nu(x) = -x^\nu J_{\nu+1}(x).$$

Multiply the first equation by $x^{-\nu}$ and the second by x^ν to obtain

$$J_\nu'(x) + \frac{\nu}{x}J_\nu(x) = J_{\nu-1}(x) \tag{6.15}$$

and

$$J_\nu'(x) - \frac{\nu}{x}J_\nu(x) = -J_{\nu+1}(x). \tag{6.16}$$

Upon subtracting the second of these equations from the first, we obtain equation 6.14.

6.2.3 Zeros of Bessel Functions

We have already observed that graphs of $J_\nu(x)$ appear to cross the axis repeatedly as x increases. This suggests that $J_\nu(x)$ has infinitely many positive zeros, which tend to infinity as $x \to \infty$.

This intuition is supported by an informal argument. From equation 6.11, $J_\nu(kx)$ is a solution of

$$x^2y'' + xy' + (k^2x^2 - \nu^2)y = 0.$$

Suppose $k > 1$ and let $u(x) = \sqrt{kx}J_\nu(kx)$ to obtain

$$u''(x) + \left(k^2 - \frac{\nu^2 - \frac{1}{4}}{x^2}\right)u(x) = 0.$$

As x increases, $(\nu^2 - 1/4)/x^2 \to 0$, so this differential equation for u becomes "increasingly like" $u'' + k^2u = 0$, which has $\cos(kx)$ and $\sin(kx)$ as solutions. This suggests that, for large x, $J_\nu(x)$ is approximated by $\cos(kx)/\sqrt{kx}$ or $\sin(kx)/\sqrt{kx}$. Because these functions have infinitely many positive zeros, so must $J_\nu(kx)$, although these zeros need not be spaced equally, as they are for a periodic function such as $\sin(kx)$.

This leaves out a lot of details, but the conclusion turns out to be correct. One can prove the following theorem.

Theorem 6.5 (Zeros of $J_\nu(x)$). Let $k > 1$ and let ν be a real number. Then

(1) For α sufficiently large, there is a zero of $J_\nu(x)$ on each of the intervals; $(\alpha, \alpha + k\pi)$, $(\alpha + k\pi, \alpha + 2k\pi)$, $(\alpha + 2k\pi, \alpha + 3k\pi)$, $\cdots$.

(2) Every positive zero of $J_\nu(x)$ is simple.

(3) $J_\nu(x)$ and $J_{\nu+1}(x)$ have no positive zeros in common.

(4) $J_\nu(x)$ and $J_{\nu-1}(x)$ have no positive zeros in common.

(5) If a and b are distinct positive zeros of $J_\nu(x)$, then $J_{\nu-1}(x)$ and $J_{\nu+1}(x)$ each has a zero between a and b.

Conclusion 5 is the *interlacing property*, or *interlacing lemma*, which can be seen in Figure 6.7.

To prove 2, suppose a is a zero of $J_\nu(x)$ of multiplicity 2 or higher. Then $J_\nu(a) = J'_\nu(a) = 0$, so $y = J_\nu(x)$ is a solution of the boundary value problem

$$y'' + \frac{1}{x}y' + \left(k^2 - \frac{\nu^2}{x^2}\right) y = 0; y(a) = y'(a) = 0.$$

But $y = 0$ is also a solution of this problem, so this would make $J_\nu(x)$ identically zero, a contradiction.

For 3, suppose $J_\nu(a) = J_{\nu+1}(a) = 0$. Then by equation 6.15, we would have $J'_\nu(a) = 0$, impossible because any zero of $J_\nu(x)$ must be simple. A similar argument proves part 4.

For part 5, let $f(x) = x^\nu J_\nu(x)$. If $f(a) = f(b) = 0$, then by Rolle's theorem, there is some c between a and b such that $f'(c) = 0$. But, by equation 6.12,

$$f'(x) = \frac{d}{dx}[x^\nu J_\nu(x)] = x^\nu J_{\nu-1}(x),$$

so $J_{\nu-1}(c) = 0$. Similar reasoning, using equation 6.13, shows that $J_{\nu+1}(d) = 0$ for some d between a and b.

6.2.4 Fourier-Bessel Expansions

We will prepare to expand functions in series of Bessel functions. To begin, divide Bessel's equation 6.8 by x^2 to write it as

$$y'' + \frac{1}{x}y' + \left(\lambda - \frac{\nu^2}{x^2}\right) y = 0.$$

We can match this with equation 6.11, with

$$a = 0, c = 1, \lambda = b^2$$

to obtain solutions $J_\nu(\sqrt{\lambda}x)$.

Now suppose we want to solve Bessel's equation on $(0, 1)$, subject to the boundary condition $y(1) = 0$. We need

$$J_\nu(\sqrt{\lambda}) = 0.$$

This means that $\sqrt{\lambda}$ must be chosen as a positive root of $J_\nu(x)$. Let these positive zeros be, in increasing order,

$$0 < j_1 < j_2 < \cdots .$$

Then $\sqrt{\lambda_n} = j_n$, so the eigenvalues and eigenfunctions of this problem are

$$\lambda_n = j_n^2 \text{ and } J_\nu(j_n x).$$

From Bessel's equation of order ν in standard Sturm-Liouville form, these eigenfunctions are orthogonal on $(0, 1)$ with weight function $p(x) = x$:

$$\int_0^1 xJ_\nu(j_n x)J_\nu(j_m x)\,dx = 0 \text{ if } n \neq m.$$

If f is piecewise smooth on $(0, 1)$, we can expand $f(x)$ in a *Fourier-Bessel* series

$$\sum_{n=1}^{\infty} c_n J_\nu(j_n x)$$

for $0 < x < 1$. The coefficients are

$$c_n = \frac{\int_0^1 xf(x)J_\nu(j_n x)\,dx}{\int_0^1 xJ_\nu^2(j_n x)\,dx}.$$

These are called the *Fourier-Bessel coefficients of* $f(x)$ on $(0, 1)$. As a simplification, we will show that the integral in the denominator of c_n can be evaluated in terms of $J_{\nu+1}(j_n)$:

$$\int_0^1 xJ_\nu^2(j_n x)\,dx = \frac{1}{2}(J_{\nu+1}(j_n))^2. \tag{6.17}$$

To verify this, begin with Bessel's equation in the form

$$x^2y'' + xy' + (j_n^2x^2 - \nu^2)y = 0.$$

One solution is $y = J_\nu(j_n x)$. Multiply the differential equation by $2y'$ to get

$$2x^2y'y'' + 2x(y')^2 + 2(j_n^2x^2 - \nu^2)yy' = 0,$$

which can be written

$$[x^2(y')^2 + (j_n^2x^2 - \nu^2)y^2]' - 2j_n^2xy^2 = 0.$$

Integrate this equation to obtain

$$\begin{aligned}
&\left[x^2(y')^2 + (j_n^2x^2 - \nu^2)y^2\right]_0^1 - 2j_n^2\int_0^1 xy^2\,dx \\
&= (y'(1))^2 - 2j_n^2\int_0^1 xy^2\,dx \\
&= j_n^2(J_\nu'(j_n))^2 - 2j_n^2\int_0^1 (J_\nu(j_n x))^2\,dx.
\end{aligned}$$

Then

$$\int_0^1 xJ_\nu(j_n x)\,dx = \frac{1}{2}(J_\nu'(j_n))^2.$$

But $J_\nu(j_n) = 0$, so equation 6.15 gives us

$$J'_\nu(j_n) = -J_{\nu+1}(j_n).$$

Therefore

$$\int_0^1 xJ_\nu^2(j_n x)\,dx = \frac{1}{2}J_{\nu+1}^2(j_n).$$

Now the Fourier-Bessel coefficients of $f(x)$ on $(0,1)$ are

$$c_n = \frac{2}{J_{\nu+1}^2(j_n)}\int_0^1 xf(x)J_\nu(j_n x)\,dx. \tag{6.18}$$

Example 6.6 Let

$$f(x) = \begin{cases} x & \text{for } 0 \le x \le 1/2, \\ 4 & \text{for } 1/2 < x \le 1. \end{cases}$$

We will examine the Fourier-Bessel expansion of $f(x)$ on $[0,1]$, using zero-order Bessel functions (so $\nu = 0$). The coefficients are

$$c_n = \frac{2}{J_1^2(j_n)}\int_0^1 xf(x)J_0(j_n x)\,dx,$$

where j_n is the nth positive zero of $J_0(x)$, written in increasing order.

We can use software to compare graphs of $f(x)$ and the 100th partial sum of its Fourier-Bessel expansion on $(0,1)$ (Figure 6.9). Details of coding the instructions for MAPLE are given in Section 9.2.

Problems for Section 6.2

In each of problems 1–5, find a solution of the differential equation in terms of a Bessel function.

1.
$$y'' - \frac{3}{x}y' + \left(1 + \frac{3}{x^2}\right)y = 0.$$

2.
$$y'' - \frac{7}{x}y' + \left(16x^2 + \frac{4}{x^2}\right)y = 0.$$

3.
$$y'' - \frac{11}{x}y' + \left(9x^4 + \frac{6}{x^2}\right)y = 0.$$

4.
$$y'' - \frac{9}{x}y' + \left(64x^2 + \frac{20}{x}\right)y = 0.$$

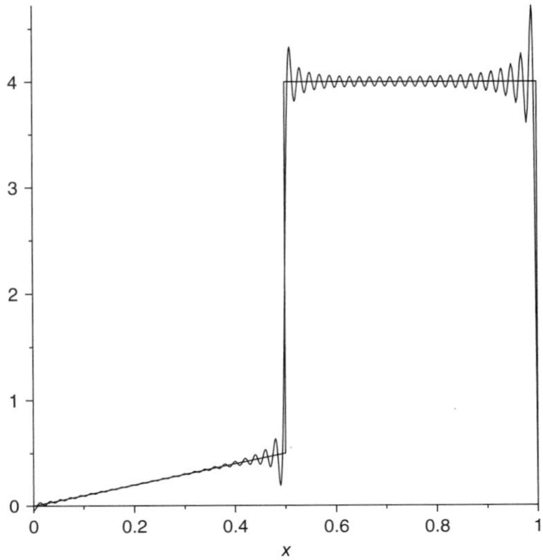

Figure 6.9: $f(x)$ and the 100th partial sum of the Fourier-Bessel expansion in example 6.6.

5.
$$y'' - \frac{23}{x}y' + \left(4x^4 + \frac{44}{x^2}\right)y = 0.$$

6. Use the transformation $u = x^{3/2}$ to solve:
$$4x^2y'' + 4xy' + (9x^3 - 36)y = 0.$$

7. Use the transformation $u = yx^{-2}$ to solve
$$9x^2y'' - 27xy' + (9x^2 + 35)y = 0.$$

8. Use the transformation $u = yx^{-2/3}$ to solve
$$36x^2y'' - 12xy' + (36x^2 + 7)y = 0.$$

9. Let a be a positive zero of $J_0(x)$. Show that
$$\int_0^1 J_1(ax)\,dx = \frac{1}{a}.$$

10. Let $u(x) = J_0(ax)$ and $v(x) = J_0(bx)$, with a and b positive constants.

(a) Show that
$$xu'' + u' + a^2xu = 0.$$

with a similar equation for $v(x)$.

(b) Multiply the differential equation for u by v, and the differential equation for v by u, and subtract to show that

$$[x\,(u'v - v'u)]' = (b^2 - a^2)xuv.$$

(c) Use the conclusion of part b to show that

$$(b^2 - a^2)\int xJ_0(ax)J_0(bx)\,dx$$
$$= x[aJ_0'(ax)J_0(bx) - bJ_0'(bx)J_0(ax)].$$

This is one of a class of integrals called *Lommel's integrals.*

11. Show that, for any positive integer n,

$$\int x^n J_{n-1}(x)\,dx = x^n J_n(x)$$

and

$$\int \frac{1}{x^n} J_{n+1}(x)\,dx = -\frac{J_n(x)}{x^n}.$$

Constants of integration have been omitted.

In each of problems 12–15, generate a graph comparing $f(x)$ with the indicated partial sum of the Fourier-Bessel expansion of $f(x)$ on $(0,1)$, using zero-order Bessel functions. These problems require the use of computer software.

12. $f(x) = -2$ (20 terms).

13. $f(x) = 3x$ (50 terms).

14. $f(x) = x^2 - x$ (10 terms).

15. $f(x) = \sin(\pi x)$ (10 terms).

16. Show that

$$J_0(x) = \frac{1}{\pi}\int_0^{\pi} \cos(x\sin(\theta))\,d\theta.$$

Hint: Use a power series expansion of $\cos(t)$ to write $\cos(x\sin(\theta))$ as a series of powers of $x\sin(\theta)$. Integrate this series term by term.

17. Use the result of problem 16 to show that $J_0(n\pi)$ is positive if n is an even positive integer, and negative if n is an odd positive integer. Use this fact to show that $J_0(x)$ has infinitely many positive zeros.

6.3 Applications of Bessel Functions

In this section we will look at a selection of problems in which Bessel functions play a role in the solution.

6.3.1 Temperature Distribution in a Solid Cylinder

Imagine an infinitely long, solid, homogeneous cylinder of radius R, with axis along the $z-$axis in $3-$space.

In cylindrical coordinates, the heat equation for the temperature distribution $U(r, \theta, z, t)$ is

$$U_t = k\left(U_{rr} + \frac{1}{r}U_r + \frac{1}{r^2}U_{\theta\theta} + U_{zz}\right).$$

We will look at the special case that the temperature at any point in the cylinder depends only on the time, t, and the radial distance, r, from the $z-$axis. In this case, U depends only on r and t, so $U = U(r,t)$ and $U_\theta = U_z = 0$. The heat equation is

$$U_t = k\left(U_{rr} + \frac{1}{r}U_r\right) \text{ for } 0 \leq r < R, t > 0.$$

Assume an initial condition

$$U(r, 0) = f(r) \text{ for } 0 \leq r < R$$

and a boundary condition

$$U(R, t) = 0 \text{ for } t \geq 0,$$

so the outer surface of the cylinder is maintained at temperature zero.

Substitute $U(r,t) = F(r)T(t)$ into the heat equation to obtain

$$\frac{T'}{kT} = \frac{F'' + \frac{1}{r}F'(r)}{F(r)} = -\lambda.$$

Then

$$T' + \lambda kT = 0 \text{ and } F'' + \frac{1}{r}F' + \lambda F = 0.$$

Since $U(R,t) = F(R)T(t) = 0$ for $t > 0$,

$$F(R) = 0.$$

Write the differential equation for $F(r)$ as

$$(rF')' + \lambda rF = 0,$$

so the problem for $F(r)$ is a singular Sturm-Liouville problem on $[0, R]$. To determine the eigenvalues and eigenfunctions, take cases on λ.

Case 1. $\lambda = 0$

Now $rF' = c$, so

$$F' = \frac{c}{r}$$

and

$$F(r) = c\ln(r) + d.$$

Since $\ln(r) \to -\infty$ as $r \to 0$ (the center of the cylinder), $c = 0$ and we are left with $F(r) = d$. But we also have $T'(t) = 0$ in this case, so $U(r,t) =$ constant. This constant must be zero because $U(R,t) = 0$. This case offers only a trivial solution.

Case 2. $\lambda < 0$

Say $\lambda = -\omega^2$ with $\omega > 0$. Now $T(t) = ce^{\omega^2 kt}$, which is unbounded as $t \to \infty$, unless $c = 0$. If we allow only bounded solutions, this case has only the trivial solution.

Case 3. $\lambda > 0$

Say $\lambda = \omega^2$ with $\omega > 0$. Now $T' + k\omega^2 T = 0$ has bounded solutions $T(t) = ce^{-k\omega^2 t}$. The problem for F becomes

$$(rF')' + \omega^2 rF = 0.$$

This is Bessel's equation of order zero on $[0, R]$, with bounded solutions that are constant multiples of $J_0(\omega r)$,

$$F(r) = c_\omega J_0(\omega r).$$

Now

$$F(R) = c_\omega J_0(\omega R) = 0,$$

so for a nontrivial solution we need to choose ω so that ωR is a positive zero of J_0. List these zeros as

$$j_1 < j_2 < \cdots .$$

Then $\omega R = j_n$, so ω can take on the values

$$\omega_n = \frac{j_n}{R} \text{ for } n = 1, 2, \cdots .$$

The eigenvalues of this problem are

$$\lambda_n = \omega_n^2 = \frac{j_n^2}{R^2}$$

and corresponding eigenfunctions are

$$J_0\left(\frac{j_n}{R} r\right) \text{ for } n = 1, 2, \cdots .$$

This leads us to attempt a solution

$$U(r,t) = \sum_{n=1}^{\infty} c_n J_0\left(\frac{j_n}{R}r\right) e^{-j_n^2 kt/R^2}.$$

To satisfy the initial condition, we need

$$U(r,0) = f(r) = \sum_{n=1}^{\infty} c_n J_0\left(\frac{j_n}{R}r\right).$$

This is a Fourier-Bessel series for $f(r)$. Because we wrote the formula 6.18 for the coefficients on the interval $(0,1)$, change variables by putting $\xi = r/R$ to write

$$f(R\xi) = \sum_{n=1}^{\infty} c_n J_0(j_n \xi).$$

The Fourier-Bessel coefficients are

$$c_n = \frac{2}{(J_1(j_n))^2} \int_0^1 \xi f(R\xi) J_0(j_n \xi)\, d\xi.$$

With these coefficients, we have the solution of the problem.

As an example, let $k = R = 1$ and $f(r) = r(1-r)$ to obtain the solution

$$U(r,t) = \sum_{n=1}^{\infty} \frac{2}{(J_1(j_n))^2} \left(\int_0^1 \xi^2(1-\xi) J_0(j_n\xi)\, d\xi \right) J_0(j_n r) e^{-j_n^2 t}.$$

Figure 6.10 shows a surface plot of $u(r,t)$ for $0 \le r \le 1$ and $0 \le t \le 3/4$.

Problems for Section 6.3.1

In each of problems 1–5, write the solution of the problem of heat transfer in the cylinder for the given initial temperature function and values of R and k. Graph the solution for the given values of t.

1. $f(r) = r(1-r), k = 2, R = 1, t = 0, 1/20,\ 1/10, 1/5.$

2. $f(r) = \sin(\pi r), R = 3, k = 2, t = 0, 1/20, 1/10, 1/5.$

3. $f(r) = r^2(3-r), R = 3, k = 2, t = 0, 1/10, 1/2, 1.$

4. $f(r) = 1 + \cos(\pi r), R = 1, k = 2, t = 0, 1/20, 1/10, 1/2.$

5. $f(r) = e^r(1-r), R = 1, k = 3, t = 0, 1/20, 1/10.$

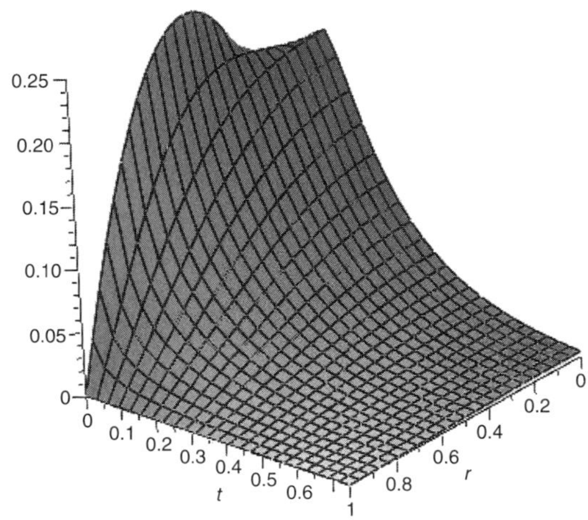

Figure 6.10: Surface graph of $U(r,t)$ for the solid cylinder.

6.3.2 Vibrations of a Circular Drum

We will solve for the wave function of a drum attached to a circular frame, lifted to some initial position, and released from rest to vibrate in 3−space. Use cylindrical coordinates and consider the special case that the amplitude $z(r,t)$ of the oscillations of the particle originally at $(r,\theta,0)$ depend only on r and t.

In this case the wave equation is

$$z_{tt} = c^2\left(z_{rr} + \frac{1}{r}z_r\right).$$

If the circular frame has radius κ, then

$$z(\kappa,t) = 0 \text{ for all } t.$$

The initial conditions are

$$z(r,0) = f(r), z_t(r,0) = 0.$$

Separate variables by putting $z(r,t) = R(r)T(t)$ to obtain

$$T'' + \lambda T = 0, (rR')' + \frac{\lambda}{c^2}R = 0.$$

The equation for $R(r)$ is a Bessel equation of order $\nu = 0$, with λ/c^2 in place of λ in the usual form of this equation.

Because the solution must be defined at $r = 0$, the center of the drum, solutions are

$$R(r) = \alpha J_0\left(\frac{\sqrt{\lambda}}{c}r\right).$$

The equation for T has solutions

$$T(t) = a\cos(\sqrt{\lambda}t) + b\sin(\sqrt{\lambda}t).$$

At this point we have functions

$$z_\lambda(r,t) = a_\lambda J_0\left(\frac{\sqrt{\lambda}}{c}r\right)\cos(\sqrt{\lambda}t) + b_\lambda J_0\left(\frac{\sqrt{\lambda}}{c}r\right)\sin(\sqrt{\lambda}t).$$

Now use the initial and boundary conditions. First,

$$z_t(r,0) = b_\lambda\sqrt{\lambda}J_0\left(\frac{\sqrt{\lambda}}{c}r\right) = 0,$$

so choose each $b_\lambda = 0$. Next,

$$z_\lambda(\kappa,t) = a_\lambda J_0\left(\frac{\sqrt{\lambda}}{c}\kappa\right)\cos(\sqrt{\lambda}t) = 0$$

for $t \geq 0$. To obtain a nontrivial solution satisfying this condition, we must choose λ so that

$$J_0\left(\frac{\sqrt{\lambda}}{c}\kappa\right) = 0.$$

If the positive zeros are J_0 are

$$j_1 < j_2 < j_3 < \cdots.$$

then we must choose, for $n = 1, 2, 3, \cdots$

$$\frac{\sqrt{\lambda}}{c}\kappa = j_n.$$

The eigenvalues of this problem are

$$\lambda_n = \left(\frac{j_n c}{\kappa}\right)^2$$

and corresponding eigenfunctions are

$$J_0(j_n r/\kappa).$$

For each positive integer n, we have found a function

$$z_n(r,t) = c_n J_0\left(\frac{j_n}{\kappa}r\right)\cos\left(\frac{j_n c}{\kappa}t\right)$$

that satisfies the wave equation and the condition that the membrane is fixed to the circular frame. Attempt a solution

$$z(r,t) = \sum_{n=1}^{\infty} c_n J_0\left(\frac{j_n}{\kappa}r\right)\cos\left(\frac{j_n c}{\kappa}t\right).$$

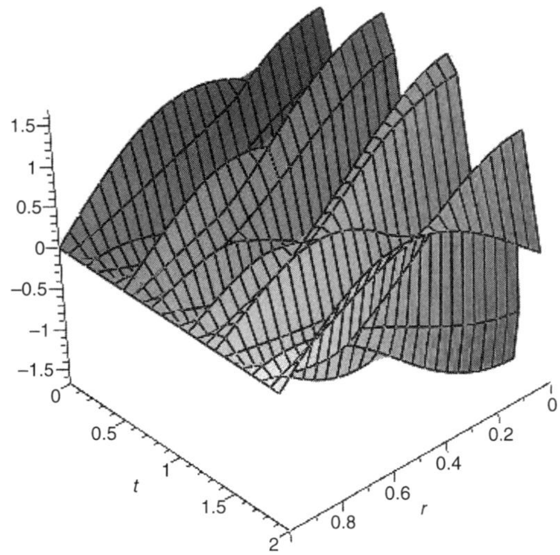

Figure 6.11: Graph of $z(r, t)$ for the circular drum.

In this sum, $z_n(r, t)$ is the *nth normal mode of vibration* of the membrane. The number $j_n c/\kappa$ is the frequency of the nth normal mode, and its period is $2\pi\kappa/j_n c$. The motion is a superposition of the normal modes.

For the initial condition, we require that

$$z(r, 0) = f(r) = \sum_{n=1}^{\infty} c_n J_0\left(\frac{j_n}{\kappa} r\right).$$

This is an eigenfunction expansion of $f(r)$ in terms of Bessel functions of order zero, so the coefficients must be chosen as the Fourier-Bessel coefficients on $(0, \kappa)$.

For example, suppose $\kappa = 1$, $c = 4$ and $f(r) = \sin(\pi r)$. The coefficients are

$$c_n = \frac{2}{J_1^2(j_n)} \int_0^1 r \sin(\pi r) J_0(j_n r)\, dr$$

and the solution is

$$z(r, t) = \sum_{n=1}^{\infty} c_n J_0(j_n r) \cos(4 j_n t).$$

Figure 6.11 is a surface graph of $z(r, t)$.

Problems for Section 6.3.2

In each of problems 1–5, write the solution $z(r, t)$ for the given κ, c and $f(r)$. Graph the solution as a function of r for the indicated values of t.

1. $\kappa = 1, c = 4, f(r) = r^2(1-r), t = 0, 1/20, 1/5, 1/10.$

2. $\kappa = 2, c = 3, f(r) = \sin(\pi r), t = 0, 1/20, 1/10, 1/2.$

3. $\kappa = 5, c = 2, f(r) = r, t = 0, 1/2, 1.$

4. $\kappa = c = 2, f(r) = r(2-r)^2, t = 0, 1/20, 1/10, 1/2.$

5. $\kappa = 1, c = 3, f(r) = \cos(\pi r/2), t = 0, 1/3, 1/6, 1/8, 3/2.$

6.3.3 Oscillations of a Hanging Chain

Imagine a flexible hanging chain or cable fastened at its upper end and free to move at its lower end. What motion will result from a small displacement at the free end?

Assume that each particle of the chain oscillates in a horizontal line. Let m be the constant mass of the cable per unit length and L its length. Let $y(x,t)$ be the displacement at time t of the particle at distance x from the fixed end.

Look at a small element of chain of length Δx, and let the forces acting on the ends of this segment have magnitudes T and $T + \Delta T$. Apply Newton's second law of motion (total external force equals the mass times the acceleration) to this segment. From the horizontal component of this vector equation,

$$m\Delta x \frac{\partial^2 y}{\partial t^2} = \frac{\partial}{\partial x}\left(T\frac{\partial y}{\partial x}\right)\Delta x.$$

Then

$$my_{tt} = (Ty_x)_x.$$

Experiment has shown that $T = mg(L-x)$ is a good approximation for small disturbances of the lower end. Then

$$y_{tt} = -gy_x + g(L-x)y_{xx}.$$

To solve this wave equation, change variables by setting

$$z = L - x \text{ and } u(z,t) = y(L-z,t).$$

The partial differential equation for u is

$$u_{tt} = gu_x + gzu_{zz}.$$

Use separation of variables. However, in anticipation of oscillatory motion in the solution, attempt a separation of the form

$$u(z,t) = f(z)\cos(\omega t - \delta).$$

Substitute this into the partial differential equation for $u(z,t)$ to obtain

$$-\omega^2 f(z)\cos(\omega t - \delta) = gf'(z)\cos(\omega t - \delta) + gzf''(z)\cos(\omega t - \delta).$$

Upon dividing by $\cos(\omega t - \delta)$, we obtain

$$-\omega^2 f(z) = gf'(z) + gzf''(z).$$

Write this as

$$f''(z) + \frac{1}{z}f'(z) + \frac{\omega^2}{gz}f(z) = 0.$$

This is equation 6.11 if we can choose a, b, c, and ν so that

$$-(2a-1) = 1, 2c-2 = -1, b^2c^2 = \frac{\omega^2}{g}, \text{ and } a^2 - \nu^2c^2 = 0.$$

Solve these to get

$$a = \nu = 0, c = \frac{1}{2}, \text{ and } b = \frac{2\omega}{g}.$$

Bounded solutions for $f(z)$ are

$$f(z) = \alpha J_0\left(2\omega\sqrt{\frac{z}{g}}\right),$$

so

$$u_\omega(z,t) = \alpha J_0\left(2\omega\sqrt{\frac{z}{g}}\right)\cos(\omega t - \delta).$$

Up to this point, $y(x,t)$ has the form

$$y_\omega(x,t) = \alpha J_0\left(2\omega\sqrt{\frac{L-x}{g}}\right)\cos(\omega t - \delta).$$

The frequencies of the oscillations are determined by the fact that the upper end of the cable does not move. This means that $y_\omega(0,t) = 0$; therefore ω must be chosen so that

$$J_0\left(2\omega\sqrt{\frac{L}{g}}\right) = 0.$$

If the positive zeros of J_0 are j_n, with

$$0 < j_1 < j_2 < \cdots$$

then ω must satisfy

$$2\omega\sqrt{\frac{L}{g}} = j_n \text{ for some positive integer } n.$$

Then

$$\omega_n = \frac{1}{2}\sqrt{\frac{g}{L}}j_n \text{ for } n = 1, 2, 3, \cdots.$$

These are the frequencies of the normal modes of vibration of the end of the chain, one normal mode for each positive zero of J_0. The periods of the oscillations are

$$\frac{4\pi}{j_n}\sqrt{\frac{L}{g}}.$$

With these values of ω_n, the fundamental modes of vibration are

$$y_n(x,t) = \alpha_n J_0\left(\frac{\sqrt{L-x}}{L}j_n\right)\cos\left(\frac{1}{2}\sqrt{\frac{g}{L}}j_n t - \delta\right).$$

The constants α_n must be determined from information about the magnitude of the original displacement.

Problems for Section 6.3.3

1. Taking $L = 10$ ft. and $\delta = 0$, graph the first three fundamental modes of vibration on the same set of axes.

2. What effect would lengthening the cable have on the motion, given the same initial displacement?

6.3.4 Did Poe Get His Pendulum Right?

Edgar Allen Poe's classic tale "The Pit and the Pendulum" is told through the eyes of a prisoner, who is tied on his back on a level surface while watching a razor-sharp pendulum swing back and forth above him, at first at a great height, but lowering toward him as the pendulum arm increases in length. He also observes that the pendulum's sweep back and forth is increasing and that the blade's swinging motion is faster as it descends.

Mathematicians and scientists do not always take great literature at face value, and the question has been posed as to whether a descending pendulum would actually behave this way.

This question was analyzed by Roger L. Borelli, Courtney S. Coleman and Dana D. Hobson of Harvey Mudd College in *Mathematics Magazine* (March 1985). We will discuss their conclusion.

First, the authors derived a differential equation involving the length $L(t)$ of the descending pendulum arm and the angle $\theta(t)$ the arm makes with the vertical at time t. Using vector calculus and Newton's second law of motion, they find that

$$L\theta'' + 2L'\theta' + g\theta = 0,$$

in which g is the constant acceleration due to gravity near the surface of the earth.

It appears from Poe's description of the motion that the pendulum descends at a constant rate. If this is the case, then $L'(t)$ is constant, so

$$L(t) = a + bt$$

for some constants a and b. Now the differential equation is

$$(a + bt)\theta'' + 2b\theta' + g\theta = 0.$$

Define

$$x = \frac{2}{b}\sqrt{(a + bt)g} \text{ and } y = \sqrt{a + bt}\theta.$$

With some computation, we obtain

$$x^2 y''(x) + xy'(x) + (x^2 - 1)y(x) = 0.$$

This is Bessel's equation of order 1, with general solution

$$y(x) = c_1 J_1(x) + c_2 Y_1(x).$$

The Bessel function of the second kind of order one is included in this general solution because x does not approach zero here. Then

$$\begin{aligned}\theta(t) &= \frac{1}{\sqrt{a + bt}} y(x) \\ &= \frac{1}{\sqrt{a + bt}} \left[c_1 J_1 \left(\frac{2}{b}\sqrt{(a + bt)g} \right) + c_2 Y_1 \left(\frac{2}{b}\sqrt{(a + bt)g} \right) \right].\end{aligned}$$

As suggested by Figures 6.5–6.8, $J_1(s)$ and $Y_1(s)$ tend to zero as $s \to \infty$. Therefore, $\theta(t) \to 0$ as t increases. By contrast, the sweep $L(t)\theta(t)$ behaves like $(a + bt)^{1/4}$ as t increases, and, therefore, increases without bound. And, because $L(t) = a + bt$, the curvilinear velocity $[L(t)\theta(t)]'$ decreases to zero as t increases. This is not the way Poe's unfortunate hero describes what he is seeing, because he apparently sees the pendulum's sweep become faster as it lowers. It can be shown by a numerical analysis that, if the pendulum began with a length of 1 ft., then the sweep is no more than 10 ft. by the time the length has reached 40 ft. This contradicts the prisoner's claim that the sweep at one point was at least 30 ft.

The authors of this analysis have asked whether there is a smooth length function $L(t)$ that yields the pendulum motion described by the prisoner. They conjecture that there is no such function. This problem is apparently still open.

6.4 Legendre Polynomials and Applications

Legendre's differential equation is

$$((1 - x^2)y')' + \lambda y = 0. \tag{6.19}$$

This equation is already in Sturm-Liouville form, with $r(x) = 1 - x^2, q(x) = 0$ and $p(x) = 1$. Because $r(1) = r(-1) = 0$, Legendre's equation forms a singular Sturm-Liouville problem on $[-1, 1]$, with no boundary conditions given. However, we will seek bounded solutions on this interval.

Substitute a proposed solution

$$y(x) = \sum_{n=0}^{\infty} a_n x^n$$

into Legendre's equation and rearrange terms to collect coefficients of each power of x and write

$$2a_2 + \lambda a_0 + [6a_3 - 2a_1 + \lambda a_1]x + \sum_{n=1}^{\infty}[(n+2)(n+1)a_{n+2} - (n^2+n-\lambda)a_n]x^n = 0.$$

This power series can be zero on an interval about zero only if the coefficient of each power of x, including the constant term, is zero. Therefore

$$2a_2 + \lambda a_0 = 0,$$
$$6a_3 - 2a_1 + \lambda a_1 = 0, \text{ and}$$
$$(n+1)(n+2)a_{n+2} - [n(n+1) - \lambda]a_n = 0 \text{ for } n = 2, 3, \cdots .$$

Then

$$a_2 = -\frac{\lambda}{2}a_0,$$
$$a_3 = \frac{2-\lambda}{6}a_1 = \frac{2-\lambda}{3!}a_1,$$

and

$$a_{n+2} = \frac{n(n+1)-\lambda}{(n+1)(n+2)}a_n \text{ for } n = 2, 3, \cdots . \tag{6.20}$$

This is a recurrence relation from the coefficients. The form of this relation tells us that every a_{2n+1} can be written in terms of a_1, and every a_{2n} in terms of a_0. It is instructive to write out a few of the coefficients in this way:

$$a_4 = \frac{6-\lambda}{(3)(4)}a_2 = -\frac{\lambda}{2}\frac{6-\lambda}{(3)(4)} = -\frac{\lambda(6-\lambda)}{4!}a_0,$$
$$a_5 = \frac{12-\lambda}{(4)(5)}a_3 = \frac{(2-\lambda)(12-\lambda)}{5!}a_1,$$
$$a_6 = \frac{20-\lambda}{(5)(6)}a_4 = \frac{-\lambda(6-\lambda)(20-\lambda)}{6!}a_0,$$
$$a_7 = \frac{30-\lambda}{(6)(7)}a_5 = \frac{(2-\lambda)(12-\lambda)(30-\lambda)}{7!}a_1,$$

and so on. The fact that the even- and odd-indexed terms separate themselves into two classes (multiples of a_0 or a_1) means that the solution separates into even and odd power terms:

$$
\begin{aligned}
y(x) &= \sum_{n=0}^{\infty} a_n x^n \\
&= a_0 \left(1 - \frac{\lambda}{2}x^2 - \frac{\lambda(6-\lambda)}{4!}x^4 - \frac{\lambda(6-\lambda)(20-\lambda)}{6!}x^6 + \cdots\right) \\
&+ a_1 \left(x + \frac{2-\lambda}{3!}x^3 + \frac{(2-\lambda)(12-\lambda)}{5!}x^5 + \frac{(2-\lambda)(12-\lambda)(30-\lambda)}{7!}x^7 + \cdots\right).
\end{aligned}
$$

Because a_0 and a_1 are arbitrary constants, this expression displays the general solution of Legendre's equation as a_0 times an even function (all even powers of x) plus a_1 times an odd function (all odd powers of x).

Now recall that we want bounded solutions on $[-1, 1]$. Polynomial solutions would be ideal. Further, from the coefficients, we get even or odd polynomial solutions of degree n by choosing $\lambda = n(n+1)$ for $n = 0, 1, 2, \cdots$, and either a_1 or a_0 equal to 0:

With $n = \lambda = a_1 = 0$:

$$y(x) = a_0 = \text{ constant.}$$

With $n = 1, \lambda = 2$ and $a_0 = 0$:

$$y(x) = a_1 x.$$

With $n = 2, \lambda = 6$ and $a_1 = 0$:

$$y(x) = a_0(1 - 3x^2).$$

With $n = 3, \lambda = 12$ and $a_0 = 0$:

$$y(x) = a_1 \left(x - \frac{5}{3}x^3\right).$$

With $n = 4, \lambda = 20$ and $a_1 = 0$:

$$y(x) = a_0 \left(1 - 10x^2 + \frac{35}{3}x^4\right),$$

and so on.

To standardize these polynomial solutions, in each case, choose the constant so that the graph passes through $(1, 1)$. In this way we obtain the *Legendre polynomials* $P_n(x)$, the first six of which are

$$
\begin{aligned}
& P_0(x) = 1, P_1(x) = x, P_2 = \frac{1}{2}(3x^2 - 1), \\
& P_3(x) = \frac{1}{2}(5x^3 - 3x), P_4(x) = \frac{1}{8}(35x^4 - 30x^2 + 3), \\
& P_5(x) = \frac{1}{8}(60x^5 - 70x^3 + 15x).
\end{aligned}
$$

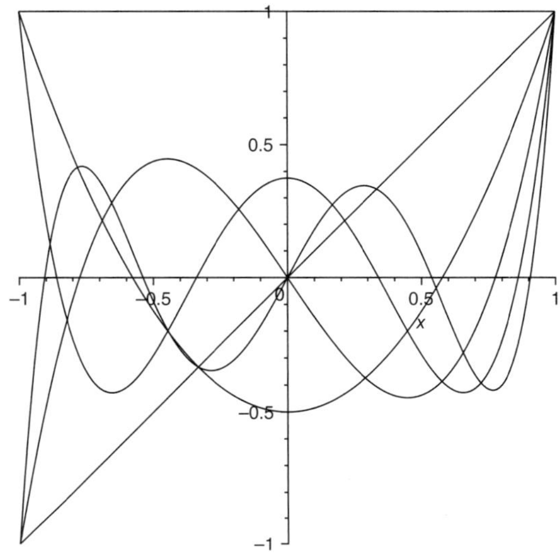

Figure 6.12: Legendre polynomials $P_0(x)$ through $P_5(x)$ on $[-1, 1]$.

Graphs of these Legendre polynomials are shown in Figure 6.12.

In the language of Sturm-Liouville theory, the singular Sturm-Liouville problem given by equation 6.19 has eigenvalues $\lambda = n(n+1)$ and corresponding eigenfunctions $P_n(x)$ for $= 0, 1, 2, \cdots$. These eigenfunctions are orthogonal on $[-1, 1]$ with respect to the weight function $p(x) = 1$:

$$\int_{-1}^{1} P_n(x)P_m(x)\,dx = 0 \text{ if } n \neq m.$$

We will outline some properties of Legendre polynomials.

6.4.1 A Generating Function

The function

$$L(x,t) = \frac{1}{\sqrt{1 - 2xt + t^2}}$$

is a *generating function* for the Legendre polynomials. This means that, if $L(x,t)$ is expanded in a power series in t about 0, the coefficient of t^n is $P_n(x)$:

$$L(x,t) = \sum_{n=0}^{\infty} P_n(x)t^n.$$

Problem 1 at the end of this section suggests some insight into this result.

Many useful facts about Legendre polynomials can be derived by using this generating function. As an example, we will show that

$$P_n(-1) = (-1)^n \text{ for } n = 0, 1, 2, \cdots.$$

Put $x = -1$ into the generating function to get

$$L(-1,t) = \frac{1}{\sqrt{1+2t+t^2}} = \frac{1}{\sqrt{(1+t)^2}} = \frac{1}{1+t} = \sum_{n=0}^{\infty} P_n(-1)t^n$$

for $-1 < t < 1$. Compare this with the well-known geometric series

$$\frac{1}{1+t} = \sum_{n=0}^{\infty}(-1)^n t^n$$

to conclude that

$$\sum_{n=0}^{\infty} P_n(-1)t^n = \sum_{n=0}^{\infty}(-1)^n t^n$$

for $-1 < t < 1$. By the uniqueness of the coefficients of a Maclaurin expansion, the coefficients of t^n on the left and right must be equal, hence $P_n(-1) = (-1)^n$.

6.4.2 A Recurrence Relation

It is possible to express $P_{n+1}(x)$ in terms of $P_n(x)$ and $P_{n-1}(x)$ for $n = 1, 2, \cdots$. This is often useful in deriving properties of Legendre polynomials by inductive arguments.

Theorem 6.6 (Recurrence Relation)

$$(n+1)P_{n+1}(x) - (2n+1)xP_n(x) + nP_{n-1}(x) = 0 \tag{6.21}$$

for $n = 1, 2, 3, \cdots$.

This recurrence relation can be derived by differentiating the generating function to obtain

$$(1 - 2xt + t^2)\frac{\partial L}{\partial t} - (x - t)L(x,t) = 0.$$

Substitute $L(x,t) = \sum_{n=0}^{\infty} P_n(x)t^n$ into this equation and collect coefficients of like powers of t to obtain

$$\begin{aligned}
&P_1(x) - xP_1(x) \\
&+ (2P_2(x) - 2xP_1(x) - xP_1(x) + P_0(x))t \\
&+ \sum_{n=2}^{\infty}[(n+1)P_{n+1}(x) - (2n+1)xP_n(x) + nP_{n-1}(x)]t^n = 0.
\end{aligned}$$

For this power series to vanish in an open interval about 0, the coefficient of each power of n must be zero:

$$\begin{aligned}
&P_1(x) - xP_0(x) = 0,\\
&2P_2(x) - (2x+1)P_1(x) + P_0(x) = 0,\\
&\text{ and, in general}\\
&(n+1)P_{n+1}(x) - (2n+1)xP_n(x) + nP_{n-1}(x) = 0.
\end{aligned}$$

This yields the recurrence relation 6.21, with the first line the case $n = 0$, the second line $n = 1$, and the last line $n = 2, 3, \cdots$.

When we consider eigenfunction expansions in terms of Legendre polynomials, we will find it useful to know the coefficient of x^n in $P_n(x)$. We can obtain this from the recurrence relation.

Theorem 6.7 (Coefficient of x^n) Let A_n be the coefficient of x^n in $P_n(x)$. Then

$$A_n = \frac{1 \cdot 3 \cdots (2n-3)(2n-1)}{n!}. \tag{6.22}$$

Proof Let A_n be the coefficient of x^n in $P_n(x)$. Now look at the highest power of x that occurs in equation 6.21. This is x^{n+1}, occurring only in $P_{n+1}(x)$, and in the $xP_n(x)$ term. These terms involving x^{n+1} must cancel, because the entire left side of equation 6.21 adds up to zero for all x. Therefore

$$(n+1)A_{n+1} - (2n+1)A_n = 0,$$

so

$$A_{n+1} = \frac{2n+1}{n+1}A_n.$$

This is a recurrence relation for A_n. Work down from this:

$$\begin{aligned}
A_{n+1} &= \frac{2n+1}{n+1}A_n\\
&= \frac{2n+1}{n+1}\frac{2(n-1)+1}{(n-1)+1}A_{n-1} = \frac{2n+1}{n+1}\frac{2n-1}{n}A_n\\
&= \cdots\\
&= \frac{2n+1}{n+1}\frac{2n-1}{n}\frac{2n-3}{n-1}\cdots\frac{3}{2}A_0\\
&= \frac{2n+1}{n+1}\frac{2n-1}{n}\frac{2n-3}{n-1}\cdots\frac{3}{2}
\end{aligned}$$

because, with $P_0(x) = 1$, $A_0 = 1$. This proves the conclusion of the theorem, with $n + 1$ in place of n.

6.4.3 Fourier-Legendre Eigenfunction Expansions

Legendre's differential equation 6.19 is in standard Sturm-Liouville form, with $r(x) = 1 - x^2$, $q(x) = 0$ and $p(x) = 1$. On $[-1, 1]$, this constitutes a singular Sturm-Liouville problem because $r(1) = r(-1) = 0$. We have noted that the numbers $\lambda_n = n(n+1$ are eigenvalues for $n = 0, 1, 2, \cdots$, with $P_n(x)$ as corresponding eigenfunctions which are bounded on $[-1, 1]$. These eigenfunctions are orthogonal on $[-1, 1]$ with weight function $p(x) = 1$.

If $f(x)$ is piecewise smooth on $[-1, 1]$, then $f(x)$ has a *Fourier-Legendre expansion*

$$\frac{1}{2}(f(x-) + f(x+)) = \sum_{n=0}^{\infty} c_n P_n(x), \tag{6.23}$$

for $-1 < x < 1$, with *Fourier-Legendre coefficients*

$$c_n = \frac{\int_{-1}^{1} f(x)P_n(x)\,dx}{\int_{-1}^{1} P_n^2(x)\,dx}. \tag{6.24}$$

One immediate consequence of Fourier-Legendre expansions is that the integral, over $[-1, 1]$, of any product of $P_n(x)$ with a polynomial of degree less than n, is zero.

Lemma 6.1 If $q(x)$ is a polynomial of degree m and $n > m$, then

$$\int_{-1}^{1} q(x)P_n(x)\,dx = 0. \tag{6.25}$$

Proof On $(-1, 1)$, $q(x)$ has the Fourier-Legendre expansion

$$q(x) = \sum_{n=0}^{\infty} c_n P_n(x) \text{ for } -1 < x < 1.$$

Because $q(x)$ has degree m, the expansion on the right cannot contain any terms containing x^n for $n > m$. Therefore, in this expansion,

$$c_n = \frac{\int_{-1}^{1} q(x)P_n(x)\,dx}{\int_{-1}^{1} P_n^2(x)\,dx} = 0 \text{ if } n > m.$$

The numerator must be zero, proving the theorem.

Using Lemma 6.1 and the recurrence relation 6.21, we can simplify the expression for the Fourier-Legendre coefficients of a function.

Lemma 6.2 For $n = 0, 1, 2, \cdots$,

$$\int_{-1}^{1} P_n^2(x)\,dx = \frac{2}{2n+1}.$$

Proof Let

$$p_n = \int_{-1}^{1} P_n^2(x)\,dx$$

and let A_n be the coefficient of x^n in $P_n(x)$. as before. Because P_{n-1} has degree $n-1$, $xP_{n-1}(x)$ has degree n. The coefficient of x^n in $xP_{n-1}(x)$ is A_{n-1}. Therefore x^n has coefficient 1 in the polynomial

$$\frac{1}{A_{n-1}}xP_{n-1}(x),$$

so

$$\frac{A_n}{A_{n-1}}xP_{n-1}(x)$$

is of degree n, and x^n has coefficient A_n. This means that x^n cancels in the polynomial

$$q(x) = P_n(x) - \frac{A_n}{A_{n-1}}xP_{n-1}(x),$$

so $q(x)$ is a polynomial of degree $\leq n-1$. By Lemma 6.1,

$$\int_{-1}^{1} q(x)P_n(x)\,dx = 0.$$

Further,

$$P_n(x) = q(x) + \frac{A_n}{A_{n-1}}xP_{n-1}(x).$$

Then

$$\begin{aligned}
p_n &= \int_{-1}^{1} P_n^2(x)\,dx \\
&= \int_{-1}^{1} P_n(x)P_n(x)\,dx \\
&= \int_{-1}^{1} \left[q(x) + \frac{A_n}{A_{n-1}}xP_{n-1}(x)\right] P_n(x)\,dx \\
&= \int_{-1}^{1} q(x)P_n(x)\,dx + \frac{A_n}{A_{n-1}}\int_{-1}^{1} xP_{n-1}(x)P_n(x)\,dx \\
&= \frac{A_n}{A_{n-1}}\int_{-1}^{1} xP_{n-1}(x)P_n(x)\,dx.
\end{aligned}$$

But, by the recurrence relation,

$$xP_{n-1}(x) = \frac{n+1}{2n+1}P_{n+1}(x) + \frac{n}{2n+1}P_{n-1}(x),$$

so

$$xP_{n-1}(x)P_n(x) = \frac{n+1}{2n+1}P_{n+1}(x)P_n(x) + \frac{n}{2n+1}P_{n-1}^2(x).$$

Using Theorem 6.7,

$$\begin{aligned}
p_n &= \frac{A_n}{A_{n-1}} \int_{-1}^{1} \left[\frac{n+1}{2n+1} P_{n+1}(x)P_n(x) + \frac{n}{2n+1} P_{n-1}^2(x) \right] dx \\
&= \frac{A_n}{A_{n-1}} \frac{n}{2n+1} \int_{-1}^{1} P_{n-1}^2(x)\, dx \\
&= \frac{A_n}{A_{n-1}} \frac{n}{2n+1} p_{n-1} \\
&= \frac{(1 \cdot 3 \cdots (2n-3)(2n-1))/n!}{1 \cdot 3 \cdots (2n-3)/(n-1)!} \frac{n}{2n+1} p_{n-1} \\
&= \frac{2n-1}{n} \frac{n}{2n+1} p_{n-1} = \frac{2n-1}{2n+1} p_{n-1}.
\end{aligned}$$

This is a recurrence relation relating p_n and p_{n-1}. The rest is a straightforward induction argument. First, by direct calculation,

$$p_0 = \int_{-1}^{1} P_0^2(x)\, dx = 2 = \frac{2}{2(0)+1},$$

so the conclusion of the lemma is true for $n = 0$. Next, if the conclusion holds for $n - 1$, then

$$p_{n-1} = \int_{-1}^{1} P_{n-1}^2(x)\, dx = \frac{2}{2(n-1)+1} = \frac{2}{2n-1}.$$

Then

$$p_n = \frac{2n-1}{2n+1} p_{n-1} = \frac{2n-1}{2n+1} \frac{2}{2n-1} = \frac{2}{2n+1},$$

completing the proof of the lemma.

The Fourier-Legendre coefficient 6.24 of $f(x)$ is therefore

$$c_n = \frac{2n+1}{2} \int_{-1}^{1} f(x)P_n(x)\, dx. \tag{6.26}$$

Example 6.7 Let $f(x) = \cos(3x) - e^{-2x} + \sin(\pi x)$. We will compare graphs of $f(x)$ and a partial sum of the Fourier-Legendre expansion of $f(x)$ on $[-1, 1]$. The Fourier-Legendre coefficients are

$$c(n) = \frac{2n+1}{2} \int_{-1}^{1} f(x)P_n(x)\, dx.$$

We will not compute these integrals numerically, but instead insert the integrals into MAPLE code to compute the graphs of the fifth partial sum $S_5(x)$ and $f(x)$. Details of this use of MAPLE are given in Section 9.2. The graphs are shown in Figure 6.13. This particular Fourier-Legendre expansion converges very rapidly to the function on the interval.

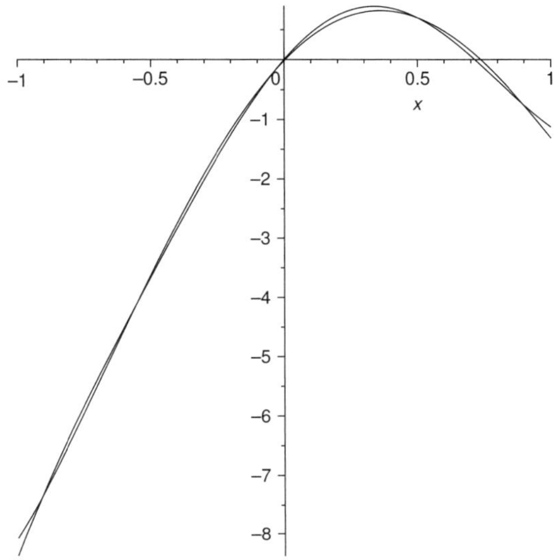

Figure 6.13: $f(x)$ and the fifth partial sum of its Fourier-Legendre expansion.

6.4.4 Zeros of Legendre Polynomials

If $n > 0$, $P_n(x)$ has n roots (zeros), which are solutions of the equation $P_n(x) = 0$. As with Bessel functions, any zero of a Legendre polynomial must be simple. For suppose x_0 is a repeated root of $P_n(x)$. Then

$$P_n(x_0) = P_n'(x_0) = 0.$$

Then $P_n(x)$ would be the unique solution of the problem consisting of Legendre's differential equation, and the boundary conditions

$$y(x_0) = y'(x_0) = 0.$$

But the identically zero function is also a solution of this problem. This makes $P_n(x)$ identically zero, a contradiction.

More than this, we claim that all of the roots of $P_n(x)$ are real, and lie between -1 and 1.

Theorem 6.8 (Zeros of Legendre Polynomials) If n is a positive integer, then $P_n(x)$ has n simple real zeros, all lying in the interval $(-1, 1)$.

Proof First we will show that $P_n(x)$ has at least one zero in $(-1, 1)$.
By the orthogonality of the Legendre polynomials, if $n \geq 1$,

$$\int_{-1}^{1} P_n(x)\, dx = \int_{-1}^{1} P_0(x) P_n(x)\, dx = 0.$$

If $P_n(x)$ were strictly positive or negative for $-1 < x < 1$, then this integral would be positive or negative. We conclude that $P_n(x)$ must change signs on $(-1, 1)$. By the intermediate value theorem, for some c in $(-1, 1)$, $P_n(c) = 0$.

Next we must show that $P_n(x)$ has n zeroes in $(-1, 1)$. Suppose instead that P_n has m zeros $-1 < x_1 < x_2 < \cdots < x_m < 1$ in this interval, with $m < n$. Define a polynomial $q(x)$ of degree $m < n$ by setting

$$q(x) = (x - x_1)(x - x_2) \cdots (x - x_m).$$

By design, $q(x)$ and $P_n(x)$ have exactly the same zeros in $(-1, 1)$, and these are all simple. Therefore $P_n(x)$ and $q_n(x)$ change sign at exactly the same points. This means that $P_n(x)$ and $q(x)$ are either of the same sign on each interval

$$(-1, x_1), (x_1, x_2), \cdots, (x_m, 1)$$

or of opposite sign on each of these intervals. But then $q(x)P_n(x)$ is either strictly positive or strictly negative on each of these subintervals. This makes

$$\int_{-1}^{1} q(x)P_n(x)\, dx$$

either positive or negative, contradicting the fact that this integral is zero if $m < n$. We conclude that $m = n$ and the proof is complete.

Legendre polynomials occur in many contexts. For example, suppose n beads, each having an electrical charge of $+2$, are distributed on the $x-$axis between -1 and 1. There is a charge of $+1$ at the end points. The charges on the beads repel each other, and the charges at the ends also repel the beads. If left alone, the beads will eventually settle into equilibrium positions at which all of these forces balance, resulting in no further motion. It can be shown that the charges distribute themselves at exactly the zeros of $P_n(x)$.

Here is an application involving Laplace's equation.

6.4.5 Steady-State Temperature in a Solid Sphere

Suppose we have a solid, homogeneous ball of radius R, and we want the steady-state temperature distribution if the temperature on the surface is given.

Use spherical coordinates, with the sphere's center at the origin, and assume axial symmetry, so the solution at a point is independent of θ and depends only on the distance ρ of the point from the origin, and the azimuthal angle φ. With $u_\theta = 0$, the steady-state heat equation is

$$\nabla^2 u(\rho, \varphi) = u_{\rho\rho} + \frac{2}{\rho} u_\rho + \frac{1}{\rho^2} u_{\varphi\varphi} + \frac{\cot(\varphi)}{\rho^2} u_\varphi = 0.$$

Assume a boundary condition

$$u(R, \varphi) = f(\varphi),$$

giving the temperature on the surface of the sphere. We, therefore, have a Dirichlet problem.

Let $u(\rho, \varphi) = X(\rho)\Phi(\varphi)$ to obtain

$$X''\Phi + \frac{2}{\rho}X'\Phi + \frac{1}{\rho^2}X\Phi'' + \frac{\cot(\varphi)}{\rho^2}X\Phi' = 0.$$

Multiply this equation by $\rho^2/X\Phi$ to obtain

$$\frac{\Phi''}{\Phi} + \cot(\varphi)\frac{\Phi'}{\Phi} = -\rho^2\frac{X''}{X} - 2\rho\frac{X'}{X} = -\lambda.$$

Then

$$\rho^2 X'' + 2\rho X' - \lambda X = 0 \text{ and } \Phi'' + \cot(\varphi)\Phi' + \lambda\Phi = 0.$$

First consider the problem for Φ. Rewrite the differential equation for Φ as

$$\frac{1}{\sin(\varphi)}[\Phi' \sin(\varphi)]' + \lambda\Phi = 0. \tag{6.27}$$

Change variables by setting $x = \cos(\varphi)$. Then $\varphi = \arccos(x)$. Let

$$G(x) = \Phi(\arccos(x)).$$

Because $0 \le \varphi \le \pi$, then $-1 \le x \le 1$. Compute

$$\begin{aligned}
\Phi' \sin(\varphi) &= \sin(\varphi)\frac{dG}{dx}\frac{dx}{d\varphi} \\
&= \sin(\varphi)G'(x)[-\sin(\varphi)] \\
&= -\sin^2(\varphi)G'(x) = -[1 - \cos^2(\varphi)]G'(x) \\
&= -(1 - x^2)G'(x).
\end{aligned}$$

Then

$$\begin{aligned}
\frac{d}{d\varphi}[\Phi'(\varphi)\sin(\varphi)] &= -\frac{d}{d\varphi}[(1 - x^2)G'(x)] \\
&= -\frac{d}{dx}[(1 - x^2)G'(x)]\frac{dx}{d\varphi} \\
&= -\frac{d}{dx}[(1 - x^2)G'(x)][-\sin(\varphi)].
\end{aligned}$$

We conclude that

$$\frac{1}{\sin(\varphi)}\frac{d}{d\varphi}[\Phi'(\varphi)\sin(\varphi)] = \frac{d}{dx}[(1 - x^2)G'(x)].$$

The point to these calculations is that the differential equation 6.27 transforms to Legendre's equation

$$[(1 - x^2)G'(x)]' + \lambda G(x) = 0.$$

We want solutions on $[-1, 1]$. The eigenvalues and eigenfunctions are

$$\lambda_n = n(n+1) \text{ and } G_n(x) = P_n(x) \text{ for } n = 0, 1, 2, \cdots.$$

In terms of φ,

$$\Phi_n(\varphi) = G_n(\cos(\varphi)) = P_n(\cos(\varphi))$$

for $n = 0, 1, 2, \cdots$.

Now that we know the values of λ, the differential equation for X becomes

$$\rho^2 X'' + 2\rho X' - n(n+1)X = 0.$$

This is an Euler equation, with solutions of the form $X(\rho) = \rho^r$. Substitute this into the equation and solve for r to obtain $r = n$ and $r = -n-1$. The general solution for X is therefore

$$X(\rho) = c\rho^n + d\rho^{-n-1}$$

for $n = 0, 1, 2, \cdots$.

Because the solution must be bounded at the center of the sphere $\rho = 0$, let $d = 0$ to obtain

$$X_n(x) = c_n \rho^n.$$

Then, for $n = 0, 1, 2, \cdots$,

$$u_n(\rho, \varphi) = c_n \rho^n P_n(\cos(\varphi)).$$

To satisfy the boundary condition, we must generally (depending on $f(\varphi)$) attempt a superposition:

$$u(\rho, \varphi) = \sum_{n=0}^{\infty} c_n \rho^n P_n(\cos(\varphi)).$$

Choose the coefficients so that

$$u(R, \varphi) = f(\varphi) = \sum_{n=0}^{\infty} c_n R^n P_n(\cos(\varphi)).$$

To put this into the context of a Fourier-Legendre expansion on $[-1, 1]$, recall that $x = \cos(\varphi)$, so $\varphi = \arccos(x)$ and this expansion is

$$f(\arccos(x)) = \sum_{n=0}^{\infty} c_n R^n P_n(x).$$

Then

$$c_n R^n = \frac{2n+1}{2} \int_{-1}^{1} f(\arccos(\xi)) P_n(\xi)\, d\xi$$

so

$$c_n = \frac{2n+1}{2R^n} \int_{-1}^{1} f(\arccos(\xi)) P_n(\xi)\, d\xi.$$

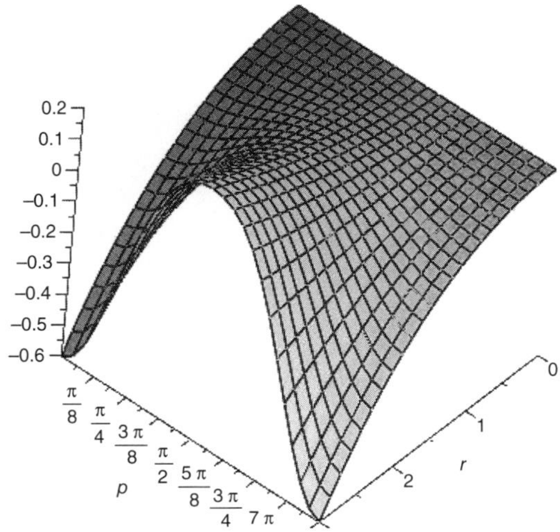

Figure 6.14: Graph of of $u(\rho, \varphi)$.

The steady-state temperature distribution is

$$u(\rho, \varphi) = \sum_{n=0}^{\infty} \frac{2n+1}{2} \left(\int_{-1}^{1} f(\arccos(\xi)) P_n(\xi)\, d\xi \right) \left(\frac{\rho}{R} \right)^n P_n(\cos(\varphi)).$$

For example, suppose $R = 3$ and $f(\varphi) = \sin(\varphi)$. The solution is

$$u(\rho, \varphi) = \sum_{n=0}^{\infty} \frac{2n+1}{2} \left(\int_{-1}^{1} \sin(\arccos(\xi)) P_n(\xi)\, d\xi \right) \left(\frac{\rho}{3} \right)^n P_n(\cos(\varphi)).$$

Figure 6.14 is a surface plot of $u(\rho, \varphi)$. This graph is something of a hybrid, with ρ and φ as two axes of a rectangular system. It provides a sense of how $u(\rho, \varphi)$ varies as ρ and φ increases.

Bessel functions and Legendre polynomials are just two examples of special functions, and there are many others. There is a large class of *orthogonal polynomials*, including Legendre, Hermite, Laguerre, Chebyschev, and Jacobi polynomials. Each type has features in common with Legendre polynomials, including their appearance as eigenfunctions of a Sturm-Liouville problem (endowing them with weighted orthogonality on an interval), a generating function and recurrence relations.

6.4.5 Spherical Harmonics

We will look for solutions of Laplace's equation in spherical coordinates (ρ, θ, φ). The relation between spherical and rectangular coordinates is given by

$$x = \rho \cos(\theta) \sin(\varphi), y = \rho \sin(\theta) \sin(\varphi), z = \rho \cos(\varphi),$$

in which $\rho \geq 0$, $0 \leq \theta \leq 2\pi$, and $0 \leq \varphi \leq \pi$.

Suppose the sphere is centered at the origin. Laplace's equation in spherical coordinates is

$$\frac{\partial}{\partial \varphi}\left(\rho^2 \frac{\partial u}{\partial \rho}\right) + \frac{1}{\sin(\varphi)}\frac{\partial}{\partial \varphi}\left(\sin(\varphi)\frac{\partial u}{\partial \varphi}\right) + \frac{1}{\sin^2(\varphi)}\frac{\partial^2 u}{\partial \theta^2} = 0.$$

To separate the variables, let $u(\rho, \theta, \varphi) = R(\rho)\Theta(\theta)\Phi(\varphi)$ in Laplace's equation and then divide by $R\Theta\Phi$ to obtain

$$\frac{\rho^2 R'' + 2\rho R'}{R} + \frac{\Phi'' + \cot(\varphi)\Phi'}{\Phi} + \frac{1}{\sin^2(\varphi)}\frac{\Theta''}{\Theta} = 0.$$

Then

$$\frac{\rho^2 R'' + 2\rho R'}{R} = -\left(\frac{\Phi'' \cot(\varphi)\Phi'}{\Phi}\right) - \left(\frac{1}{\sin^2(\varphi)}\frac{\Theta''}{\Theta}\right).$$

Both sides must equal some constant λ, because the left side depends only on ρ, which is independent of θ and φ. Then, for some λ

$$\frac{\rho^2 R'' + 2\rho R'}{R} = \lambda$$

and

$$\frac{\Phi'' + \cot(\varphi)\Phi'}{\Phi} + \frac{1}{\sin^2(\varphi)}\frac{\Theta''}{\Theta} = -\lambda. \tag{6.28}$$

The equation for R is

$$\rho^2 R'' + 2\rho R' - \lambda R = 0. \tag{6.29}$$

This is an Euler equation with solutions $R(\rho) = \rho^k$. Substitute $R(\rho) = \rho^k$ into the differential equation for R and divide out ρ^k to obtain

$$k(k-1) + 2k - \lambda = 0,$$

so

$$k = \frac{1}{2}(-1 \pm \sqrt{1 + 4\lambda})$$

and

$$\lambda = k^2 + k = k(k+1).$$

This leaves a lot of possibilities for k and λ. However, notice something we have seen before. Suppose we look for solutions $R(\rho) = \rho^k$ with k a nonnegative integer, say $k = m$ for $m = 0, 1, 2, \cdots$. Now λ must be of the form

$$\lambda = m(m+1).$$

These are the eigenvalues of Legendre's equation! With $\lambda = m(m+1)$,

$$\begin{aligned} k &= \frac{1}{2}(-1 + \sqrt{1 + 4\lambda}) = \frac{-1 + \sqrt{1 + 4m(m+1)}}{2} \\ &= \frac{-1 \pm \sqrt{(2m+1)^2}}{2}, \end{aligned}$$

so k must be m or $-m-1$. This means that $R(\rho)$ must be:

$$R(\rho) = \alpha\rho^m + \beta\rho^{-m-1}.$$

For a solution that is bounded near $\rho = 0$ we must choose $\beta = 0$, so

$$\lambda = m(m+1), R_m(\rho) = \rho^m \text{ for } m = 0, 1, 2, \cdots .$$

Now equation 6.28 is

$$m(m+1) + \frac{\Phi'' + \cot(\varphi)\Phi'}{\Phi} + \frac{1}{\sin^2(\varphi)}\frac{\Theta''}{\Theta} = 0,$$

or

$$-\frac{\Theta''}{\Theta} = m(m+1)\sin^2(\varphi) + \sin^2(\varphi)\frac{\Phi'' + \cot(\varphi)\Phi'}{\Phi} = 0.$$

Because θ and φ are independent, we can separate variables in this equation. Further, Θ must be periodic of period 2π, so we expect Θ to have the form

$$\Theta(\theta) = A_n\cos(n\theta) + B_n\sin(n\theta).$$

Then

$$\frac{\Theta''}{\Theta} = -n^2$$

so

$$\Phi'' + \cot(\varphi)\Phi' + \left(m(m+1) - \frac{n^2}{\sin^2(\varphi)}\right)\Phi = 0. \tag{6.30}$$

There is a clever device for writing solutions of this equation in terms of known functions. First let $u = \cos(\varphi)$. This differential equation for Φ transforms to

$$(1-u^2)\frac{d^2\Phi}{du^2} - 2u\frac{d\Phi}{du} + \left(m(m+1) - \frac{n^2}{1-u^2}\right)\Phi = 0. \tag{6.31}$$

This is called *Legendre's associated equation*, and it reduces to Legendre's differential equation when $n = 0$. Now make another change of variables,

$$\Phi = (1-u^2)^{n/2}y.$$

Substitute this into equation 6.28 to find that y satisfies the following equation:

$$(1-u^2)\frac{d^2y}{du^2} - 2(n+1)u\frac{dy}{du} + (m(m+1) - n(n+1))y = 0. \tag{6.32}$$

Now recall that $P_m(x)$ is a solution of

$$(1-u^2)y'' - 2uy' + m(m+1)y = 0. \tag{6.33}$$

Therefore the nth derivative P_m, denoted $P_m^{(n)}$, is a solution of the equation obtained by differentiating equation 6.32 n times. If this is done, we obtain

$$(1-u^2)y^{(n+2)} - (n+1)(2u)y^{(n+1)} + [m(m+1) - n(n+1)]y^{(n)} = 0. \tag{6.34}$$

This means that the function

$$\Phi_{nm}(x) = \sin^n(\varphi) P_m^{(n)}(\cos(\varphi))$$

is a solution of equation 6.30 if $n \leq m$.

This gives us functions

$$u_{mn}(\theta, \varphi) = \cos(n\theta) \sin^m(\varphi) P_m^{(n)}(\cos(\varphi))$$

and

$$v_{mn}(\theta, \varphi) = \sin(n\theta) \sin^m(\varphi) P_m^{(n)}(\cos(\varphi)),$$

which are called *spherical harmonics*. In terms of these we have a solution of Laplace's equation

$$u(\rho, \theta, \varphi) = \sum_{m=1}^{\infty} \sum_{n \leq m} \rho^m [a_{mn} u_{mn}(\theta, \varphi) + b_{mn} v_{mn}(\theta, \varphi)].$$

Problems for Section 6.4

1. This problem deals with the generating function for Ledgendre polynomials. Begin with the Maclaurin expansion

$$\frac{1}{\sqrt{1-w}} = 1 + \frac{1}{2}w + \frac{3}{8}w^2 + \frac{15}{48}w^3 + \frac{105}{384}w^4 + \frac{945}{3840}w^5 + \cdots.$$

   for $-1 < w < 1$. Let $w = 2xt - t^2$ in this expansion. Collect the coefficients of t, t^2, and t^3 in this Maclaurin series, and show that they are, respectively, $P_1(x), P_2(x)$, and $P_3(x)$.

2. Use the generating function $L(x,t)$ to show that

$$\sum_{n=0}^{\infty} \left(\frac{1}{2^{n+1}}\right) P_n(1/2) = \frac{1}{\sqrt{3}}.$$

3. The gravitational potential at a point $P : (x, y, z)$ due to a unit mass at (x_0, y_0, z_0) is

$$\varphi(x, y, z) = \frac{1}{\sqrt{(x-x_0)^2 + (y-y_0)^2 + (z-z_0)^2}}.$$

   In some areas, such as astronomy, it is convenient to expand $\varphi(x, y, z)$ in powers of r or $1/r$, where

$$r = \sqrt{x^2 + y^2 + z^2}.$$

   To derive such an expansion, let

$$r = \sqrt{x_0^2 + y_0^2 + z_0^2}$$

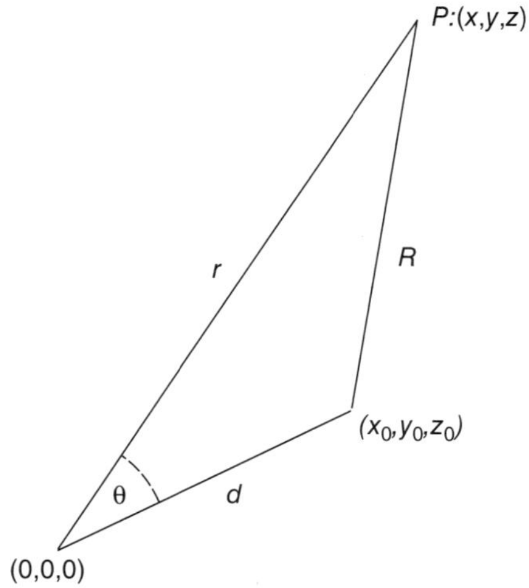

Figure 6.15: Problem 3.

and

$$R = \sqrt{(x - x_0)^2 + (y - y_0)^2 + (z - z_0)^2}.$$

Refer to Figure 6.15 for the following derivation.

(a) Use the law of cosines to write

$$\varphi(x, y, z) = \frac{1}{d\sqrt{1 - 2(r/d)\cos(\theta) + (r/d)^2}}.$$

(b) If $r < d$, use the generating function for Legendre polynomials to show that

$$\varphi(r) = \sum_{n=0}^{\infty} \frac{1}{d^{n+1}} P_n(\cos(\theta)) r^n.$$

(c) If $r > d$, use the generating function to show that

$$\varphi(r) = \frac{1}{r} \sum_{n=0}^{\infty} d^n P_n(\cos(\theta)) r^{-n}.$$

4. Let n be a nonnegative integer. Show that

$$P_{2n+1}(0) = 0 \text{ and } P_{2n}(0) = \frac{(2n)!}{2^{2n}(n!)^2}(-1)^n.$$

In problems 5–7, write $f(x)$ as a (finite) sum of Legendre polynomials.

5. $f(x) = x^2 - x + 1$.

6. $f(x) = -4x^3 + x^2 + 6x$.

7. $f(x) = 1 - x^3$.

In each of problems 8–13, graph $f(x)$ and the sum of the first eleven terms of the Fourier-Legendre expansion of the function on $[-1, 1]$.

8. $f(x) = |x|$.

9. $f(x) = \cos(2x)$.

10. $f(x) = x\sin(x)$.

11. $f(x) = e^{-x}$.

12. $f(x) = x\cos(x)$.

13. $f(x) = \sin(\pi x)$.

In each of problems 14–20, write the solution for the steady-state temperature distribution for a solid sphere, given the radius and the temperature function $f(\varphi)$ on the surface. Compute (approximately) the coefficients of the first 10 terms of this solution and graph $u(\rho, \varphi)$ as a surface in 3−space.

14. $R = 1, f(\varphi) = \varphi$.

15. $R = 2, f(\varphi) = 2 + \varphi^2$.

16. $R = 1/2, f(\varphi) = \varphi^2 - \varphi$.

17. $R = 4, f(\varphi) = \cos(2\varphi)$.

18. $R = 3, f(\varphi) = \sin^2(\varphi)$.

19. $R = 5, f(\varphi) = \varphi\cos(\varphi)$.

20. $R = 2, f(\varphi) = \varphi\cos(\varphi)$.

Problems 21–23 refer to spherical harmonics.

21. Derive equation 6.31.

22. Verify the assertion made that, by setting $\Phi = (1-u^2)^{n/2}y$, then y satisfies equation 6.32, for the cases that $n = 3$ and $n = 4$.

23. Derive equation 6.34.

24. Determine the spherical harmonics $u_{52}(\theta, \varphi)$, $u_{62}(\theta, \varphi)$, $v_{41}(\theta, \varphi)$, and $v_{53}(\theta, \varphi)$.

Chapter 7

Integral Transform Methods of Solution

In mathematics, a transform is often some expression or process that is used to convert one type of problem into another that is in some sense easier to solve. It must then be possible to reverse the process to obtain the solution of the original problem from the solution of the transformed problem.

In this chapter we define four transforms (Fourier, Fourier sine, Fourier cosine, and Laplace), which are defined in terms of integrals and are well suited to the solution of some types of initial-boundary value problems.

7.1 The Fourier Transform

The *Fourier transform* $\mathcal{F}[f]$ of a function f is defined to be the function of ω defined by

$$\mathcal{F}[f](\omega) = \int_{-\infty}^{\infty} f(\xi)e^{-i\omega\xi}\,d\xi. \tag{7.1}$$

Because it is awkward to write $\mathcal{F}[f](\omega)$ in performing calculations, we often denote the Fourier transform of f as $\widehat{f}$. In this notation,

$$\widehat{f}(\omega) = \int_{-\infty}^{\infty} f(\xi)e^{-i\omega\xi}\,d\xi.$$

We rarely compute a Fourier transform by integration, but instead refer to a set of tables or a software package.

Example 7.1 Let a be a positive constant and let

$$f(t) = \begin{cases} e^{-at} & \text{for } t \geq 0, \\ 0 & \text{for } t < 0. \end{cases}$$

When working with the Fourier transform, piecewise continuous functions are often written in terms of the Heaviside function $H(t)$, defined by

$$H(t) = \begin{cases} 1 & \text{for } t \geq 0, \\ 0 & \text{for } t < 0. \end{cases}$$

This enables us to simplify the definition of $f(t)$

$$f(t) = H(t)e^{-at}.$$

A routine integration, or a software routine will return the transform

$$\widehat{f}(\omega) = \frac{1}{a + i\omega}.$$

We need a way to retrieve a function from its Fourier transform. To understand how to do this, as well as where the definition of the transform comes from, begin with the Fourier integral representation of $f(t)$:

$$\frac{1}{2}(f(t-) + f(t+)) = \frac{1}{\pi}\int_0^\infty \int_{-\infty}^\infty f(\xi)\cos(\omega(\xi - t))\, d\xi\, d\omega.$$

From Euler's formula,

$$e^{it} = \cos(t) + i\sin(t) \text{ and } e^{-it} = \cos(t) - i\sin(t).$$

Solve these for $\cos(t)$ and $\sin(t)$:

$$\cos(t) = \frac{1}{2}\left(e^{it} + e^{-it}\right) \text{ and } \sin(t) = \frac{1}{2i}\left(e^{it} - e^{-it}\right).$$

Then

$$\cos(\omega(\xi - t)) = \frac{1}{2}\left(e^{i\omega(\xi - t)} + e^{-i\omega(\xi - t)}\right).$$

Substitute this into the Fourier representation to obtain

$$\frac{1}{2}(f(t-) + f(t+)) = \frac{1}{\pi}\int_0^\infty \int_{-\infty}^\infty \frac{1}{2}f(\xi)\left(e^{i\omega(\xi - t)} + e^{-i\omega(\xi - t)}\right) d\xi\, d\omega$$

$$= \frac{1}{2\pi}\int_0^\infty \int_{-\infty}^\infty f(\xi)e^{i\omega(\xi - t)}\, d\xi\, d\omega + \frac{1}{2\pi}\int_0^\infty \int_{-\infty}^\infty f(\xi)e^{-i\omega(\xi - t)}\, d\xi\, d\omega. \quad (7.2)$$

In the first integral on the second line of this equation, put $\omega = -w$:

$$\int_0^\infty \int_{-\infty}^\infty f(\xi)e^{i\omega(\xi - t)}\, d\xi\, d\omega = \int_{-\infty}^0 \int_{-\infty}^\infty f(\xi)e^{-iw(\xi - t)}\, d\xi\, dw.$$

Substitute this into equation 7.2, but write ω in place of w to have the same variable of integration:

$$\begin{aligned}\frac{1}{2}(f(t-)+f(t+)) &= \frac{1}{2\pi}\int_{-\infty}^{0}\int_{-\infty}^{\infty} f(\xi)e^{-i\omega(\xi-t)}\,d\xi\,d\omega \\ &+ \frac{1}{2\pi}\int_{0}^{\infty}\int_{-\infty}^{\infty} f(\xi)e^{-i\omega(\xi-t)}\,d\xi\,d\omega \\ &= \frac{1}{2\pi}\int_{-\infty}^{\infty}\int_{-\infty}^{\infty} f(\xi)e^{-i\omega(\xi-t)}\,d\xi\,d\omega. \qquad (7.3)\end{aligned}$$

This is the *complex Fourier integral representation of* $f(x)$. Thus far we have just rewritten something we already knew. However, we will find some value in this reformulation. Reverse the order of integration in equation 7.3 and put one integral in large parentheses for emphasis:

$$\begin{aligned}\frac{1}{2}(f(t-)+f(t+)) &= \frac{1}{2\pi}\int_{-\infty}^{\infty}\left(\int_{-\infty}^{\infty} f(\xi)e^{-i\omega\xi}\,d\xi\right)e^{i\omega t}\,d\omega \\ &= \frac{1}{2\pi}\int_{-\infty}^{\infty} \widehat{f}(\omega)e^{i\omega t}\,d\omega,\end{aligned}$$

because the term in large parentheses is $\widehat{f}(\omega)$. In summary,

$$\frac{1}{2}(f(t-)+f(t+)) = \frac{1}{2\pi}\int_{-\infty}^{\infty} \widehat{f}(\omega)e^{i\omega t}\,d\omega. \qquad (7.4)$$

This is exactly what we wanted! Equation 7.4 enables us to recover a function from its transform. For this reason, equation 7.4 is called the *inversion formula*.

Definition 7.1, coupled with the inversion formula 7.4, form a *transform pair* for the Fourier transform. In applications, we will transform a problem using 7.1, solve the transformed problem, then invert the transform of the solution using 7.4, obtaining the solution of the original problem.

When we use the Fourier transform to solve an initial-boundary value problem, we have to be able to transform not just a function, but its derivatives. For this we need an *operational rule*. In its statement, $f^{(n)}$ denotes the nth derivative of f if n is a positive integer. It is convenient to let $n=0$ in this notation, where $f^{(0)}$ means the function itself: $f^{(0)}(t) = f(t)$.

Theorem 7.1 (Operational Rule for the Fourier Transform) Let n be a positive integer. Suppose $f^{(n)}(t)$ is piecewise continuous on the real line and that $|f^{(n-1)}(t)|$ is absolutely integrable. Assume also that

$$\lim_{t\to-\infty} f^{(k)}(t) = \lim_{t\to\infty} f^{(k)}(t) = 0$$

for $k = 0, 1, \cdots, n-1$. Then

$$\mathcal{F}[f^{(n)}](\omega) = (i\omega)^n \widehat{f}(\omega). \qquad (7.5)$$

We may also write equation 7.5 as

$$\widehat{f^{(n)}}(\omega) = (i\omega)^n \widehat{f}(\omega).$$

This gives the transform of the nth derivative of f in terms of the transform of f. Two frequently used cases arise. When $n = 1$:

$$\widehat{f'}(\omega) = i\omega \widehat{f}(\omega),$$

and, when $n = 2$

$$\widehat{f''}(\omega) = (i\omega)^2 \widehat{f}(\omega) = -\omega^2 \widehat{f}(\omega).$$

Notice that the operational formula removes derivatives from the game, because the Fourier transform of a derivative involves only a power of $i\omega$ and the transform of the original function.

It is straightforward to prove the theorem by induction on n, using integration by parts to reduce an integral involving the nth derivative of f to an expression involving the $n-1$ order derivative, and so on.

We will list some other properties of the Fourier transform. Their short proofs are left as exercises. We will not repeat hypotheses each time, but generally it is assumed that the function f is piecewise smooth on every interval, and absolutely integrable.

1. **scaling** If p is a positive number, then $\mathcal{F}[pf(pt)](\omega) = \widehat{f}(\omega/p)$.

2. **time shifting** For any real number t_0, $\mathcal{F}[f(t-t_0)](\omega) = e^{-i\omega t_0}\widehat{f}(\omega)$.

3. **frequency shifting** If k is a real number, $\mathcal{F}[e^{ikt}f(t)](\omega) = \widehat{f}(\omega - k)$.

4. **another form of scaling** If k is any nonzero real number, $\mathcal{F}[f(kt)](\omega) = \frac{1}{|k|}\widehat{f}(\omega/k)$.

5. **time reversal** $\mathcal{F}[f(-t)](\omega) = \widehat{f}(-\omega)$.
The effect of replacing t by $-t$ in the function being transformed, is to replace ω by $-\omega$ in the transformed function.

6. **symmetry** $\mathcal{F}[\widehat{f}(t)](\omega) = 2\pi f(-\omega)$.
If we transform f and replace ω by t in $\widehat{f}$ to obtain the function $\widehat{f}(t)$ of t, and then transform this function of t, we obtain 2π times the original function f, evaluated at $-\omega$.

7. **modulation** $\mathcal{F}[f(t)\cos(kt)](\omega) = \frac{1}{2}\left[\widehat{f}(\omega + k) + \widehat{f}(\omega - k)\right]$.
The transform of $f(t)\cos(kt)$ is the average of the transform of f, evaluated at $\omega + k$, and the transform of f, evaluated at $\omega - k$.

7.1.1 Convolution

Throughout this section functions are assumed to be absolutely integrable and piecewise smooth on every interval.

The transform $\widehat{fg}$ of a product is in general not equal to the product $\widehat{f}\widehat{g}$ of the transforms. However, it is sometimes useful to know what function has the product $\widehat{f}\widehat{g}$ as its transform. The convolution of f with g provides the answer to this question. This is particularly useful when we want to invert a function that can be recognized as the product of two transformed functions.

To define the convolution operation, let f and g be real- or complex-valued functions defined on the entire real line. We say that f *has a convolution with* g if the following hold:

1. $\int_a^b f(\xi)\,d\xi$ and $\int_a^b g(\xi)\,d\xi$ exist for every interval $[a,b]$.

2. For every real t, $\int_{-\infty}^{\infty} |f(t-\xi)g(\xi)|\,d\xi$ converges.

In this event, the *convolution* $f*g$ of f with g is the function defined for all real t by

$$(f*g)(t) = \int_{-\infty}^{\infty} f(t-\xi)g(\xi)\,d\xi.$$

It follows from properties of integrals that the convolution operation is linear:

$$(\alpha f + \beta g) * h = \alpha(f*h) + \beta(g*h),$$

for any real numbers α and β. It may be less obvious that the convolution operation is commutative.

Theorem 7.2 If f has a convolution with g, then g has a convolution with f, and

$$f*g = g*f.$$

Proof

$$\begin{aligned}
(f*g)(t) &= \int_{-\infty}^{\infty} f(t-\xi)g(\xi)\,d\xi \\
&= \lim_{r\to\infty,R\to\infty} \int_{-r}^{R} f(t-\xi)g(\xi)\,d\xi \\
&(\text{ let } w = t-\xi) \\
&= \lim_{r\to\infty,R\to\infty} \int_{t+r}^{t-R} f(w)g(t-w)(-1)\,dw \\
&= \lim_{r\to\infty,R\to\infty} \int_{t-R}^{t+R} g(t-w)f(w)\,dw \\
&= \int_{-\infty}^{\infty} g(t-w)f(w)\,dw \\
&= (g*f)(t).
\end{aligned}$$

We will also use the following property of convolution.

Theorem 7.3

$$\int_{-\infty}^{\infty} (f * g)(t)\, dt = \int_{-\infty}^{\infty} f(t)\, dt \int_{-\infty}^{\infty} g(t)\, dt.$$

Proof Manipulate the integrals as follows:

$$\begin{aligned}\int_{-\infty}^{\infty} (f * g)(t)\, dt &= \int_{-\infty}^{\infty} \int_{-\infty}^{\infty} \left(f(t-\xi) g(\xi)\, d\xi \right) dt \\ &= \int_{-\infty}^{\infty} \left(\int_{-\infty}^{\infty} f(t-\xi) g(\xi)\, dt \right) d\xi \\ &= \int_{-\infty}^{\infty} \left(\int_{-\infty}^{\infty} f(t-\xi)\, dt \right) g(\xi)\, d\xi.\end{aligned}$$

But, for any real x,

$$\int_{-\infty}^{\infty} f(t-\xi)\, dt = \int_{-\infty}^{\infty} f(t)\, dt.$$

Therefore

$$\begin{aligned}\int_{-\infty}^{\infty} (f * g)(t)\, dt &= \int_{-\infty}^{\infty} \left(\int_{-\infty}^{\infty} f(t)\, dt \right) g(\xi)\, d\xi \\ &= \int_{-\infty}^{\infty} f(t)\, dt \int_{-\infty}^{\infty} g(\xi)\, d\xi \\ &= \int_{-\infty}^{\infty} f(t)\, dt \int_{-\infty}^{\infty} g(t)\, dt.\end{aligned}$$

In the middle of this argument a change of order of integration was slipped in without notice. Under the assumptions we are making about functions in this section, it can be shown that this interchange is justified.

Theorem 7.4 (Convolution Theorem) For any real ω,

$$\widehat{f * g}(\omega) = \widehat{f}(\omega)\widehat{g}(\omega).$$

We may also write

$$\mathcal{F}(f * g) = \mathcal{F}[f]\mathcal{F}[g].$$

Proof The proof starts with a trick. Let

$$F(t) = e^{-i\omega t} f(t) \text{ and } G(t) = e^{-i\omega t} g(t).$$

Then

$$\begin{aligned}\widehat{f*g}(\omega) &= \int_{-\infty}^{\infty} (f*g)(t)e^{-i\omega t}\,dt \\ &= \int_{-\infty}^{\infty} \left(\int_{-\infty}^{\infty} f(t-\xi)g(\xi)\,d\xi\right) e^{-i\omega t}\,dt \\ &= \int_{-\infty}^{\infty} \left(\int_{-\infty}^{\infty} e^{-i\omega t} f(t-\xi)g(\xi)\,d\xi\right) dt \\ &= \int_{-\infty}^{\infty} \left(\int_{-\infty}^{\infty} e^{-i\omega(t-\xi)} f(t-\xi)e^{-i\omega\xi} g(\xi)\,d\xi\right) dt.\end{aligned}$$

Recognize that the integral within the large parentheses is $(F*G)(t)$ and use the conclusion of Theorem 7.3:

$$\begin{aligned}\widehat{f*g}(\omega) &= \int_{-\infty}^{\infty} (F*G)(t)\,dt \\ &= \int_{-\infty}^{\infty} F(t)\,dt \int_{-\infty}^{\infty} G(t)\,dt \\ &= \int_{-\infty}^{\infty} f(t)e^{-i\omega t}\,dt \int_{-\infty}^{\infty} g(t)e^{-i\omega t}\,dt \\ &= \widehat{f}(\omega)\widehat{g}(\omega).\end{aligned}$$

7.1.2 Fourier Sine and Cosine Transforms

The Fourier transform of a function is the coefficient in a Fourier integral representation of the function. In similar fashion, we will obtain the Fourier sine (cosine) transform as a coefficient in the Fourier sine (cosine) integral representation.

For the sine transform, first recall the integral sine representation:

$$\frac{1}{2}(f(t-)+f(t+)) = \frac{2}{\pi}\int_0^{\infty} \left(\int_0^{\infty} f(\xi)\sin(\omega\xi)\,d\xi\right) \sin(\omega t)\,d\omega.$$

The integral in large parenthesis is the *Fourier sine transform* $\mathcal{F}_S[f]$, or $\widehat{f}_S$, of f:

$$\mathcal{F}_S[f](\omega) = \widehat{f}_S(\omega) = \int_0^{\infty} f(\xi)\sin(\omega\xi)\,d\xi. \tag{7.6}$$

Because of the way this transform was defined, we also have the inverse transform

$$\frac{1}{2}(f(t-)+f(t+)) = \frac{2}{\pi}\int_0^{\infty} \widehat{f}_S(\omega)\sin(\omega t)\,d\omega. \tag{7.7}$$

This enables us to recover f from its transform, and equations 7.6 and 7.7 form a *transform pair for the Fourier sine transform.*.

Using the Fourier cosine integral

$$\frac{1}{2}(f(t-)+f(t+)) = \frac{2}{\pi}\int_0^\infty \left(\int_0^\infty f(\xi)\cos(\omega\xi)\,d\xi\right)\cos(\omega t)\,d\omega$$

we have the *Fourier cosine transform* defined by

$$\mathcal{F}_C[f](\omega) = \widehat{f}_C(\omega) = \int_0^\infty f(\xi)\cos(\omega\xi)\,d\xi \tag{7.8}$$

and the corresponding inverse given by

$$\frac{1}{2}(f(t-)+f(t+)) = \frac{2}{\pi}\int_0^\infty \widehat{f}_C(\omega)\cos(\omega t)\,d\omega. \tag{7.9}$$

.

Equations 7.8 and 7.9 form the *transform pair for the Fourier cosine transform.*

Example 7.2 Let k be a positive number, and

$$f(t) = \begin{cases} 0 & \text{for } 0 \le t < 4, \\ k & \text{for } 4 \le t \le 6, \\ 0 & \text{for } t > 6. \end{cases}$$

The Fourier sine transform of f is

$$\begin{aligned}\widehat{f}_S(\omega) &= \int_0^\infty f(\xi)\sin(\omega\xi)\,d\xi \\ &= \int_4^6 k\sin(\omega\xi)\,d\xi \\ &= \frac{k}{\omega}(\cos(4\omega)-\cos(6\omega)).\end{aligned}$$

In terms of the Heaviside function,

$$f(t) = k[H(t-4)-H(t-6)].$$

The Fourier cosine transform of f is

$$\begin{aligned}\widehat{f}_C(\omega) &= \int_0^\infty f(\xi)\cos(\omega\xi)\,d\xi \\ &= \int_4^6 k\cos(\omega\xi)\,d\xi \\ &= \frac{k}{\omega}(\sin(6\omega)-\sin(4\omega)).\end{aligned}$$

We will state operational rules for the second derivative of a Fourier sine and cosine transform.

Theorem 7.5 (Operational Rules) Let f and f' be continuous on $[0, \infty)$. Suppose $f(t) \to 0$ and $f'(t) \to 0$ as $t \to \infty$. Suppose also that f'' is piecewise continuous on $[0, \infty)$. Then

$$\mathcal{F}_S[f''](\omega) = -\omega^2 \widehat{f}_S(\omega) + \omega f(0). \tag{7.10}$$

and

$$\mathcal{F}_C[f''](\omega) = -\omega^2 \widehat{f}_C(\omega) - f'(0). \tag{7.11}$$

Both conclusions follow from integrations by parts.

The next section illustrates the use of a transform to solve wave and diffusion problems.

Problems for Section 7.1

In each of problems 1–11, determine $\widehat{f}(\omega)$ for the given function $f(t)$. Wherever they appear, k, a and b are positive numbers, with $a < b$.

1. $f(t) = \begin{cases} \cos(t) & \text{for } -a \le t \le a, \\ 0 & \text{for } |t| > a. \end{cases}$

2. $f(t) = e^{-|t|}$.

3. $f(t) = te^{-|t|}$.

4. $f(t) = \begin{cases} |t| & \text{for } -a \le t \le a, \\ 0 & \text{for } |t| > a. \end{cases}$

5. $f(t) = \begin{cases} k & \text{for } 0 \le t \le a, \\ -k & \text{for } -a \le t < 0, \\ 0 & \text{for } |t| > a. \end{cases}$

6. $f(t) = \begin{cases} k & \text{for } -a \le t < 0, \\ 2k & \text{for } 0 \le t \le a, \\ 0 & \text{for } t < -a \text{ and for } t > a. \end{cases}$

7. $f(t) = \begin{cases} k & \text{for } a \le t \le b, \\ 0 & \text{for } t < a \text{ and } t > b. \end{cases}$

8. $f(t) = \begin{cases} \sin(t) & \text{for } -a \le t \le 2a, \\ 0 & \text{for } t < -a \text{ and for } t > 2a. \end{cases}$

9. $f(t) = e^{-k|t|}$.

10. $f(t) = e^{-|t|}\cos(t)$.

11. $f(t) = \begin{cases} e^{-t} & \text{for } t \geq 0, \\ 0 & \text{for } t < 0. \end{cases}$

12. The *amplitude spectrum* of f is a graph of the magnitude of its Fourier transform, $|\widehat{f}(\omega)|$. Generate the amplitude spectrum of each of the functions in problems 6–11.

13. Write out the details of a proof of the operational formula for the Fourier transform.

14. Prove the scaling property for the Fourier transform.

15. Prove the time shifting property of the Fourier transform.

16. Prove the frequency shifting property of the Fourier transform.

17. Prove property 4 of the Fourier transform.

18. Prove time reversal for the Fourier transform.

19. Prove the symmetry property of the Fourier transform.

In each of problems 2–22, compute $f * g$.

20. $f(t) = e^{-t}$ and $g(t) = \begin{cases} t & \text{for } |t| \leq 1, \\ 0 & \text{for } |t| > 1. \end{cases}$ Determine $f * g$.

21. Let k be a positive number. Let $f(t) = e^{-|t|}$ and $g(t) = e^{-k|t|}$. Determine $f * g$.

22. Let k be a positive constant. Let $f(t) = \begin{cases} t & \text{for } |t| \leq k, \\ 0 & \text{for } |t| > k. \end{cases}$ Determine $f * f$.

23. Prove that

$$\int_{-\infty}^{\infty} |(f * g)(t)|\, dt \leq \int_{-\infty}^{\infty} |f(t)|\, dt \int_{-\infty}^{\infty} |g(t)|\, dt.$$

Assume that f and g satisfy the conditions properties assumed throughout this section.

24. Suppose that $f * g$, $f' * g$, and $f * g'$ exist. Is it true that

$$(f * g)' = f' * g + f * g'?$$

25. (Convolution and the Dirac Delta Function) Define the *Heaviside function* H by

$$H(t) = \begin{cases} 1 & \text{for } t \geq 0, \\ 0 & \text{for } t < 0. \end{cases}$$

The *Dirac delta function* δ is defined by

$$\delta(t) = \lim_{\epsilon \to 0+} \frac{1}{2\epsilon}(H(t+\epsilon) - H(t-\epsilon)).$$

(Actually, δ is not a function, but an object called a *distribution*).

Show that

$$H(t+\epsilon) - H(t-\epsilon) = \begin{cases} 1 & \text{for } -\epsilon \leq t < \epsilon, \\ 0 & \text{for } t \geq \epsilon \text{ and for } t < -\epsilon. \end{cases}$$

Assuming that this limit can be interchanged with the operation of taking the Fourier transform, define

$$\widehat{\delta}(\omega) = \lim_{\epsilon \to 0}[\widehat{H}(t+\epsilon) - \widehat{H}(t-\epsilon)].$$

(a) Show that

$$\mathcal{F}[H(t+\epsilon) - H(t-\epsilon)](\omega) = 2\frac{\sin(\omega)}{\omega}.$$

(b) Conclude from part a and the definition of $\widehat{\delta}(\omega)$ that

$$\widehat{\delta}(\omega) = 1.$$

(c) Show that, if f has a convolution with δ, then

$$f * \delta = \delta * f = f.$$

This means that the delta function behaves like a group identity element under the convolution operation.

In each of problems 26–30, determine the Fourier sine transform and Fourier cosine transform of the function.

26. $f(t) = e^{-t}$.

27. $f(t) = te^{-at}$, with a any positive number.

28. $f(t) = \begin{cases} \cos(t) & \text{for } 0 \leq t \leq a, \\ 0 & \text{for } t > a. \end{cases}$

29. $f(t) = \begin{cases} k & \text{for } 0 \le t \le a, \\ -k & \text{for } a < t \le 2a, \\ 0 & \text{for } t > 2a. \end{cases}$ Here a and k are positive constants.

30. $f(t) = e^{-t}\cos(at)$, with a a nonzero constant.

31. Prove the operational rule for the Fourier sine transform.

32. Prove the operational rule for the Fourier cosine transform.

7.2 Heat and Wave Equations

7.2.1 The Heat Equation on the Real Line

We will solve the problem

$$\begin{aligned} u_t &= ku_{xx} \text{ for } -\infty < x < \infty, t > 0, \\ u(x,0) &= f(x) \text{ for } -\infty < x < \infty. \end{aligned}$$

Because x varies over the entire real line in this problem, we can try a Fourier transform in x. Begin with the transform of the heat equation:

$$\mathcal{F}[u_t] = \mathcal{F}[ku_{xx}] = k\mathcal{F}[u_{xx}].$$

On the left, the transform in x and the partial differentiation with respect to t pass through each other:

$$\begin{aligned} \mathcal{F}[u_t](\omega) &= \int_{-\infty}^{\infty} \frac{\partial u(x,t)}{\partial t} e^{-i\omega x}\, dx \\ &= \frac{\partial}{\partial t} \int_{-\infty}^{\infty} u(x,t) e^{-i\omega x}\, dx \\ &= \frac{\partial}{\partial t} \hat{u}(\omega, t). \end{aligned}$$

where $\widehat{u}(\omega, t)$ is the transform of $u(x,t)$ in the $x-$variable.

For the transform of u_{xx} we must use the operational formula 7.5, with $n = 2$, because now the transform and the derivatives are in the same variable:

$$\mathcal{F}[u_{xx}](\omega) = (i\omega)^2 \widehat{u}(\omega, t) = -\omega^2 \widehat{u}(\omega, t).$$

The Fourier transform of the heat equation is

$$\frac{\partial}{\partial t}\widehat{u} = -k\omega^2 \widehat{u},$$

or

$$\widehat{u}_t + k\omega^2 \widehat{u} = 0.$$

Think of this as an ordinary differential equation in the variable t, with ω carried along as a parameter. The solution is

$$\widehat{u}(\omega, t) = A_\omega e^{-\omega^2 kt}.$$

in which A_ω may depend on ω, and must be determined to suit the initial condition.

To determine A_ω, apply $\mathcal{F}$ to the initial condition $u(x, 0) = f(x)$ to obtain

$$\widehat{u}(\omega, 0) = \widehat{f}(\omega) = A_\omega.$$

Now

$$\widehat{u}(\omega, t) = \widehat{f}(\omega)e^{-\omega^2 kt}.$$

We have obtained the Fourier transform of the solution $u(x, t)$. Apply the inverse transform 7.4 to obtain $u(x, t)$:

$$u(x,t) = \mathcal{F}^{-1}[\widehat{f}(\omega)e^{-\omega^2 kt}] = \frac{1}{2\pi}\int_{-\infty}^{\infty} \widehat{f}(\omega)e^{-\omega^2 kt}e^{i\omega x}\, d\omega.$$

We want to write this solution in terms of $f(x)$, not $\widehat{f}(\omega)$. To do this, insert the definition of the transform into this integral for $\widehat{f}(\omega)$ and perform some manipulations:

$$\begin{aligned}
u(x,t) &= \frac{1}{2\pi}\int_{-\infty}^{\infty}\left(\int_{-\infty}^{\infty} f(\xi)e^{-i\omega\xi}\, d\xi\right) e^{-\omega^2 kt}e^{i\omega x}\, d\omega \\
&= \frac{1}{2\pi}\int_{-\infty}^{\infty}\int_{-\infty}^{\infty} f(\xi)e^{-i\omega(\xi - x)}e^{-\omega^2 kt}\, d\xi\, d\omega \\
&= \frac{1}{2\pi}\int_{-\infty}^{\infty}\int_{-\infty}^{\infty} f(\xi)[\cos(\omega(\xi - x)) - i\sin(\omega(\xi - x))]e^{-\omega^2 kt}\, d\xi\, d\omega \\
&= \frac{1}{2\pi}\int_{-\infty}^{\infty}\int_{-\infty}^{\infty} f(\xi)\cos(\omega(\xi - x))e^{-\omega^2 kt}\, d\xi\, d\omega \\
&- \frac{i}{2\pi}\int_{-\infty}^{\infty}\int_{-\infty}^{\infty} f(\xi)\sin(\omega(\xi - x))e^{-\omega^2 kt}\, d\xi\, d\omega.
\end{aligned}$$

Because $u(x, t)$ is a temperature distribution and must be real-valued, take the real part of this expression to obtain

$$u(x,t) = \frac{1}{2\pi}\int_{-\infty}^{\infty}\int_{-\infty}^{\infty} f(\xi)\cos(\omega(\xi - x))e^{-\omega^2 kt}\, d\xi\, d\omega. \tag{7.12}$$

We can simplify this expression by using the integral 5.8 to obtain

$$u(x,t) = \frac{1}{2\sqrt{\pi kt}}\int_{-\infty}^{\infty} e^{-(x-\xi)^2/4kt} f(\xi)\, d\xi, \tag{7.13}$$

in agreement with the solution obtained in Chapter 5 using the Fourier integral.

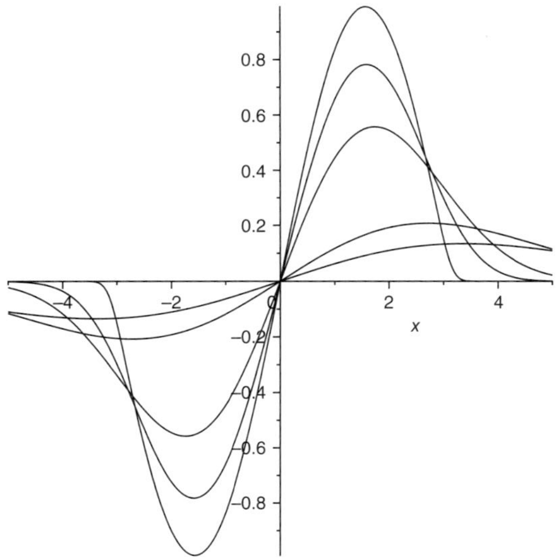

Figure 7.1: Temperature function decreasing at selected times in example 7.3.

Example 7.3 For the problem just solved, let $k = 1$ and

$$f(x) = \begin{cases} \sin(x) & \text{for } -\pi \leq x \leq \pi, \\ 0 & \text{for } |x| > \pi. \end{cases}$$

The solution is

$$u(x,t) = \frac{1}{2\sqrt{\pi t}} \int_{-\pi}^{\pi} e^{-(x-\xi)^2/4t} \sin(\xi)\, d\xi.$$

Figure 7.1 shows $u(x,t)$ at times $t = 1/100, 1/4, 2/3, 3$, and 5. Figure 7.2 is a plot of the surface $u(x,t)$ for $0 \leq t \leq 5$ and $-12 \leq x \leq 12$.

7.2.2 Solution by Convolution

It is also possible to solve the problem of Section 7.2.1 using the convolution for the Fourier transform. Begin by applying the transform in x to the heat equation, as we have just done, to obtain

$$\widehat{u}(\omega,t) = \widehat{f}(\omega)e^{-\omega^2 kt}.$$

The solution $u(x,t)$ is the inverse transform of the right side of this equation, which is a product of a transformed function with an exponential function. If we can write $e^{-\omega^2 kt}$ as the transform of some function, then $\widehat{u}(\omega,t)$ will be a product of two transformed functions, and this is the transform of the convolution of the individual functions. This convolution will be the solution of the problem.

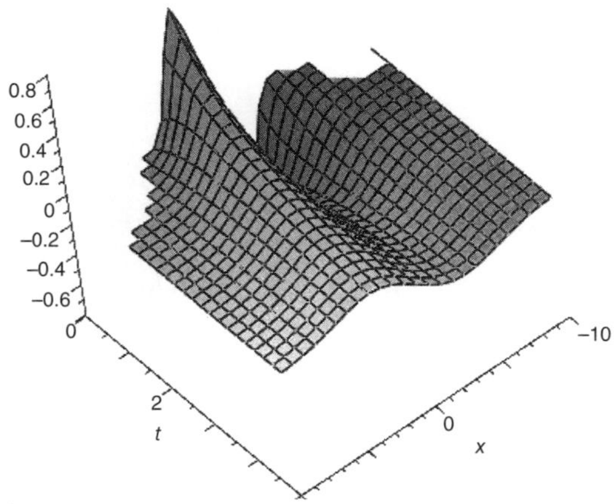

Figure 7.2: Temperature surface in example 7.3.

To carry out this strategy, use the following two results. First, use the scaling property of the Fourier transform. This was listed as item 1 in Section 7.1, and it says, if $p > 0$ and $g(x)$ is given, and if we set $g_p(x) = pg(px)$, then

$$\widehat{g_p}(\omega) = \widehat{g}(\omega/p).$$

Now use a table or software to find a function whose transform looks like the exponential term in $\widehat{u}(\omega, t)$. We find that the transform of $g(x) = e^{-x^2/2}$ is

$$\widehat{g}(\omega) = \sqrt{2\pi} e^{-\omega^2/2} = \sqrt{2\pi} g(\omega),$$

where

$$g(\omega) = e^{-\omega^2/2}.$$

Now set $p = 1/\sqrt{2kt}$ to obtain

$$\begin{aligned} e^{-\omega^2 kt} &= g(\sqrt{2kt}\omega) = \frac{1}{\sqrt{2\pi}} \widehat{g}(\sqrt{2kt}\omega) = \frac{1}{\sqrt{2\pi}} \widehat{g}(\omega/p) \\ &= \frac{1}{\sqrt{2\pi}} \widehat{g_p}(\omega) = \frac{1}{\sqrt{2\pi}} \widehat{g_{1/\sqrt{2kt}}}(\omega). \end{aligned}$$

This enables us to write $\widehat{u}(\omega, t)$ as the product of two transformed functions:

$$\begin{aligned} \widehat{u}(\omega, t) &= \widehat{f}(\omega) \widehat{g_{1/\sqrt{2kt}}}(\omega) \\ &= \frac{1}{2\pi} \widehat{g_{1/\sqrt{2kt}}}(\omega) \widehat{f}(\omega), \end{aligned}$$

in which we have interchanged the order of the factors to obtain the particular solution given in equation 7.13. By the convolution theorem,

$$\begin{aligned} u(x,t) &= \frac{1}{\sqrt{2\pi}}\left(g_{1/\sqrt{2kt}} * f\right)(x) \\ &= \frac{1}{\sqrt{2\pi}}\int_{-\infty}^{\infty} g_{1/\sqrt{2kt}}(x-\xi)f(\xi)\,d\xi \\ &= \frac{1}{2\pi}\int_{-\infty}^{\infty} g((x-\xi)/\sqrt{2kt})f(\xi)\,d\xi \\ &= \frac{1}{2\sqrt{\pi kt}}\int_{-\infty}^{\infty} e^{-(x-\xi)^2/4kt}f(\xi)\,d\xi, \end{aligned}$$

and this is the solution obtained previously.

7.2.3 The Heat Equation on a Half Line

We will use a transform to solve a problem on a half line:

$$\begin{aligned} u_t &= ku_{xx} \text{ for } x > 0, \\ u(0,t) &= 0 \text{ for } t > 0, \\ u(x,0) &= f(x) \text{ for } x > 0. \end{aligned} \tag{7.14}$$

Previously we solved this problem using the Fourier sine integral representation of the initial temperature function. Here we will illustrate the use of a Fourier transform, chosen to suit the initial and boundary conditions. This is an example of fitting the transform to be used to the information of the problem.

Because $t \geq 0$ and $x \geq 0$, the problem is not suitable for use of the Fourier transform in the space or time variable. With $x \geq 0$ we might try a Fourier sine or cosine transform in x. The operational formula for the cosine transform requires that we know $u_x(0,t)$, and we are not given this. However, the operational formula for the sine transform requires $u(0,t)$, which is given. This leads us to try the sine transform in the space variable. Apply $\mathcal{F}_S$ to the heat equation, getting

$$\mathcal{F}_S[u(x,t)](\omega,t) = \widehat{u}_S(\omega,t).$$

The partial derivative with respect to t passes through the transform, and by using the operational formula for $\mathcal{F}_S$ we obtain

$$\begin{aligned} \frac{\partial}{\partial t}\widehat{u}_S(\omega,t) &= k\mathcal{F}_S\left(\frac{\partial^2 u}{\partial x^2}\right) \\ &= k[-(i\omega)^2\widehat{u}_S(\omega,t) + \omega u(0,t)] \\ &= k[-\omega^2\widehat{u}_S(\omega,t)], \end{aligned}$$

because $u(0,t) = 0$. Rewrite this equation as

$$\frac{\partial}{\partial t}\widehat{u}_S(\omega,t) + \omega^2 k\widehat{u}_S(\omega,t) = 0.$$

Think of this as a first-order differential equation in t. The solution has the form

$$\widehat{u}_S(\omega,t) = b_\omega e^{-\omega^2 kt}.$$

To choose b_ω, take the sine transform of the initial condition:

$$\widehat{u}_S(\omega,0) = b_\omega = \mathcal{F}_S[u(x,0)](\omega) = \mathcal{F}_S[f(x)](\omega) = \widehat{f}_S(\omega).$$

Therefore

$$\widehat{u}_S(\omega,t) = \widehat{f}_S(\omega)e^{-\omega^2 kt}.$$

This is the Fourier sine transform of the solution, and it is the product of the sine transform of the initial temperature with an exponential function of ω and t. Using equation 7.7 for the inverse transform, the solution is

$$\begin{aligned} u(x,t) &= \frac{2}{\pi}\int_0^\infty \widehat{f}_S(\omega)e^{-\omega^2 kt}\sin(\omega x)\,d\omega \\ &= \frac{2}{\pi}\int_0^\infty \int_0^\infty f(\xi)\sin(\omega\xi)\sin(\omega x)e^{-\omega^2 kt}\,d\xi\,d\omega. \end{aligned} \tag{7.15}$$

If we write

$$\sin(\omega\xi)\sin(\omega x) = \frac{1}{2}[\cos(\omega(x-\xi)) - \cos(\omega(x+\xi))],$$

the solution 7.15 can be written as

$$\begin{aligned} u(x,t) &= \frac{1}{\pi}\int_0^\infty \left(\int_0^\infty e^{-\omega^2 kt}\cos(\omega(x-\xi))\,d\omega\right) f(\xi)\,d\xi \\ &- \frac{1}{\pi}\int_0^\infty \left(\int_0^\infty e^{-\omega^2 kt}\cos(\omega(x+\xi))\,d\omega\right) f(\xi)\,d\xi. \end{aligned} \tag{7.16}$$

Now use the fact that

$$\frac{1}{\pi}\int_0^\infty e^{-\omega^2 kt}\cos(\alpha\omega)\,d\omega = \frac{1}{2\sqrt{\pi kt}}e^{-\alpha^2/4kt}$$

to write the solution of problem 7.14 as

$$u(x,t) = \frac{1}{2\sqrt{\pi kt}}\int_0^\infty \left[e^{-(x-\xi)^2/4kt} - e^{-(x+\xi)^2/4kt}\right] f(\xi)\,d\xi. \tag{7.17}$$

Example 7.4 In the problem just solved, let $k = 1/4$ and

$$f(x) = \begin{cases} 0 & \text{for } x = 0, \\ 1 & \text{for } 0 < x < 1, \\ -1 & \text{for } 1 \le x \le 2, \\ 0 & \text{for } x > 2. \end{cases}$$

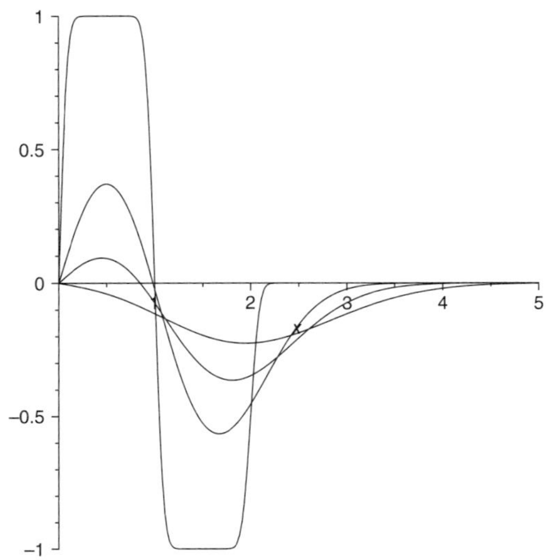

Figure 7.3: Temperature function in example 7.4 at specific times.

Now the solution is

$$u(x,t) = \frac{1}{\sqrt{\pi t}} \int_0^1 \left(e^{-(x-\xi)^2/t} - e^{-(x+\xi)^2/t} \right) d\xi$$
$$+ \frac{1}{\sqrt{\pi t}} \int_1^2 \left(-e^{-(x-\xi)^2/t} + e^{-(x+\xi)^2/t} \right) d\xi.$$

Figure 7.3 shows graphs of $u(x, 1/100), u(x, 1/2)$, $u(x, 1)$ and $u(x, 2)$. These graphs are decreasing over this time frame. Figure 7.4 is a graph of the temperature surface for $0 \le x \le 4$ and $1/20 \le t \le 6$.

7.2.4 The Wave Equation by Fourier Transform

We will use the Fourier transform to solve the Cauchy problem 3.7 for the wave equation:

$$u_{tt} = c^2 u_{xx} \text{ for } -\infty < x < \infty, t > 0,$$
$$u(x,0) = \varphi(x), u_t(x,0) = \psi(x) \text{ for } -\infty < x < \infty.$$

Transform the wave equation with respect to x. Because t is transparent to the transform in x, we have

$$\mathcal{F}[u(x,t)](\omega) = \widehat{u}_{tt}(\omega, t).$$

For the transform of u_{xx}, we must use the operational formula:

$$\mathcal{F}[u(x,t)](\omega) = (i\omega)^2 \widehat{u}(\omega, t) = -\omega^2 \widehat{u}_{xx}(\omega, t).$$

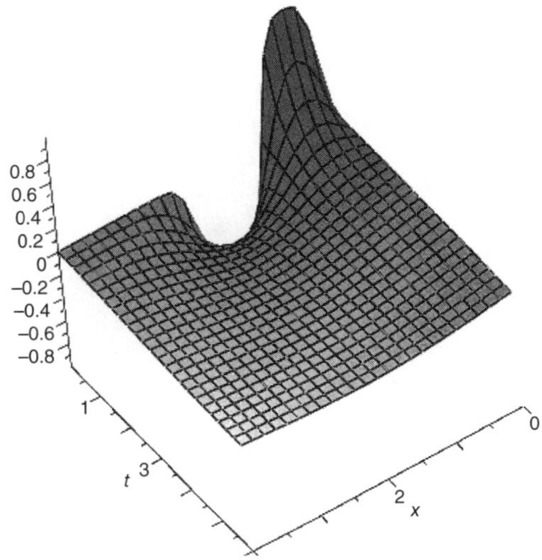

Figure 7.4: Temperature surface in example 7.4.

The wave equation transforms to

$$\widehat{u}_{tt}(\omega, t) = -c^2\omega^2\widehat{u}(\omega, t).$$

Treat this as a second-order differential equation in t. The general solution is

$$\widehat{u}(\omega, t) = a_\omega \cos(\omega ct) + b_\omega \sin(\omega ct).$$

Now,

$$a_\omega = \widehat{u}(\omega, 0) = \mathcal{F}[u(x, 0)](\omega) = \mathcal{F}[\varphi(x)](\omega) = \widehat{\varphi}(\omega)$$

and

$$\omega c b_\omega = \widehat{u}_t(\omega, 0) = \mathcal{F}[u_t(x, 0)](\omega) = \widehat{\psi}(\omega).$$

Therefore,

$$b_\omega = \frac{1}{\omega c}\widehat{\psi}(\omega).$$

The transform of the solution is

$$\widehat{u}(\omega, t) = \widehat{\varphi}(\omega)\cos(\omega ct) + \frac{1}{\omega c}\widehat{\psi}(\omega, t)\sin(\omega ct).$$

The solution is

$$\begin{aligned} u(x, t) &= \mathcal{F}^{-1}[u(\omega, t)](x, t) \\ &= \frac{1}{2\pi}\int_{-\infty}^{\infty} [\widehat{\varphi}(\omega)\cos(\omega ct) + \frac{1}{\omega c}\widehat{\psi}(\omega)\sin(\omega ct)]e^{i\omega x}\, d\omega. \end{aligned} \qquad (7.18)$$

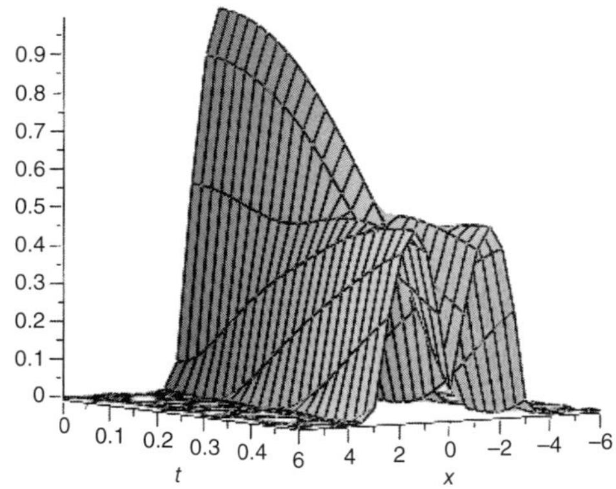

Figure 7.5: Part of the wave surface in example 7.5.

Example 7.5 We will solve this Cauchy problem for the case that $c = 3, \psi(x) = 0$ and

$$\varphi(x) = \begin{cases} \cos(x) & \text{for } -\pi/2 \leq x \leq \pi/2, \\ 0 & \text{for } |x| > \pi/2. \end{cases}$$

The solution is

$$u(x,t) = \frac{1}{2\pi} \int_{-\infty}^{\infty} \widehat{\varphi}(\omega) \cos(3\omega t) e^{i\omega x} \, d\omega.$$

We find that

$$\widehat{\varphi}(\omega) = \frac{2\cos(\pi\omega/2)}{1-\omega^2}.$$

Therefore the solution is

$$u(x,t) = \frac{1}{\pi} \int_{-\infty}^{\infty} \frac{\cos(\pi\omega/2)}{1-\omega^2} \cos(3\omega t) e^{i\omega x} \, d\omega.$$

Because $e^{i\omega x} = \cos(x) + i\sin(x)$ and the solution $u(x,t)$ must be real valued, we can take the real part of this integral to obtain

$$u(x,t) = \frac{1}{\pi} \int_{-\infty}^{\infty} \frac{\cos(\pi\omega/2)}{1-\omega^2} \cos(3\omega t) \cos(\omega x) \, d\omega,$$

in agreement with the solution obtained using the Fourier integral. Figure 7.5 is a surface plot of this wave function for $-6 \leq x \leq 6$ and $0 \leq t \leq 1/2$.

Problems for Section 7.2

1. For the problem of the heat equation on the real line, determine a function $G(x,t)$ such that the solution has the form

$$u(x,t) = \int_{-\infty}^{\infty} G(x-\xi,t)f(\xi)\,d\xi.$$

Such a function is called the *Green's function* for the problem. Physically, it may be thought of as the temperature at point x and time t, due to a heat source concentrated at ξ at time zero. Show that

$$\int_{-\infty}^{\infty} G(x-\xi,t)\,d\xi = 1$$

for all x.

2. Use a Fourier transform to solve

$$\begin{aligned} u_t &= ku_{xx} \text{ for } x > 0, t > 0, \\ u(x,0) &= f(x) \text{ for } x > 0, \\ u_x(0,t) &= 0 \text{ for } t > 0. \end{aligned}$$

3. Solve problem 2 for the case that $f(x) = \begin{cases} e^{-x} & \text{for } 0 \le x \le 1, \\ 0 & \text{for } x > 1. \end{cases}$

4. Use a Fourier transform to solve

$$\begin{aligned} u_t &= ku_{xx} - tu \text{ for } x > 0, t < 0, \\ u(x,0) &= xe^{-x} \text{ for } x > 0, \\ u(0,t) &= 0 \text{ for } t > 0. \end{aligned}$$

5. Use a Fourier transform to solve

$$\begin{aligned} u_t &= ku_{xx} - u \text{ for } x > 0, t > 0, \\ u_x(0,t) &= f(t) \text{ for } t > 0, \\ u(x,0) &= 0 \text{ for } x > 0. \end{aligned}$$

In each of problems 6–12, use a Fourier transform to solve the wave equation on the real line. In problems 6–10, the initial position $\varphi(x)$ is given, and the initial velocity is zero. In problems 11 and 12, there is no initial displacement, and the initial velocity is $\psi(x)$.

6. $\varphi(x) = \begin{cases} 1 - x^2 & \text{for } -1 \le x \le 1, \\ 0 & \text{for } |x| > 1. \end{cases}$

7. $\varphi(x) = \begin{cases} \sin(x) & \text{for } -\pi \leq x \leq \pi, \\ 0 & \text{for } |x| > \pi. \end{cases}$

8. $\varphi(x) = \begin{cases} 4 - |x| & \text{for } -4 \leq x \leq 4, \\ 0 & \text{for } |x| > 4. \end{cases}$

9. $\varphi(x) = \begin{cases} e^{-x} & \text{for } x > 0, \\ 0 & \text{for } x < 0. \end{cases}$

10. $\varphi(x) = \begin{cases} \sin^2(x) & \text{for } -4\pi \leq x \leq 4\pi, \\ 0 & \text{for } |x| > 4\pi. \end{cases}$

11. $\psi(x) = \begin{cases} \cos(x) & \text{for } |x| \leq \pi/2, \\ 0 & \text{for } |x| > \pi/2. \end{cases}$

12. $\psi(x) = \begin{cases} x & \text{for } |x| \leq 5, \\ 0 & \text{for } |x| > 5. \end{cases}$

7.3 The Telegraph Equation

If C is the capacitance, R the resistance, L the inductance, and S the leakage of a cable, all measured per unit length, then the current $u(x,t)$ at point x and time t in the cable is modeled by

$$u_{xx} = Cu_{tt} + (RC + SL)u_t + RSu.$$

Because of this model of current flow through a cable, the wave equation

$$u_{tt} + 2bu_t + au = c^2 u_{xx} \tag{7.19}$$

is called the *telegraph equation*. The coefficient of u_t is denoted $2b$ for convenience in solving a quadratic equation we will encounter shortly. We assume that a, b, and c are constants, and that x varies over all real values, while $t \geq 0$. We will use the Fourier transform to write solutions of the telegraph equation with the initial conditions

$$u(x,0) = f(x), u_t(x,0) = g(x),$$

assuming that f and g have Fourier transforms.

Because x varies over the real line, we can take the Fourier transform of the telegraph equation with respect to x, obtaining

$$\widehat{u}''(\omega,t) + 2b\widehat{u}'(\omega,t) + a\widehat{u}(\omega,t) = -\omega^2 c^2 \widehat{u}(\omega,t).$$

Primes denote differentiation with respect to t. The operational rule for $\mathcal{F}$ was used on the right to transform the u_{xx} term.

This gives us a homogeneous ordinary differential equation (in t):

$$\widehat{u}''(\omega,t)+2b\widehat{u}'(\omega,t)+(a+\omega^2c^2)\widehat{u}(\omega,t)=0. \tag{7.20}$$

As usual, look for solutions of the form $\widehat{u}(\omega,t)=e^{rt}$. Substitute this into equation 7.20 and divide out e^{rt} to obtain the characteristic equation for r:

$$r^2+2br+(a+\omega^2c^2)=0.$$

The roots are

$$r=-b\pm\sqrt{b^2-a-\omega^2c^2}.$$

The sign of the term under the radical dictates what solutions look like. Consider cases.

Case 1. $b^2-a<0$

Now $b^2-a-(\omega c)^2<0$ and the roots of the characteristic equation are

$$r=-b\pm\sqrt{a-b^2+(\omega c)^2}i.$$

In this case

$$\widehat{u}(\omega,t)=$$
$$e^{-bt}\left[c_1(\omega)\cos\left(\sqrt{a-b^2+(\omega c)^2}t\right)+c_2(\omega)\sin\left(\sqrt{a-b^2+(\omega c)^2}t\right)\right].$$

For the "constants" $c_1(\omega)$ and $c_2(\omega)$, take the transform of the initial conditions. First,

$$c_1(\omega)=\widehat{u}(\omega,0)=\mathcal{F}[u(x,0)](\omega)=\widehat{f}(\omega).$$

Next,

$$\begin{aligned}\widehat{u}(\omega,0)&=\mathcal{F}[u_t(x,0)](\omega=\widehat{g}(\omega)\\&=-b\widehat{f}(\omega)+c_2(\omega)\sqrt{a-b^2+(\omega c)^2}.\end{aligned}$$

Then

$$c_2(\omega)=\frac{\widehat{g}(\omega)+b\widehat{f}(\omega)}{\sqrt{a-b^2+(\omega c)^2}}.$$

The solution for $\widehat{u}(\omega,t)$ is

$$\begin{aligned}\widehat{u}(\omega,t)&=\widehat{f}(\omega)e^{-bt}\cos\left(\sqrt{a-b^2+(\omega c)^2}t\right)\\&\quad+\frac{\widehat{g}(\omega)+b\widehat{f}(\omega)}{\sqrt{a-b^2+(\omega c)^2}}e^{-bt}\sin\left(\sqrt{a-b^2+(\omega c)^2}t\right).\end{aligned}$$

Apply the inversion formula to write the solution in this case:

$$u(x,t)=\frac{1}{2\pi}\int_{-\infty}^{\infty}\widehat{u}(\omega,t)e^{i\omega x}\,d\omega.$$

Case 2. $b^2 - a = 0$

Now the roots of the characteristic equation are $r = -b \pm \omega c i$. Then

$$\widehat{u}(\omega, t) = e^{-bt}[c_1(\omega)\cos(\omega c t) + c_2(\omega)\sin(\omega c t)].$$

Then

$$\widehat{u}(\omega, 0) = c_1(\omega) = \mathcal{F}[u(x,0)](\omega) = \widehat{f}(\omega).$$

To obtain $c_2(\omega)$, it is routine to compute $\widehat{u}'(\omega, t)$ and let $t = 0$:

$$\widehat{u}'(\omega, 0) = -b\widehat{f}(\omega) + c_2(\omega)\omega c = \widehat{g}(\omega).$$

Then

$$c_2(\omega) = \frac{\widehat{g}(\omega) + b\widehat{f}(\omega)}{\omega c}.$$

The transform of the solution in this case is, therefore,

$$\widehat{u}(\omega, t) = \widehat{f}(\omega)e^{-bt}\cos(\omega c t) + \frac{\widehat{g}(\omega) + b\widehat{f}(\omega)}{\omega c}e^{-bt}\sin(\omega c t).$$

The solution when $b^2 - a = 0$ is obtained by inverting this:

$$u(x,t) = \frac{1}{2\pi}\int_{-\infty}^{\infty} \widehat{u}(\omega, t)e^{i\omega x}\, d\omega.$$

Case 3. $b^2 - a > 0$

This is the most complicated case because $b^2 - a - (\omega c)^2$ can be positive or less than or equal to zero, depending on ω. We will obtain different expressions for $\widehat{u}(\omega, t)$ in each of these subcases.

Subcase 3 - 1. $b^2 - a - (\omega c)^2 > 0$

This occurs if

$$|\omega| < \frac{1}{c}\sqrt{b^2 - a}.$$

Now

$$\widehat{u}_1(\omega, t) = e^{-bt}\left[c_1(\omega)e^{\sqrt{b^2-a-(\omega c)^2}t} + c_2(\omega)e^{-\sqrt{b^2-a-(\omega c)^2}t}\right]. \tag{7.21}$$

The subscript on $\widehat{u}_1$ is a reminder that this is subcase 1 of case 3. We immediately have

$$\widehat{u}_1(\omega, 0) = \widehat{f}(\omega) = c_1(\omega) + c_2(\omega).$$

Further, a routine differentiation yields

$$\begin{aligned}\widehat{u}_1'(\omega, 0) &= -b(c_1 + c_2) + \sqrt{b^2 - a - (\omega c)^2}(c_1 - c_2) \\ &= -b\widehat{f}(\omega) + (c_1 - c_2)\sqrt{b^2 - a - (c\omega)^2} = \widehat{g}(\omega).\end{aligned}$$

This gives us two equations for c_1 and c_2. Solve these to obtain

$$c_1(\omega) = \frac{1}{2}\widehat{f}(\omega) + \frac{1}{2}\frac{\widehat{g}(\omega) + b\widehat{f}(\omega)}{\sqrt{b^2 - a - (\omega c)^2}}$$

and

$$c_2(\omega) = \frac{1}{2}\widehat{f}(\omega) - \frac{1}{2}\frac{\widehat{g}(\omega) + b\widehat{f}(\omega)}{\sqrt{b^2 - a - (\omega c)^2}}.$$

With these choices of $c_1(\omega)$ and $c_2(\omega)$, $\widehat{u}_1(\omega, t)$ is completely determined by equation 7.21.

Subcase 3 - 2. $b^2 - a - (c\omega^2) \leq 0$

This occurs if

$$|\omega| \geq \frac{1}{c}\sqrt{b^2 - a}.$$

Now the transformed solution is

$$\begin{aligned}\widehat{u}_2(\omega, t) &= e^{-bt}c_1(\omega)\cos(\sqrt{a - b^2 + (\omega c)^2}t) \\ &\quad + e^{-bt}c_2(\omega)\sin(\sqrt{a - b^2 + (\omega c)^2}t),\end{aligned} \tag{7.22}$$

where

$$c_1(\omega) = \widehat{u}_2(\omega, 0) = \widehat{f}(\omega).$$

Further,

$$\widehat{u}_2(\omega, 0) = -b\widehat{f}(\omega) + c_2(\omega)\sqrt{a - b^2 + (\omega c)^2}$$

so

$$c_2(\omega) = \frac{\widehat{g}(\omega) + b\widehat{f}(\omega)}{\sqrt{a - b^2 + (\omega c)^2}}.$$

With these choices of $c_1(\omega)$ and $c_2(\omega)$, $\widehat{u}_2(\omega, t)$ is determined by equation 7.22.

Finally, we can combine these cases and write the solution in case 3 as

$$\begin{aligned}u(x,t) &= \frac{1}{2\pi}\int_{-\infty}^{\infty} \widehat{u}(\omega, t)e^{i\omega x}\,d\omega \\ &= \frac{1}{2\pi}\left[\int_{|\omega|<\sqrt{b^2-a}/c} \widehat{u}_1(\omega, t)e^{i\omega x}\,d\omega + \int_{|\omega|\geq\sqrt{b^2-a}/c} \widehat{u}_2(\omega, t)e^{i\omega x}\,d\omega\right].\end{aligned}$$

We now have integral solutions of the telegraph equation on the real line, for all cases.

Problems for Section 7.3

In each of problems 1–3, write the solution of the telegraph equation for the given constants, with $g(x) = 0$ and $f(x) = \begin{cases} 1 - x & \text{for } 0 \leq x \leq 1, \\ 1 + x & \text{for } -1 \leq x \leq 0, \\ 0 & \text{for } |x| > 1. \end{cases}$

1. $a = 1, b = 2, c = 3$.

2. $a = 6, b = 2, c = 5$.

3. $a = 9, b = 3, c = 4$.

In each of problems 4–6, write solutions of the telegraph equation for the given constants, with $f(x) = 0$ and $g(x) = \begin{cases} 1 - x^2 & \text{for } -1 \leq x \leq 1, \\ 0 & \text{for } |x| > 1. \end{cases}$

4. $a = 4, b = c = 2$.

5. $a = 3, b = 4, c = 1$.

6. $a = 8, b = 2, c = 3$.

7. Look at the telegraph equation on the real line with $f(x) = 0$. Show that the solution for the case $b^2 - a = 0$ can be written

$$u(x,t) = \frac{1}{2c}e^{-bt}\int_{x-ct}^{x+ct} g(\xi)\,d\xi.$$

8. Suppose $u(x,t)$ satisfies the telegraph equation (just the equation, not any particular boundary conditions). Let $v(x,t) = e^{bt}u(x,t)$. Show that v satisfies

$$v_{tt} + (a - b^2)v = c^2 v_{xx}.$$

9. Solve the following problem on the half-line:

$$\begin{aligned} u_{tt} + 2bu_t + au &= c^2 u_{xx}, \\ u(x,0) &= f(x), u_t(x,0) = 0, \\ u(0,t) &= 0. \end{aligned}$$

Note the boundary condition at $x = 0$, which is missing for the problem on the entire line.

7.4 The Laplace Transform

The Laplace transform is another integral transform, defined by

$$\mathcal{L}[f](s) = \int_0^\infty e^{-st} f(t)\,dt.$$

A standard convention is to denote the transform of a function $f(t)$ by $F(s)$, with the lower case letter for the input function, and the upper case for its transform.

As with any integral transform, the Laplace transform is linear:

$$\mathcal{L}[cf + dg] = c\mathcal{L}[f] + d\mathcal{L}[g],$$

or

$$\mathcal{L}[cf + dg] = cF + dG.$$

Denote the inverse Laplace transform by $\mathcal{L}^{-1}$. In this notation,

$$\mathcal{L}^{-1}[F] = f \text{ if } \mathcal{L}[f] = F.$$

To use the Laplace transform to solve differential equations, we need an operational formula.

Theorem 7.6 (Operational Rules for the Laplace Transform) Let n be a positive integer and suppose that $f, f', \cdots, f^{(n-1)}$ are continuous for $t > 0$, while $f^{(n)}$ is piecewise continuous on $[0, k]$ for every $k > 0$. Suppose also that

$$\lim_{t\to\infty} e^{-st} f^{(j)}(k) = 0$$

for $s > 0$ and $j = 1, \cdots, n-1$. Then

$$\mathcal{L}[f^n](s) =$$
$$s^n F(s) - s^{n-1} f(0) - s^{n-2} f'(0) - \cdots - s f^{(n-2)}(0) - f^{(n-1)}(0). \tag{7.23}$$

This gives the transform of the nth derivative of $f(t)$ in terms of $s^n F(s)$, minus a sum of powers of s times values of the function (not the transformed function) and its first $n-1$ derivatives evaluated at 0.

We will use this most often in the cases $n = 1$ and $n = 2$:

$$\mathcal{L}[f'](s) = sF(s) - f(0)$$

and

$$\mathcal{L}[f''](s) = s^2 F(s) - sf(0) - f'(0).$$

The theorem is proved by integrations by parts.

As with the Fourier transform, the Laplace transform has a convolution operation. The *convolution* of f and g is a function defined by

$$(f * g)(t) = \int_0^t f(t-\tau) g(\tau)\, d\tau.$$

As with the Fourier transform, this convolution for the Laplace transform is defined so that the transform of the convolution of f and g is the product of the individual transforms of f and g:

$$\mathcal{L}[f * g] = \mathcal{L}[f]\mathcal{L}[g]. \tag{7.24}$$

This is the *convolution theorem* for the Laplace transform. We may also write

$$\mathcal{L}[f * g](s) = F(s)G(s).$$

One use for this result is in taking inverses of transforms. If we want the inverse Laplace transform of $H(s)$, and if we can write $H(s) = F(s)G(s)$, in which we know the inverse transforms $f(t)$ of $F(s)$ and $g(t)$ of $G(t)$, then the inverse transform of $H(s)$ is $f(t) * g(t)$:

$$\mathcal{L}^{-1}[F(s)G(s)](t) = f(t) * g(t). \tag{7.25}$$

The *Heaviside function* $H(t)$ is often used in connection with the Laplace transform. This is defined by

$$H(t) = \begin{cases} 1 & \text{for } t \geq 0, \\ 0 & \text{for } t < 0. \end{cases}$$

We may think of $H(t)$ as a signal of unit magnitude that is turned off for $t < 0$, and is then turned on at $t = 0$. When used in a product, $H(t)$ turns off the function for $t < 0$ and leaves it unaffected for $t \geq 0$:

$$H(t)f(t) = \begin{cases} f(t) & \text{for } t \geq 0, \\ 0 & \text{for } t < 0. \end{cases}$$

We may also want to shift a function as well as turn it off before a certain time. For example, $H(t-3)f(t-3)$ shifts the graph of $f(t)$, 3 units to the right, then cancels out the part of this graph to the left of 3:

$$H(t-3)f(t-3) = \begin{cases} f(t-3) & \text{for } t \geq 3, \\ 0 & \text{for } t < 3. \end{cases}$$

As with the Fourier transform, we usually consult a table or a software routine when we want to compute the Laplace transform, or inverse Laplace transform, of a function.

We will illustrate the use of the Laplace transform to solve initial-boundary value problems.

7.4.1 Temperature Distribution in a Semi-Infinite Bar

We will write a general formula for the solution of the problem

$$u_t = ku_{xx} \text{ for } x > 0, t > 0,$$
$$u(x,0) = 0, u(0,t) = f(t).$$

We can think of this initial-boundary value problem as modeling the temperature distribution in a homogeneous bar of material lying along the positive $x-$axis, with zero initial temperature and a control function on the temperature at the left end.

In the absence of a bound on increasing x, we will seek a solution such that $u(x,t) \to 0$ as $x \to \infty$, for all times t.

Apply the Laplace transform with respect to time to the heat equation to obtain

$$sU(x,s) = kU_{xx}(x,s).$$

Here $U(x,s)$ is the Laplace transform of $u(x,t)$, leaving x along and transforming in the $t-$variable. The term $sU(x,s)$ comes from the operational formula for $\mathcal{L}$, with $n = 1$, and the condition $u(x,0) = 0$. We also have

$$U(0,s) = \mathcal{L}[u(0,t)](s) = \mathcal{L}[f(t)](s) = F(s)$$

and

$$\lim_{x \to \infty} U(x,s) = 0.$$

The transformed differential equation is

$$U_{xx}(x,s) = \frac{s}{k}U(x,s),$$

with the general solution

$$U(x,s) = c_1 e^{\sqrt{s/k}x} + c_2 e^{-\sqrt{s/k}x}.$$

We require that $c_1 = 0$ to have $U(x,s) \to 0$ as $x \to \infty$. Further,

$$U(0,s) = c_2 = F(s).$$

Therefore,

$$U(x,s) = F(s)e^{-\sqrt{s/k}x}.$$

This is the Laplace transform of the solution, which is obtained by inverting $U(x,s)$:

$$u(x,t) = \mathcal{L}^{-1}[U(x,s)] = \mathcal{L}^{-1}\left[F(s)e^{-\sqrt{s/k}x}\right]$$

This is the inverse transform of a product of transformed functions. We know that the inverse of one factor, $F(s)$, is $f(t)$. If we can write $e^{-\sqrt{s/k}x}$ as the transform of $g(x,t)$, then by the convolution theorem, the solution is

$$u(x,t) = f(t) * g(x,t).$$

This inverse $g(x,t)$ is nontrivial to compute. There are methods of complex analysis that can be used, or one can consult a very good table to find that

$$\mathcal{L}^{-1}\left[e^{-\sqrt{s/k}x}\right] = g(x,t) = \frac{x}{\sqrt{\pi k}t^{3/2}}e^{-x^2/4kt}.$$

The solution is the convolution of the given function $f(x)$, with this inverse transform:

$$u(x,t) = f(t) * \frac{x}{\sqrt{\pi k}t^{3/2}}e^{-x^2/4kt}.$$

Keep in mind that this convolution is with respect to the t–variable, since we took the transform with respect to the time. The solution is

$$\begin{aligned} u(x,t) &= f(t) * g(x,t) \\ &= \int_0^t f(t-\tau)\frac{x}{\sqrt{\pi k}\tau^{3/2}} e^{-x^2/4k\tau}\, d\tau. \end{aligned}$$

7.4.2 A Diffusion Problem in a Semi-Infinite Medium

Consider again the heat equation for $x > 0$, except now use the boundary and initial conditions:

$$u(x,0) = A, u(0,t) = \begin{cases} B & \text{for } 0 \leq t \leq h, \\ 0 & \text{for } t > h. \end{cases}$$

Here $A, B,$ and h are positive constants. The temperature function at the left end has a jump discontinuity at $t = h$.

It will be convenient to write $u(0,t)$ in terms of the Heaviside function:

$$u(0,t) = B[1 - H(t-h)].$$

As a check, if $t > h$ then $t - h > 0$ so $H(t-h) = 1$ and $u(0,t) = B(1-1) = 0$. If $t < h$, then $t - h < 0$, so $H(t-h) = 0$ and we get $u(0,t) = B(1-0) = B$, as we want.

Take the Laplace transform of $u_t = ku_{xx}$, with respect to t. Now the operational rule must be applied to u_t and we obtain

$$\mathcal{L}[u_t](s) = sU(x,s) - u(x,0) = sU(x,s) - A = kU_{xx}.$$

Then

$$U_{xx} - \frac{s}{k}U = -\frac{A}{k}.$$

As a differential equation in x, this has the general solution

$$U(x,s) = c_1(s)e^{\sqrt{s/k}x} + c_2(s)e^{-\sqrt{s/k}x} + \frac{A}{s}.$$

Here the "constants" $c_1(s)$ and $c_2(s)$ may depend on s, which is thought of as constant in this differential equation with respect to x. To have a solution that remains bounded as x increases, choose $c_1 = 0$. This leaves

$$U(x,s) = c_2 e^{-\sqrt{s/k}x} + \frac{A}{s}.$$

To obtain c_2, take the Laplace transform of the boundary condition:

$$\begin{aligned} U(0,t) &\doteq \mathcal{L}[u(0,t)](s) = \mathcal{L}[B](s) - \mathcal{L}[BH(t-h)](s) \\ &= \frac{B}{s} - B\frac{e^{-hs}}{s} = c_2 + \frac{A}{s}. \end{aligned}$$

Then

$$c_2 = \frac{B-A}{s} - \frac{B}{s}e^{-hs}.$$

Finally, we have the transform of the solution:

$$U(x,s) = \left[\frac{B-A}{s} - \frac{B}{s}e^{-hs}\right]e^{-\sqrt{s/k}x} + \frac{A}{s}.$$

Again, the inversion process is nontrivial by hand, but the inverse can be read from a table in terms of the error function and the complementary error function. These functions are defined, respectively, by

$$\operatorname{erf}(t) = \frac{2}{\sqrt{\pi}}\int_0^t e^{-w^2}\,dw$$

and

$$\operatorname{erfc}(t) = 1 - \operatorname{erf}(t) = \frac{2}{\sqrt{\pi}}\int_t^\infty e^{-w^2}\,dw.$$

We obtain

$$\begin{aligned} u(x,t) &= A\operatorname{erf}\left(\frac{x}{2\sqrt{kt}}\right) + B\operatorname{erfc}\left(\frac{x}{2\sqrt{kt}}\right) \\ &\quad - B\operatorname{erfc}\left(\frac{x}{2\sqrt{k(t-h)}}\right)H(t-h). \end{aligned}$$

7.4.3 Vibrations in an Elastic Bar

We will use the Laplace transform to solve the following problem:

$$\begin{aligned} u_{tt} &= c^2 u_{xx} \text{ for } 0 < x < L, t > 0, \\ u(x,0) &= u_t(x,0) = 0 \text{ for } 0 < x < L, \\ u(0,t) &= 0, Eu_x(0,t) = f(t) \text{ for } t > 0. \end{aligned}$$

Here $u(x,t)$ is the wave function for vibrations in an elastic bar of length L and constant density ρ. $c^2 = E/\rho$, where E is Young's modulus for the material of the bar. This is the ratio of stress to strain for linear compression or elongation. A force $f(t)$ is applied to the right end of the bar, directed along its length.

In this problem, x is bounded, but we can try using the Laplace transform in the t variable, denoting $\mathcal{L}[u(x,t)](s) = U(x,s)$. First, write the wave equation as

$$u_{xx} = \frac{1}{c^2}u_{tt}.$$

The transform in t passes through the differentiations with respect to x, and we use the operational formula on the right side to obtain

$$\begin{aligned} U''(x,s) &= \frac{1}{c^2}\left(s^2U(x,s) - su(x,0) - u_t(x,0)\right) \\ &= \frac{s^2}{c^2}U(x,s) \end{aligned}$$

because $u(x,0) = u_t(x,0) = 0$. The primes denote differentiation with respect to x. Now

$$U'' - \frac{s^2}{c^2}U = 0,$$

with the general solution

$$U(x,s) = ae^{sx/c} + be^{-sx/c}.$$

Transform the condition $u(0,t) = 0$ to obtain

$$U(x,0) = a + b = 0,$$

so $b = -a$ and

$$U(x,s) = k\sinh(sx/c).$$

To determine k, transform the condition $Eu_x(L,t) = f(t)$ to obtain

$$EU'(L,s) = F(s),$$

where $F(s) = \mathcal{L}[f](s)$. Then

$$EU'(L,s) = E\frac{ks}{c}\cosh(sL/c) = F(s).$$

Then

$$k = \frac{c}{E}\frac{F(s)}{s\cosh(sL/c)}$$

so

$$U(x,s) = \frac{c}{E}\frac{F(s)\sinh(sx/c)}{s\cosh(sL/c)}.$$

This is the transform of the solution, so

$$u(x,t) = \mathcal{L}^{-1}\left[\frac{c}{E}\frac{F(s)\sinh(sx/c)}{s\cosh(sL/c)}\right](t).$$

Given $f(t)$, we can attempt to carry out this calculation and solve for $u(x,t)$. Depending on $f(t)$, this could be a formidable task. We will illustrate with $f(t) = K$, constant. Now $F(s) = K/s$ and

$$U(x,s) = \frac{Kc}{E}\frac{\sinh(sx/c)}{s^2\cosh(sL/c)}.$$

Inverting this is nontrivial, but can be done with some manipulation. Begin with

$$\begin{aligned}
\frac{\sinh(sx/c)}{\cosh(sL/c)} &= \frac{e^{sx/c} - e^{-sx/c}}{e^{sL/c} + e^{-sL/c}} \\
&= \frac{(e^{sx/c} - e^{-sx/c})e^{-sL/c}}{1 + e^{-2sL/c}} \\
&= \left(e^{-(L-x)s/c} - e^{-(L+x)sx/c}\right)\frac{1}{1 + e^{-2sL/c}}.
\end{aligned}$$

Now recall the geometric series

$$\sum_{n=0}^{\infty}(-1)^n r^n = \frac{1}{1+r} \text{ if } |r| < 1$$

and let $r = e^{-2sL/c}$, which is less than 1 in magnitude. Then

$$\frac{1}{1+e^{-2sL/c}} = \sum_{n=0}^{\infty}(-1)^n e^{-2nsL/c}.$$

Substitute this into the equation for $U(x,s)$ to obtain, after some manipulation,

$$\begin{aligned} U(s,x) &= \frac{Kc}{E}\sum_{n=0}^{\infty}(-1)^n \frac{1}{s^2} e^{((-(2n+1)L-x)/c)s} \\ &- \frac{Kc}{E}\sum_{n=0}^{\infty}(-1)^n \frac{1}{s^2} e^{((-(2n+1)L+x)/c)s}. \end{aligned}$$

This expression allows us to invert $U(x,s)$ because

$$\mathcal{L}^{-1}\left[\frac{1}{s^2}e^{-\alpha s}\right] = (t-\alpha)H(t-\alpha),$$

in which H is the Heaviside function. Finally, we can write the solution as an infinite series

$$\begin{aligned} u(x,t) &= \frac{Kc}{E}\sum_{n=0}^{\infty}(-1)^n \left(t - \frac{(2n+1)L-x}{c}\right) H\left(t - \frac{(2n+1)L-x}{c}\right) \\ &- \frac{Kc}{E}\sum_{n=0}^{\infty}(-1)^n \left(t - \frac{(2n+1)L+x}{c}\right) H\left(t - \frac{(2n+1)L+x}{c}\right). \end{aligned}$$

Figure 7.6 shows a graph of the surface $u(x,t)$ corresponding to $K = 1, L = 4$, $c = 7$ and $E = 1$, for $0 \leq x \leq 4$, $0 \leq t \leq 20$.

Problems for Section 7.4

In each of problems 1 and 2, use the Laplace transform to solve the heat equation on the half line $x > 0$, subject to the given conditions.

1. $u(x,0) = e^{-x}, u(0,t) = 0, \lim_{x\to\infty} u(x,t) = 0.$

2. $u(x,0) = 0, u(0,t) = t^2, \lim_{x\to\infty} u(x,t) = 0.$

In each of problems 3 and 4, solve the heat equation on $[0, L]$ subject to the given conditions.

3. $u(x,0) = 0, u(0,t) = 0, u(L,t) = K =$ constant .

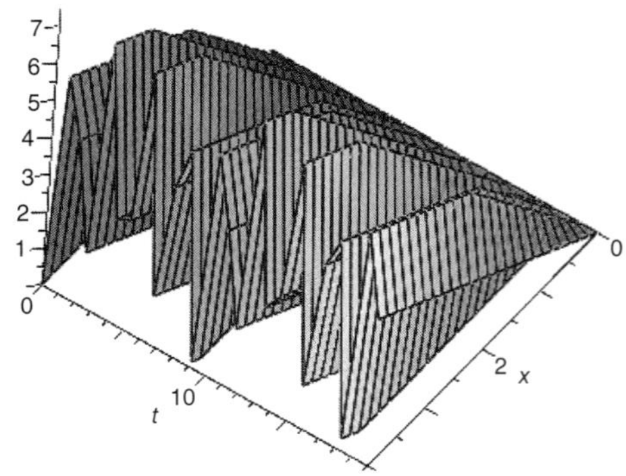

Figure 7.6: $u(x,t)$ for the elastic bar.

4. $u(x,0) = 1, u(0,t) = u(L,t) = 0.$

5. Use the Laplace transform to solve:

$$\begin{aligned} u_{tt} &= c^2 u_{xx} + Axt \text{ for } x > 0, t > 0, \\ u(x,0) &= u_t(x,0) = 0, \\ u(0,t) &= e^{-t}, \\ \lim_{x\to\infty} u(x,t) &= 0 \text{ for } t > 0. \end{aligned}$$

Chapter 8

First-Order Equations

Up to this point we have concentrated on problems involving second-order partial differential equations. These see wide application and also have a rich theory that leads naturally to further topics in partial differential equations.

Higher order differential equations occur, for example, in modeling buckling of beams in elasticity theory. In the other direction, first-order equations are also seen in many contexts. As an illustration, we will derive an equation known as a conservation law.

Imagine a fluid, or perhaps energy, flowing in some medium. As a " "thought device" constructed to analyze the flow, imagine a channel or tube in the medium, aligned along an $x-$axis, with uniform cross sections of area A (Figure 8.1). The fluid flows from left to right in the diagram.

Let $u(x,t)$ be the fluid density (amount per unit volume) at time t and point x. Choose any point a and any $\Delta x > 0$ and consider the segment of tube between a and $a + \Delta x$. The fluid is flowing into this segment over the cross section at a and out through the cross section at $a + \Delta x$. Let $\varphi(x,t)$ be the flux of the fluid at time t, so $A\varphi(x,t)$ is the quantity flowing across the section at x, at time t. Finally, let $S(x,t)$ account for sources or sinks (fluid being produced or lost, per unit area of the cross section at x at time t). We want a relationship between $u(x,t)$, $\varphi(x,t)$ and $S(x,t)$.

The principle of conservation that applies here is that, at any time t, the rate of change of the amount of fluid in the section of tube between a and $a + \Delta x$ at any time must be accounted for by the flow into the segment at a, out of the segment at $a + \Delta x$, and produced or lost in this segment. We may state this balancing equation as

$$\frac{d}{dt}\int_a^{a+\Delta x} Au(x,t)\,dx = A\varphi(a,t) - A\varphi(a+\Delta x,t) + \int_a^{a+\Delta x} AS(x,t)\,dx.$$

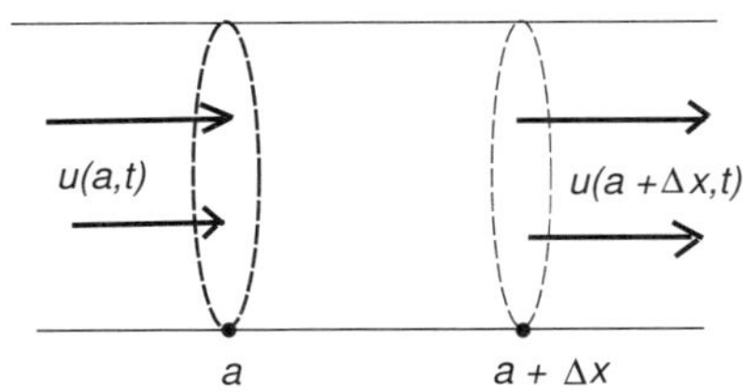

Figure 8.1: Derivation of a conservation law.

Write this equation as

$$\frac{d}{dt}\int_a^{a+\Delta x} Au(x,t)\,dx$$
$$= -\int_a^{a+\Delta x} A\varphi_x(x,t)\,dx + \int_a^{a+\Delta x} AS(x,t)\,dx.$$

Then

$$\int_a^{a+\Delta x} [u_t + \varphi_x(x,t) - S(x,t)]\,dx = 0.$$

This integral can be zero for every a and any positive Δx only if the integrand is zero:

$$u_t(x,t) + \varphi_x(x,t) = S(x,t).$$

This is a *conservation law* for the fluid.

With this as background, we will look at some of the ideas centered about linear and quasi-linear first-order partial differential equations.

8.1 Linear First-Order Equations

In two variables, the *linear first-order partial differential equation* has the form

$$a(x,y)u_x + b(x,y)u_y + c(x,y)u = f(x,y). \tag{8.1}$$

It is generally assumed that a, b, c, and f are continuous in some region of the plane, and that $a(x,y)$ and $b(x,y)$ are not both zero for the same point (x,y).

One approach to solving equation 8.1 is to find a change of variables

$$(x,y) \to (\xi,\eta),$$

which converts equation 8.1 to the simpler form

$$w_\xi + h(\xi,\eta)w = F(\xi,\eta), \tag{8.2}$$

where

$$w(\xi,\eta) = u(x(\xi,\eta), y(\xi,\eta)),$$

is obtained by substituting for x and y in terms of ξ and η in $u(x,y)$.

The reason equation 8.2 is easier to work with than 8.1 is that we can solve it as a linear ordinary differential equation in ξ, thinking of η as "constant". We will illustrate this point, then look for a transformation of equation 8.1 to equation 8.2.

Example 8.1 We will solve

$$w_\xi + 2\xi\eta w = \xi.$$

Treat η as a constant, because the only differentiation is with respect to ξ. Multiply the differential equation by the integrating factor

$$e^{\int 2\xi\eta\, d\xi} = e^{\xi^2\eta}$$

to obtain

$$w_\xi e^{\xi^2\eta} + 2\xi\eta w e^{\xi^2\eta} = \xi e^{\xi^2\eta}.$$

The left side is the partial derivative of a product, so the equation is

$$\frac{\partial}{\partial\xi}\left(e^{\xi^2\eta}w\right) = \xi e^{\xi^2\eta}.$$

Integrate with respect to ξ:

$$we^{\xi^2\eta} = \int \xi e^{\xi^2\eta}\, d\xi = \frac{1}{2\eta}e^{\xi^2\eta} + g(\eta),$$

in which $g(\eta)$ is the "constant of the integration" with respect to ξ. g can be any differentiable function, because the partial derivative of $g(\eta)$ with respect to ξ is zero. Finally, solve for w:

$$w(\xi,\eta) = \frac{1}{2\eta} + g(\eta)e^{-\xi^2\eta}.$$

It is routine to check by substitution that this is a solution for $\eta \neq 0$ and any differentiable function g.

Now we want to find a change of variables that produces equation 8.2 from 8.1. Begin the search with a general transformation relating (x, y) to (ξ, η):

$$\xi = \alpha(x, y), \eta = \beta(x, y).$$

for (x, y) in some region D of the plane.

In addition to transforming 8.1 to 8.2, we will require that this transformation be one to one, meaning that, at least in some part D of the plane, each (x, y) is associated with exactly one (ξ, η), and each (ξ, η) with exactly one (x, y). In this case it is possible, at least in theory, to invert the transformation and write (x, y) in terms of (ξ, η):

$$x = x(\xi, \eta), y = y(\xi, \eta).$$

For example, the mapping

$$\xi = x^2 - y^2, \eta = x + y^4$$

is not one to one because it sends $(1, 1)$ and $(1, -1)$ to the same point $(0, 2)$.

A convenient criterion for a mapping to be one to one is that

$$J(\alpha, \beta) = \begin{vmatrix} \alpha_x & \alpha_y \\ \beta_x & \beta_y \end{vmatrix} \neq 0$$

for all (x, y) in the region D of definition of the transformation. $J(\alpha, \beta)$ is the *Jacobian* of the transformation.

Now we can get to the details of choosing the transformation to get equation 8.2 from 8.1. Use chain rule differentiation to compute

$$\begin{aligned} u_x &= w_\xi \xi_x + w_\eta \eta_x, \\ u_y &= w_\xi \xi_y + w_\eta \eta_y. \end{aligned}$$

Substitute these into equation 8.1 to write it in terms of ξ and η:

$$a(w_\xi \xi_x + w_\eta \eta_x) + b(w_\xi \xi_y + w_\eta \eta_y) + cw = f.$$

Rearrange terms in this equation to obtain

$$(a\xi_x + b\xi_y)w_\xi + (a\eta_x + b\eta_y)w_\eta + cw = f. \tag{8.3}$$

This is nearly in the form of equation 8.2 if we can choose η so that

$$a\eta_x + b\eta_y = 0$$

for (x, y) in D. Assuming that $\eta_y \neq 0$, this requires that

$$\frac{\eta_x}{\eta_y} = -\frac{b}{a}.$$

If there is such a function $\eta(x, y)$, then setting $\eta(x, y) = k$, with k an arbitrary constant, gives us

$$d\eta = \eta_x \, dx + \eta_y \, dy = 0,$$

so

$$\frac{dy}{dx} = -\frac{\eta_x}{\eta_y} = \frac{b}{a}.$$

This means that the place to look for the function $\eta(x, y)$ is in solutions of the ordinary differential equation

$$\frac{dy}{dx} = \frac{b(x, y)}{a(x, y)}. \tag{8.4}$$

This differential equation is called the *characteristic equation* of the partial differential equation 8.1. If $\eta(x, y) = k$ defines solutions of this equation we call the curves $\eta(x, y) = c$ in the $x, y-$plane *characteristic curves* of 8.1.

Thus far we have a way of looking for $\eta(x, y)$. For ξ, we will make a simple choice

$$\xi = \alpha(x, y) = x.$$

This may appear arbitrary, but the idea is to try something simple that will work, and here we observe that, with $\xi = x$,

$$J = \begin{vmatrix} 1 & 0 \\ \eta_x & \eta_y \end{vmatrix} = \eta_y \neq 0,$$

so this transformation is one to one.

Up to this point we have set $\xi = x$, and η is obtained by finding the general solution $\eta(x, y) = k$ of the characteristic equation

$$\frac{dy}{dx} = \frac{b(x, y)}{a(x, y)}.$$

With these choices, equation 8.3 becomes

$$a(x, y)w_\xi + c(x, y)w = f(x, y).$$

All we remains is to substitute for x and y in terms of ξ and η in $a(x, y), c(x, y)$ and $f(x, y)$ to obtain:

$$A(\xi, \eta)w_\xi + C(\xi, \eta)w = F(\xi, \eta).$$

Finally, restricting the transformation to points (ξ, η) such that $A(\xi, \eta) \neq 0$, divide the last equation by $A(\xi, \eta)$ to obtain the reduced form of equation 8.2.

In summary, given equation 8.1, find the solution $\eta(x, y) = k$ of the characteristic equation

$$\frac{dy}{dx} = \frac{b(x, y)}{a(x, y)}.$$

Use this function to define the transformation

$$\xi = x, \eta = \eta(x, y).$$

Example 8.2 We will solve

$$u_x + \cos(x)u_y + u = xy.$$

Begin with the characteristic equation to find a transformation of this equation to the form 8.2. Solve

$$\frac{dy}{dx} = \frac{b}{a} = \cos(x).$$

This has a general solution defined by

$$y - \sin(x) = k.$$

Graphs of $y - \sin(x) = k$ are the characteristics of this partial differential equation. The transformation we want is

$$\xi = x, \eta = y - \sin(x).$$

This has Jacobian

$$\begin{vmatrix} 1 & 0 \\ -\cos(x) & 1 \end{vmatrix} = 1,$$

so this transformation is one to one over the entire plane.

Now

$$u_x = w_\xi \xi_x + w_\eta \eta_x = w_\xi - w_\eta \cos(x)$$

and

$$u_y = w_\xi \xi_y + w_\eta \eta_y = w_\eta.$$

Substitute these into the differential equation to obtain

$$w_\xi - \cos(x)w_\eta + \cos(x)w_\eta + w = xy,$$

or

$$w_\xi + w = xy.$$

We must put xy in terms of ξ and η. Put $x = \xi$ and $y = \eta + \sin(x) = \eta + \sin(\xi)$ to get:

$$w_\xi + w = \xi[\eta + \sin(\xi)].$$

This is the transformed differential equation. Because the coefficient of w is 1, multiply the differential equation by

$$e^{\int d\xi} = e^\xi$$

to obtain

$$e^\xi w_\xi + e^\xi w = \xi e^\xi \eta + \xi e^\xi \sin(\xi).$$

This is

$$\frac{\partial}{\partial \xi}\left(e^{\xi} w\right) = \xi e^{\xi} \eta + \xi e^{\xi} \sin(\xi).$$

Think of this as a differential equation in ξ and integrate with respect to this variable:

$$\begin{aligned} e^{\xi} w &= \eta \int \xi e^{\xi}\, d\xi + \int \xi e^{\xi} \sin(\xi)\, d\xi + g(\eta) \\ &= \eta e^{\xi}(\xi - 1) + \frac{1}{2} \xi e^{\xi}(\sin(\xi) - \cos(\xi)) + \frac{1}{2} e^{\xi} \cos(\xi) + g(\eta). \end{aligned}$$

Here g is any differentiable function of one variable. Divide this equation by e^{ξ} to obtain the solution $w(\xi, \eta)$ of the transformed equation. For the solution $u(x, y)$ of the original partial differential equation, substitute $\xi = x$ and $\eta = y - \sin(x)$:

$$\begin{aligned} u(x, y) = (y - \sin(x))(x - 1) + \frac{1}{2} x(\sin(x) - \cos(x)) \\ + \frac{1}{2} \cos(x) + e^{-x} g(y - \sin(x)). \end{aligned}$$

Example 8.3 We will solve the general constant coefficient homogeneous linear first-order partial differential equation

$$a u_x + b u_y + c u = 0.$$

The characteristic equation is

$$\frac{dy}{dx} = \frac{b}{a},$$

with general solution defined by the equation

$$bx - ay = k,$$

with k any number. Define the transformation

$$\xi = x, \eta = bx - ay.$$

A routine calculation transforms the partial differential equation to

$$w_{\xi} + \frac{c}{a} w = 0.$$

Multiply this equation by

$$e^{\int (c/a)\, d\xi} = e^{c\xi/a}$$

to get

$$e^{c\xi/a} w_{\xi} + \frac{a}{c} e^{c\xi/a} w = 0,$$

which is the same as

$$\frac{\partial}{\partial \xi}\left(e^{c\xi/a} w\right) = 0.$$

Integrate with respect to ξ to get

$$e^{c\xi/a}w = g(\eta),$$

so

$$w(\xi,\eta) = e^{-c\xi/a}g(\eta),$$

in which g is an arbitrary differentiable function of one variable. Finally, transform this solution for $w(\xi,\eta)$ to the solution $u(x,y)$:

$$u(x,y) = e^{-cx/a}g(bx - ay).$$

In these examples, we found a solution containing an arbitrary function (arising from the integration with respect to one of the variables of the transformed equation). Such a solution is called a *general solution* of the partial differential equation. This is analogous to general solutions of first-order ordinary differential equations, which contain an arbitrary constant.

Problems for Section 8.1

In each of problems 1–12, find the general solution of the partial differential equation.

1. $3u_x + 5u_y - xyu = 0.$
2. $u_x - u_y + yu = 0.$
3. $u_x + 4u_y - xu = x.$
4. $-2u_x + u_y - yu = 0.$
5. $xu_x + u = x.$
6. $x^2u_x - 2u_y - xu = x^2.$
7. $u_x - xu_y = 4.$
8. $x^2u_x + xyu_y + xu = x - y.$
9. $u_x + u_y - u = y.$
10. $u_x - y^2u_y - yu = 0.$
11. $u_x + 7u_y + xu = 0.$
12. $xu_x + yu_y + 2 = 0.$
13. Find the general solution of

$$u_x + \alpha(y-1)u_y = \frac{1}{2}\beta f(x)(y-1)u,$$

in which α and β are real numbers and f is continuous on the real line. Use the general solution to find the solution satisfying

$$u(0,y) = y^n,$$

in which n is a nonnegative integer.

8.2 The Significance of Characteristics

In Section 8.1 we used the characteristic equation of of a linear, first-order partial differential equation to find a transformation to a simpler form that we could solve. In this section we will look more closely at characteristics and their significance in specifying solutions, or in the existence of a solution satisfying a given condition.

An example will be instructive. The characteristic equation of

$$2u_x + 3u_y + 8u = 0$$

is

$$\frac{dy}{dx} = \frac{3}{2},$$

so the characteristics are graphs of $3x - 2y = k$. We find that the differential equation has the general solution

$$u(x, y) = e^{-4x} g(3x - 2y).$$

Now suppose we want to pick out a unique solution satisfying a given condition. Specifying the value of $u(x, y)$ at a particular point will not be enough, because this cannot completely determine the function g.

For a first-order partial differential equation, we will consider instead the problem of finding a solution that takes on prescribed values along a curve Γ in the plane.

See how this works for the case that Γ is the x−axis. Say we specify that

$$u(x, 0) = \sin(x) \text{ if } x \text{ is real.}$$

Now we need g so that

$$u(x, 0) = e^{-4x} g(3x) = \sin(x),$$

and we can achieve this by putting

$$g(x) = e^{4x/3} \sin(x/3).$$

With this g, we have

$$\begin{aligned} u(x, y) &= e^{-4x} g(3x - 2y) = e^{-4x} e^{4(3x-2y)/3} \sin((3x - 2y)/3) \\ &= e^{-8y/3} \sin\left(x - \frac{2}{3}y\right). \end{aligned}$$

This function satisfies the partial differential equation, and also the condition that $u(x, 0) = \sin(x)$. Figure 8.2 shows part of the surface $u = u(x, y)$.

Specifying values of the solution along this curve has completely determined g, hence a unique solution.

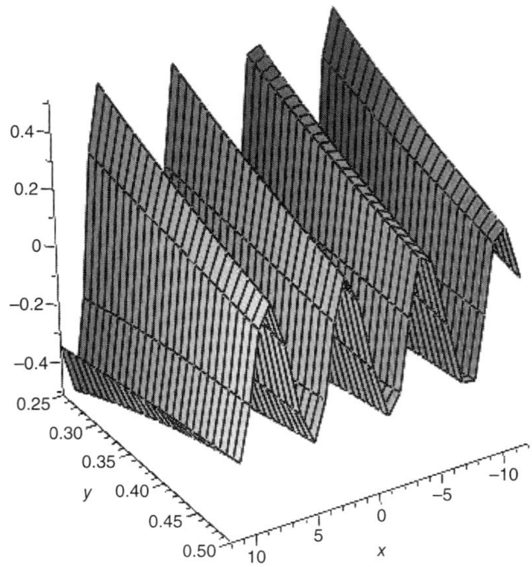

Figure 8.2: Surface $u(x, y)$ with $u(x, 0) = \sin(x)$.

Try another curve. Let Γ be the line $y = x$. We will try to find a solution taking on the value x^4 at points on this curve:

$$u(x, x) = x^4.$$

Now we must choose g so that

$$u(x, x) = e^{-4x} g(3x - 2x) = e^{-4x} g(x) = x^4,$$

so choose

$$g(x) = x^4 e^{4x}.$$

The solution for this g is

$$\begin{aligned} u(x, y) &= e^{-4x} g(3x - 2y) = e^{-4x}(3x - 2y)^4 e^{4(3x-2y)} \\ &= e^{-4x}(3x - 2y)^4 e^{12x} e^{-8y} = e^{8(x-y)}(3x - 2y)^4. \end{aligned}$$

This is the unique solution of the differential equation satisfying $u(x, y) = x^4$ along the curve $\Gamma : y = x$. Figure 8.3 shows part of a graph of this solution as a surface in $x, y, u-$space.

Can we find a unique solution assuming given values on any (reasonable) curve in the plane? The answer is no. Take Γ to be the straight line $3x - 2y = 1$. Suppose we require that, for (x, y) on this curve, $u(x, y) = x^2$.

On Γ, $y = (3x - 1)/2$, so we must choose g so that,

$$u\left(x, \frac{1}{2}(3x - 1)\right) = x^2.$$

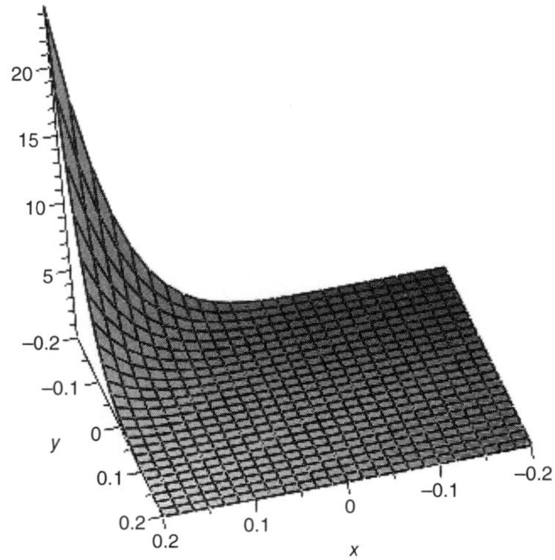

Figure 8.3: Solution surface with $u(x, y) = x^4$ along $\Gamma : y = x$.

This requires that

$$u\left(x, \frac{1}{2}(3x-1)\right) = e^{-4x} g\left(3x - (3x-1)\right) = e^{-4x} g(1) = x^2.$$

But then we must choose g so that

$$g(1) = x^2 e^{4x},$$

with a variable quantity equal to a constant. This is impossible.

Why did some choices of Γ allow us to solve the problem, while another choice led to an absurdity? In general, if information is specified along a non-characteristic, the problem has a unique solution (assuming reasonable conditions on the curve, such as piecewise smoothness). If the curve is a characteristic, then there may be no solution, or there may be many solutions. As the next example shows, either existence or uniqueness may fail with data given along a characteristic.

Example 8.4 We will examine solutions of

$$xu_x + 2x^2 u_y - u = x^2 e^x.$$

The characteristic equation is

$$\frac{dy}{dx} = \frac{2x^2}{x} = 2x,$$

so the characteristics are graphs of $y = x^2 + k$. This is a family of parabolas. Use the transformation

$$\xi = x, \eta = y - x^2$$

to obtain, after some manipulation,

$$w_\xi - \frac{1}{\xi}w = \xi e^\xi.$$

Multiply this by $e^{\int(-1/\xi)\,d\xi}$, which is $1/\xi$, to obtain

$$\frac{1}{\xi}w_\xi - \frac{1}{\xi^2}w = e^\xi$$

or

$$\frac{\partial}{\partial\xi}\left(\frac{1}{\xi}w\right) = e^\xi.$$

Upon integrating with respect to ξ, we have

$$\frac{1}{\xi}w = e^\xi + g(\eta).$$

The general solution for $w(\xi,\eta)$ is

$$w(\xi,\eta) = \xi e^\xi + \xi g(\eta).$$

Therefore the general solution of the original partial differential equation is

$$u(x,y) = xe^x + xg(y - x^2).$$

Now we will attempt to find solutions taking on given values along given curves.

First take Γ to be the parabola $y = x^2 + 4$. This is a characteristic. Suppose we want

$$u(x,y) = \sin(x) \text{ on } \Gamma.$$

Because $y - x^2 = 4$ on Γ, we need g so that

$$u(x,y) = u(x, x^2+4) = xe^x + xg(4) = \sin(x).$$

But this would make

$$\frac{\sin(x) - xe^x}{x} = g(4),$$

impossible for any choice of g because the left side is not constant. This problem (for the given Γ and data on Γ) has no solution.

Next, suppose we want a solution such that

$$u(x,y) = xe^x - x \text{ on the characteristic } y = x^2 + 4.$$

Now we need

$$u(x, x^2+4) = xe^x + xg(4) = xe^x - x,$$

and we can do this if $g(4) = -1$. There are infinitely many choices of $g(x)$ having $g(4) = -1$, and each one produces a different solution $u(x,y)$ having the desired values on this curve. This problem has infinitely many solutions. With Γ chosen as this characteristic, existence is fine, but uniqueness fails.

Finally, suppose we want $u(x, y) = \cos(x)$ along the noncharacteristic parabola $y = x^2 + 4x$. Now we need

$$u(x, x^2 + 4x) = xe^x + xg(4x) = \cos(x),$$

so we must have

$$g(4x) = \frac{\cos(x) - xe^x}{x}$$

for $x \neq 0$. This is easy to do. Choose

$$g(x) = \frac{4}{x}\left(\cos(x/4) - \frac{x}{4}e^{x/4}\right).$$

The solution of this problem, say for $x > 0$, is

$$\begin{aligned} u(x, y) &= xe^x + xg(y - x^2) \\ &= xe^x + \frac{4x}{y - x^2}\left[\cos\left(\frac{y - x^2}{4}\right) - \frac{1}{4}(y - x^2)e^{(y-x^2)/4}\right]. \end{aligned}$$

The problem of finding a solution of the partial differential 8.1 assuming given values on a given curve, is called a *Cauchy problem* for this equation. The condition that the solution must satisfy is called *Cauchy data* for the problem. The examples suggest that we can expect a unique solution of a Cauchy problem if the curve is noncharacteristic (and, say, smooth). If the curve is a characteristic, there may be no solution or infinitely many solutions.

Problems for Section 8.2

In each of problems 1–6, determine the characteristics and the general solution of the partial differential equation. For each set of Cauchy data given, attempt to find a particular solution of the Cauchy problem. Determine if there is a unique solution, no solution, or infinitely many solutions.

1. $3yu_x - 2xu_y = 0$
   (a) $u(x, y) = x^2$ on the line $y = x$.
   (b) $u(x, y) = 1 - x^2$ on the line $y = -x$.
   (c) $u(x, y) = 2x$ on the ellipse $2x^2 + 3y^2 = 4$.

2. $u_x - 6u_y = y$
   (a) $u(x, y) = e^x$ on the line $y = -6x + 2$.
   (b) $u(x, y) = 1$ on the line $y = -x^2$.
   (c) $u(x, y) = -4x$ on the line $y = -6x$.

3. $4yu_x + 8u_y - u = 1$
   (a) $u(x, y) = \cos(x)$ on the line $y = 3x$.
   (b) $u(x, y) = x$ on the line $y = 2x$.
   (c) $u(x, y) = 1 - x$ on the parabola $y = x^2$.

4. $-4yu_x + u_y - yu = 0$
   (a) $u(x, y) = x^3$ on the line $x + 2y = 3$.
   (b) $u(x, y) = -y$ on the the parabola $y^2 = x$.
   (c) $u(x, y) = 2$ on the parabola $x + 2y^2 = 1$.

5. $yu_x + x^2u_y = xy$
   (a) $u(x, y) = 4x$ on the curve $y = \frac{1}{3}x^{3/2}$.
   (b) $u(x, y) = x^2$ on the curve $3y^2 = 2x^3$.
   (c) $u(x, y) = \sin(x)$ on the line $y = 0$.

6. $y^2u_x + x^2u_y = y^2$
   (a) $u(x, y) = x$ on the line $y = 4x$.
   (b) $u(x, y) = -2y$ on $y^2 = x^2 - 2$.
   (c) $u(x, y) = y^2$ on the line $y = -x$.

8.3 The Quasi-Linear Equation

The first-order partial differential equation

$$f(x, y, u)u_x + g(x, y, u)u_y = h(x, y, u) \tag{8.5}$$

is called a *quasi-linear equation.*

For this equation, the characteristics are defined as solutions of the system

$$\frac{dx}{dt} = f(x, y, u), \frac{dy}{dt} = g(x, y, u), \frac{du}{dt} = h(x, y, u). \tag{8.6}$$

Solutions of this system are curves in $x, y, u-$space. We will show that a solution $u(x, y)$ of equation 8.5 can be thought of as a surface in $u = u(x, y)$ in $x, y, u-$space, and that this surface is made up of the differential equation's characteristic curves. This idea can sometimes be exploited to solve a Cauchy problem for a quasi-linear equation.

To get some feeling for this assertion, we will need two facts.

Fact 1 Suppose $u = \varphi(x, y)$ is a solution of equation 8.5, defining a surface Σ as the graph of $u = \varphi(x, y)$. Let $P_0 : (x_0, y_0, u_0)$ be a point on Σ, so $u_0 = \varphi(x_0, y_0)$. Then the characteristic passing through P_0 lies entirely on Σ.

To see why this is true, let the characteristic through P_0 have parametric equations

$$x = x(t), y = y(t), z = z(t).$$

Because P_0 is on this curve, there is some t_0 such that

$$x_0 = x(t_0), y_0 = y(t_0), u_0 = u(t_0).$$

But because this curve is a characteristic, we can use equations 8.6 to compute

$$\frac{d}{dt}\varphi(x(t), y(t)) = \varphi_x \frac{dx}{dt} + \varphi_y \frac{dy}{dt}$$
$$= \varphi_x f(x, y, u) + \varphi_y g(x, y, u) = h(x, y, u) = \frac{du}{dt}.$$

This means that

$$u(t) = \varphi(x(t), y(t)) + k$$

for some constant k. But $k = 0$, because we know that

$$u_0 = u(t_0) = \varphi(x(t_0), y(t_0)).$$

Therefore

$$u(t) = \varphi(x(t), y(t))$$

and the characteristic lies on Σ.

Fact 2 If we start with a noncharacteristic curve Γ and construct, through each point of Γ, the characteristic through that point, the resulting surface Σ is the graph of a solution of the partial differential equation.

To see why this is true, suppose Σ is the graph of $u = \varphi(x, y)$. We want to show that φ is a solution of equation 8.5.

Suppose Γ is parametrized by

$$x = x(s), y = y(s), z = z(s).$$

At any point (x, y, u) on Σ,

$$\frac{dx}{ds} = f(x, y, u), \frac{dy}{ds} = g(x, y, u), \frac{du}{ds} = h(x, y, u)$$

because Σ is made up of characteristics. Then

$$\frac{du}{ds} = h(x, y, u) = \varphi_x \frac{du}{ds} + \varphi_y \frac{dy}{ds} = f\varphi_x + g\varphi_y,$$

and this makes $\varphi(x, y)$ a solution of the partial differential equation.

These two observations suggest the *method of characteristics* for solving a Cauchy problem when the partial differential equation is quasi-linear. Given Cauchy data along a curve Γ which is *not* a characteristic, construct the characteristic through each point of Γ. These characteristics will form a surface that is the graph of the solution (Figure 8.4).

This strategy also suggests why we do not want to specify Cauchy data along a characteristic. If Γ is a characteristic, then the characteristics through the points of Γ are all just Γ itself, and Σ is a curve, not a surface.

We will illustrate the method of characteristics.

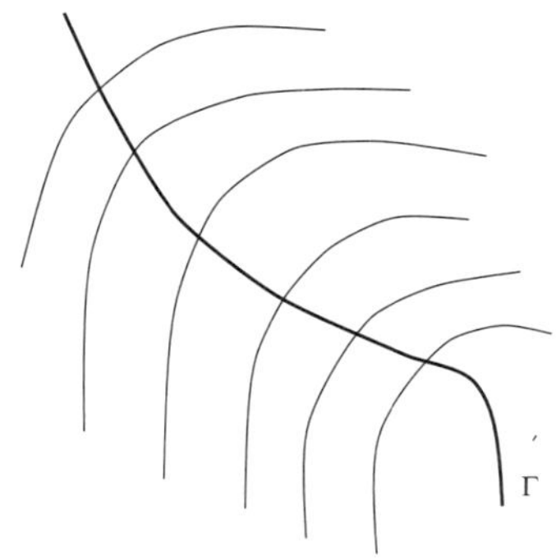

Figure 8.4: Characteristics through points of Γ form a solution surface containing Γ.

Example 8.5 We will find the solution of

$$xu_x + yu_y = \sec(u)$$

that passes through the curve

$$\Gamma : x = s^2, y = \sin(s), u = 0.$$

By this we mean that the curve is on the graph of the solution $u(x, y)$, if graphed as a surface in $x, y, u-$space. We can also write this condition as

$$u(s^2, \sin(s)) = 0.$$

The characteristics of the partial differential equation satisfy

$$\frac{dx}{dt} = x, \frac{dy}{dt} = y, \frac{du}{dt} = \sec(u).$$

These ordinary differential equations have general solutions

$$x = Ae^t, y = Be^t, \sin(u) = t + C.$$

Here we are using s as the parameter in the equations defining Γ, and t for the variable on characteristics, as a way of keeping track of these quantities as we proceed.

We want to construct the characteristic curve through each point of Γ. Suppose a characteristic passes through Γ at $P : (s^2, \sin(s), 0)$. Then

$$x = A = s^2, B = \sin(s), \text{ and } \sin(0) = 0 = 0 + C.$$

Therefore,

$$x = s^2, y = \sin(s), \text{ and } u = 0.$$

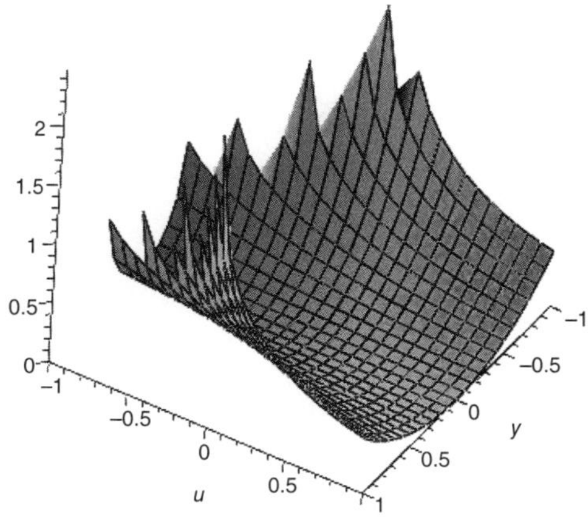

Figure 8.5: Solution surface in example 8.5.

Then

$$x = s^2 e^t, u = \sin(s)e^t \text{ and } \sin(u) = t.$$

These are the parametric equations (with t as parameter) of the characteristic curve passing through P on Γ.

We want to eliminate s and t from these equations. First, $t = \sin(u)$, so

$$y = \sin(s)e^{-\sin(u)}$$

so

$$s = \arcsin(ye^{-\sin(u)}).$$

Using the fact that $x = e^t s^2$, we have

$$x = e^{\sin(u)} \left[\arcsin\left(ye^{-\sin(u)}\right)\right]^2.$$

This equation implicitly defines the solution $u(x, y)$, and it is easy to check by substitution that $u(s^2, \sin(s)) = 0$, so this surface does contain Γ.

Figure 8.5 is a graph of part of this surface.

Example 8.6 We will find the solution of

$$yu_x - xu_y = e^u$$

that contains the curve $\Gamma : y = \sin(x), y = 0$. This means that we want a solution satisfying

$$u(x, \sin(x)) = 0.$$

The characteristics are specified by

$$\frac{dx}{dt} = y, \frac{dy}{dt} = -x, \frac{du}{dt} = e^u.$$

From the first two of these equations,

$$\frac{dy}{dx} = -\frac{x}{y}.$$

Then

$$x\,dx + y\,dy = 0,$$

so

$$x^2 + y^2 = k^2.$$

Graphs of this equation are circles about the origin, leading us to look at sine and cosine solutions. If we set

$$x(t) = a\cos(t) + b\sin(t),$$

then

$$y = \frac{dx}{dt} = b\cos(t) - a\sin(t),$$

with a and b constants. Finally, from $du/dt = e^u$ we have

$$-e^{-u} = t + c.$$

The characteristics therefore have parametric equations

$$x(t) = a\cos(t) + b\sin(t), y(t) = b\cos(t) - a\sin(t), e^{-u} = c - t.$$

Parametrize Γ as

$$x = s, y = \sin(s), u = 0,$$

using s as parameter on Γ.

We are now in a position to determine a characteristic curve through each point $P : (s, \sin(s), 0)$ of Γ. We may suppose (by scaling the parameter if necessary) that the characteristic passes through P when $t = 0$. At $t = 0$,

$$x = a = s, y = b = \sin(s) \text{ and } e^0 = 1 = c - 0.$$

Therefore the characteristic passing through Γ has parametric equations

$$x = s\cos(t) + \sin(s)\sin(t), y = \sin(s)\cos(t) - s\sin(t), e^{-u} = 1 - t.$$

Eliminate t and s from these equations. From the first two parametric equations,

$$s = x\cos(t) - y\sin(t) \text{ and } \sin(s) = y\cos(t) + x\sin(t).$$

Therefore

$$\sin(x)\cos(t) - y\sin(t) = y\cos(t) + x\sin(t). \tag{8.7}$$

But $e^{-u} = 1 - t$ implies that

$$t = 1 - e^{-u}.$$

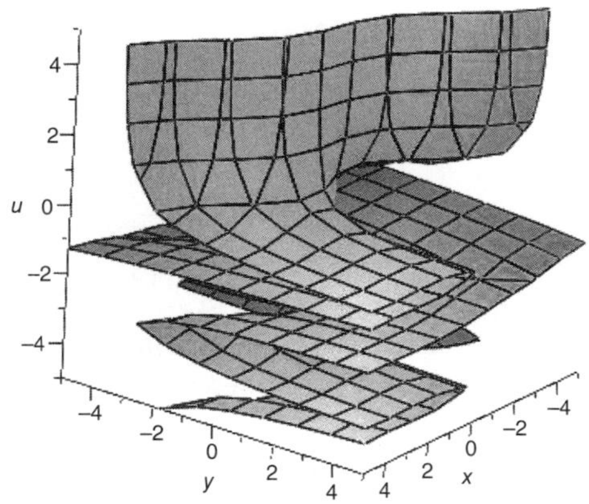

Figure 8.6: Solution surface in example 8.6, for $-3 \leq x, y, u \leq 3$.

Upon substituting this for t in equation 8.7, we have

$$\sin(x)\cos\left(1-e^{-u}\right) - y\sin\left(1-e^{-u}\right) = y\cos\left(1-e^{-u}\right) + x\sin\left(1-e^{-u}\right).$$

This equation implicitly defines $u = u(x, y)$, the solution of the Cauchy problem with data given on Γ. Figure 8.6 shows a graph of part of this surface.

Problems for Section 8.3

In each of problems 1–10, use the method of characteristics to find a solution of the Cauchy problem consisting of the partial differential equations and the curve the solution surface is to contain. This solution may be implicitly defined.

1. $xu_x + yu_y = \sec(u), \Gamma$ is given by equations $y = x^2, u = 0$.

2. $u_x - xu_y = 4, \Gamma$ is given by $y = 4x, u = 0$.

3. $u_x - y^2u_y = 4, \Gamma$ is given by $y = x^2 + 2, u = 0$.

4. $u_x - y^3u_y = \sec(u), \Gamma$ is given by $y = x^2, u = 0$.

5. $u_x + yu_y = u, \Gamma$ is given by $y = 1 - x, u = 1$.

6. $u_x - y^2u_y = \cos(u), \Gamma$ is given by $x = y^2, u = 0$.

7. $u_x - u_y = u^2, \Gamma$ is given by $y = 2x - 1, u = 4$.

8. $x^3u_x - yu_y = u, \Gamma$ is given by $x = y^2 - 1, u = 1$.

9. $u_x - y^2u_y = u, \Gamma$ is given by $y = 1 - x^2, u = 2$.

10. $xu_x + u_y = e^u, \Gamma$ is given by $y = x = 1, u = 0$.

11. Use the method of characteristics to derive the implicitly defined solution

$$u(x,y) = f(x - yu(x,y))$$

of the problem

$$uu_x + u_y = 0; u(x,0) = f(x).$$

Think of Γ as the curve $x = s, y = 0, u = f(s)$.

12. Show that, if f is a continuously differentiable function of one variable, the quasi-linear first-order partial differential equation

$$u_y = f(u)u_x$$

has a solution defined implicitly by

$$u(x,y) = \varphi(x + f(u)y)$$

in which φ can be any continuously differentiable function of one variable. Use this idea to determine (perhaps implicitly) a solution of each of the following partial differential equations.

(a) $u_y = ku_x$, with k a positive constant.

(b) $u_y = uu_x$.

(c) $u_y = \cos(u)u_x$.

(d) $u_y = e^u u_x$.

(e) $u_y = u\sin(u)u_x$.

Chapter 9

End Materials

9.1 Notation

Notation in this book is for the most part standard. This list includes notation that might be less familiar. Page references are included for convenience.

$f(c-), f(c+)$ left and right limits of $f(x)$ at $x = c$ (17)

$S(x, y, z; t)$ sphere of radius t about (x, y, z) (141)

∇ del (or gradient) operator (147)

$\nabla^2 u$ Laplacian of u (147)

R^n space of $n-$vectors (165)

$B(\mathbf{a}, r)$ open ball of radius r about $\mathbf{a}$ (166)

$x \in A$ x is an element of a set A (165)

$\{x \mid x \text{ has property } P\}$ the set of objects having property P (166)

$\overline{B(\mathbf{a}, r)}$ closed ball of radius r about $\mathbf{a}$ (167)

∂A boundary of A (167)

$\overline{A}$ closure of A (170)

$B \subset A$ B is a subset of A (166)

$\varnothing$ the empty set (167)

$B - A$ set difference consisting of all elements of B not in A (166)

$C^2(A)$ set of functions that are continuous on A with continuous first and second partial derivatives (172)

$\partial u/\partial n$ normal derivative of u (directional derivative along a unit normal vector) (172)

$f_{K,v}$ a function that is harmonic on K and agrees with v on ∂K. (203)

ω_Q barrier function (209)

$PS[a,b]$ space of piecewise smooth functions on $[a,b]$ (260)

$f \cdot g$ dot product of f and g (262)

$\| f \|$ norm of f (262)

$\Gamma(x)$ gamma function (267)

$J_\nu(x)$ Bessel function of the first kind of order ν (268)

$Y_\nu(x)$ Bessel function of the second kind of order ν (269)

$P_n(x)$ Legendre polynomial of degree n (290)

$u_{mn}(\theta,\varphi), v_{mn}(\theta,\varphi)$ spherical harmonics (304)

$\mathcal{F}[f]$, or $\widehat{f}$ Fourier transform of f (307)

$f^{(n)}(x)$ nth derivative of $f(x)$ (309)

$\mathcal{F}[f]$ Laplace transform of f (332)

$f * g$ Fourier (311) or Laplace (333) transform convolution of f with g

9.2 Use of MAPLE

In learning to use computational software, such as MAPLE, it is a good practice to experiment with it and acquire some feeling for entering commands, as well as an expectation of the responses these commands will produce. In addition, it is necessary to know some of the particular code or types of instructions that the package will recognize, as well as some of its capabilities and limitations.

9.2.1 Numerical Computations, Functions, Solving Equations, and Graphing

MAPLE performs numerical calculations as we might expect, with an asterisk denoting products and a wedge ∧ denoting powers. π is entered as Pi, and the imaginary unit i as I.

In performing numerical computations, the evalf command will return decimal approximations. For example,

(3∧4)*Pi

will return the result 81π. For a decimal expression of this product, enter

evalf((3∧4)*Pi)

to obtain 254.4690050.

Enter a function into MAPLE by using the := command. For example, to define $g(x) = 4x^3 - \sin(4x)$, enter

g:=x → 4*x∧3 - sin(4*x)

The arrow can be entered from a palette, or entered as a dash followed by a "greater than" symbol.

For a graph of a function $f(x)$ that has been entered into MAPLE, type

plot(f(x),x=-a..b,color=black)

in which a and b must be specific numbers. The command to draw the graph in black is optional.

If we want to plot several functions, say f, g, and h, define these functions in the program, then use curly brackets to enclose the functions in the plot command:

plot($\{f(x), g(x), h(x)\}$, x = a..b, color = black)

If the color instruction is omitted, the three graphs will be drawn in different colors.

We can enter the function directly into the plot command if we wish. For example

plot(x∧4 - cos(3*x) - 1,x=-2..Pi)

will graph $x^4 - \cos(3x) - 1$ for $-2 \leq x \leq \pi$.

Sometimes we want to enter a function that has one or more discontinuities. In such a case, enter the definition of the function on each of the relevant intervals as follows. Say, for example,

$$h(x) = \begin{cases} -1 & \text{for } -2 \le x < 3, \\ \sin(2x) & \text{for } 3 \le x < \pi, \\ e^x & \text{for } \pi < x < 5, \\ 0 & \text{for } x < -2 \text{ or } x > 5. \end{cases}$$

Enter

h:= x $\rightarrow$ piecewise($x < -2, 0, x < 3, -1, x <$ Pi, sin(2*x),$x < 5$, exp(x), 0)

The idea in entering the function is to imagine walking to the right along the real line, specifying the function value on each interval up to the next jump discontinuity.

The solve command can be used to solve some equations. For example, to solve $\cos(3x) + \sin(x) = 1/2$, enter

solve(cos(3*x) + sin(x) = 1/2,x)

Notice that the command includes the name of the variable for which a solution is sought. This is because some equations have other symbols in them, and MAPLE does not know your preference in naming variables. In this example, MAPLE will return a long and intimidating expression. For an evaluation of a solution in decimal form, precede the solve command with evalf:

evalf(solve(cos(3*x) + sin(x) = 1/2,x))

It is important in MAPLE commands to reconcile parentheses. If, for example, the last parenthesis is omitted, then we would have more left than right parentheses in the expression, and MAPLE will return an error message. As written, this evaluation of a solve command will return five solutions, three real and two complex:

$$0.5215987758, 2.617993878, 1.401725865, -2.149280954294 \pm 0.28576533812400i$$

In solving equations, particularly transcendental ones, it is often helpful to use a graph to get some sense of how many roots there might be, and how they are distributed. In this example, use, say,

plot(cos(3*x) + sin(x) - 1/2,x=-5..5)

Real solutions of $\cos(3x) + \sin(x) = 1/2$ occur where the graph crosses the $x-$axis. The use of $-5 \le x \le 5$ here is arbitrary, and various intervals can be used to get some sense of how many real roots there may be, and even approximate locations.

Instead of evalf and solve, we can also go directly to a numerical approximation of solutions by using the fsolve command. For example, to solve $\sqrt{2}\cos(2x) - \sin(3x) - 1/4 = 0$, enter fsolve(sqrt(2)*cos(2*x) - sin(3*x) - 1/4,x)

to obtain 0.3392673756. Is this the only root? We can try graphing to get some insight. However, we can also try to solve (approximately) for another root by dividing this function by $(x - 0.3392673756)$. If the first root found is simple, then this new function does not have this root, and the fsolve command produces the new root -0.8734862440 of the original equation. To try for another root, divide again, this time by $(x + 0.8734862440)$ and apply fsolve to this function to obtain -2.268106410.

We can be a little more precise in seeking roots of an equation by specifying a range of values in which we want roots found (if there are any there). For example, to search for solutions of $\sin(x) + \cos(x) - 1/2 = 0$ on $[\pi, 5\pi]$, enter

fsolve(sin(x) + cos(x) - 1/2,x,Pi..5*Pi)

This returns the root 8.27801673. If we change the range to, say, -8 .. -2, we obtain the root -6.707216347.

To integrate $f(x)$ over $[a, b]$, enter

int(f(x),x=a..b)

For an indefinite integral, enter

int(f(x),x)

It is necessary that the variable of integration be specified, because $f(x)$ may involve other symbols. Improper integrals can sometimes be evaluated by using infinity as one of the limits, for example,

int(exp(-3*x),x=0..infty)

will evaluate $\int_0^\infty e^{-3x}\,dx$.

MAPLE contains a sum command. If $k(n)$ is a number defined for, say, positive integers n, then

sum(k(n),n = 1..N)

will return $\sum_{n=1}^{N} k(n)$. Here N must be specified as some positive integer. If we want, say, to graph the $20th$ partial sum of the series

$$\sum_{n=1}^{\infty} \frac{2}{n}(-1)^n \sin(nx)$$

for $0 \leq x \leq \pi$, we can first enter this as a function, using the sum command, in which the $20th$ partial sum is called $S(x)$:

S:= $x \rightarrow$ sum(((2/n)*((-1)∧ n))*sin(n*x),n=1..20)

Now this partial sum can be graphed by:

plot(S(x),x=0..Pi)

For a list of all plotting (graphing) routines, enter

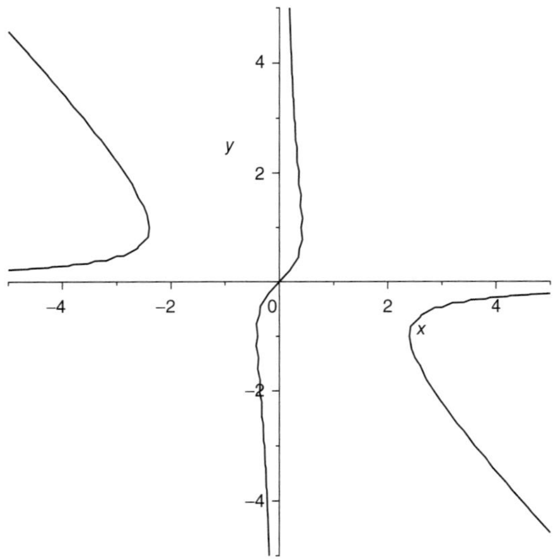

Figure 9.1: Implicit plot of $x^2y + xy^2 + x - y = 0$ for $-5 \leq x, y \leq 5$.

with(plots)

One routine that is frequently used is the implicitplot command. For example, suppose we want to graph the equation

$$x^2y + xy^2 + x - y = 0.$$

Use

implicitplot((x∧2)*y + x*(y∧2) + x - y = 0, x = a..b,y=c..d)

in which a range of values of x and of y must be specified. This may require some experimentation to specify ranges that yield informative graphs. With $x = -5..5$ and $y = -5..5$, the graph shown in Figure 9.1 results.

We can also graph a surface in 3−space as a plot of a function in two variables. As an example, suppose

$$u(x, y) = e^{-x-y}\sin(2\pi(x + y)).$$

First enter the function:

u:= (x,y) → exp(- x - y)*sin(2*Pi*(x + y))

Now use

plot3d(u(x,y),x=a..b,y=c..d)

with numbers specified for a, b, c, and d. Figure 9.2 shows the graph with $a = c = -2, b = d = 2$.

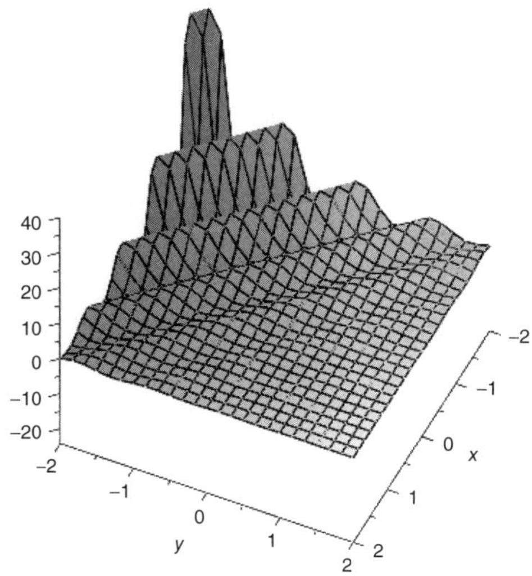

Figure 9.2: Plot of $u(x,y) = e^{-x-y}\sin(2\pi(x+y))$ for $-2 \le x, y \le 2$.

In using plot3d, a right click with the cursor on the graph will display a list of options, such as choice of axes, and also open up a line under the toolbars at the top of the screen, allowing rotations of the surface for different viewing perspectives.

There is also an implicitplot3d command, which works like implicitplot, except now there are two independent variables and the objective is to plot a given equation in three variables. Ranges must be specified for all three variables (again often requiring some experimentation).

9.2.2 Ordinary Differential Equations

Given a function $f(x)$ of x, compute $f'(x)$ by

```
diff(f(x),x)
```

The variable with respect to which the derivative is taken must be specified in this command, because an expression may have various symbols in it. For a second derivative, differentiate the first derivative:

```
diff(diff(f(x),x),x)
```

and so on.

The dsolve command will solve some ordinary differential equation. To solve $y'(t) - 5y(t) = t^2 + 1$, enter

```
dsolve(diff(y(t),t) - 5*y(t) = t∧2 + 1)
```

to obtain the general solution

$$y(t) = C_1 e^{5t} - \frac{1}{5}t^2 - \frac{2}{25}t - \frac{27}{125}.$$

In the dsolve command, the function is denoted y(t). This notation must be kept throughout the command. If, say, y(t) is written as just y in part of the command, an error message will result.

As another example, to solve

$$y'' + 3y' - y = \cos(2t) + t,$$

enter

dsolve(diff(diff(y(t),t),t) + 3*diff(y(t),t) - y(t) = cos(2*t) + t,y(t))

to obtain the general solution

$$y(t) = C_1 e^{(-3+\sqrt{13})t} + C_2 e^{(-3-\sqrt{13})t} - t - 3 - \frac{5}{61}\cos(2t) + \frac{6}{61}\sin(2t).$$

For an initial-value problem, place the initial value in the command, enclosing the differential equation and the initial value(s) in curly brackets. For example, to solve

$$y' - 4y = -4t; y(1) = 4,$$

enter

dsolve({diff(y(t),t) - 4*y(t) = -4t∧3, y(1) = 4}, y(t))

to obtain

$$y = \frac{11}{4} e^{4(t-1)} + \frac{1}{4} + t.$$

9.2.3 Integral Transforms

MAPLE has a package of integral transform routines. Enter

with(inttrans)

to obtain a list of these transforms and activate this package. Included are the Fourier, Fourier sine, Fourier cosine, and Laplace transforms, called, respectively, fourier, fouriersin, fouriercos, and laplace. To take the transform of a function, enter the MAPLE name of the transform, then the function to be transformed, then the variable of this function, and then the variable of the transformed function. For example, for the Laplace transform of $t\sin(t)$, enter

laplace(t*sin(t),t,s)

to obtain

2s/((s∧2 + 1)∧2.

For the inverse of one of these transforms, use invfourier, invfouriersin, invfouriercos, or invlaplace. For example, for the inverse Fourier transform of $1/(1+\omega)$, write this as $1/(1+w)$ in entering it into MAPLE:

invfourier(1/(1 + w),w,t)

to obtain

```
(I/2)*exp∧(-I*t)*(2*Heaviside(t) - 1).
```

Here *Heaviside*(t) is $H(t)$, the Heaviside function. Using standard notation, this result can be written

$$\mathcal{F}^{-1}\left(\frac{1}{1+\omega}\right) = \frac{i}{2}e^{-it}(2H(t)-1).$$

9.2.4 Special Functions

MAPLE has routines for some special functions that are frequently used, including Bessel functions and Legendre polynomials.

The Bessel function $J_n(x)$ of the first kind of order n is denoted BesselJ(n,x) in MAPLE. To evaluate this function at some x, enter

```
evalf(BesselJ(n,x))
```

To evaluate an integral involving a product of a function $g(x)$ and a Bessel function, say from 0 to 1, enter

```
evalf(int(g(x)*BesselJ(n,x),x=0..1)
```

BesselY(n,x) is $Y_n(x)$, the Bessel function of the second kind of order n.

For the *kth* positive zero of $J_n(x)$, enter

```
evalf(BesselJZeros(n,k))
```

For example, enter

```
evalf(BesselJZeros(4,6))
```

to obtain 24.01901952, which is the sixth (in increasing order) positive zero of $J_4(x)$.

As an example that puts many of these ideas together, suppose we want to plot the 20th partial sum of the Fourier-Bessel expansion of $g(x) = x^2$ on $[0, 1]$, in terms of Bessel functions of order 3. Although this can be done in several ways, it is perhaps easiest to break the calculation up into simple components and enter each individually. The partial sum we want to plot is

$$\sum_{n=1}^{20} c_n J_3(j_n x),$$

where j_n is the *nth* positive zero of $J_3(x)$, and

$$c_n = \frac{2}{J_4^2(j_n)} \int_0^1 x g(x) J_3(j_n x)\, dx.$$

First enter $g(x)$ into the program. Next enter the zeros as a sequence j so it can be easily referenced:

```
j:=n → evalf(BesselJZeros(3,n))
```

Now $j(n)$ is the nth zero. Next define the sequence of coefficients. We will do this in two pieces (not necessary, but easy to check). First, one term in c_n is $2/(J_4(j(n)))^2$, which we will call $d(n)$:

d:=n → 2/(BesselJ(4,j(n)))∧2

The nth coefficient in the expansion is

c:=n → d(n)*int(x*g(x)*BesselJ(3,j(n)),x=0..1)

Now enter the 20th partial sum of this Fourier-Bessel expansion as a function:

S:=x → sum(c(n)*BesselJ(3,j(n)*x),n=1..20)

Finally, graph this 20th partial sum and the given function $g(x)$:

plot({$g(x), S(x)$},x=0..1,color=black)

This will plot the function and this partial sum, with both curves drawn in black (optional coloring).
The nth Legendre polynomial $P_n(x)$ is denoted LegendreP(n,x). If we enter

LegendreP(4,x)

the return is just LegendreP(4,x), not very informative. For the polynomial to be displayed, enter

simplify(LegendreP(4,x))

This will return $P_4(x)$ as a 4th degree polynomial. For an integral of $g(x)P_n(x)$, use

int(g(x)*LegendreP(n,x),x=a..b)

9.3 Answers to Selected Problems

Chapter 1: First Ideas

1.1 Two Fundamental Partial Differential Equations

13. $B^2 - 4AC = -12 < 0$, so the equation is elliptic. Let

$$\xi = x + t, \eta = \sqrt{3}t$$

to obtain the canonical form

$$V_{\xi\xi} + V_{\eta\eta} + \frac{2}{3}V_{\xi} - \left(\xi - \frac{1}{\sqrt{3}\eta}\right)V = 0.$$

15. $B^2 - 4AC = 9 > 0$, so the equation is hyperbolic. Let $\xi = t, \eta = 3x + t$ to obtain the canonical form

$$V_{\xi\eta} - \frac{1}{27}(\eta - \xi)V = 0.$$

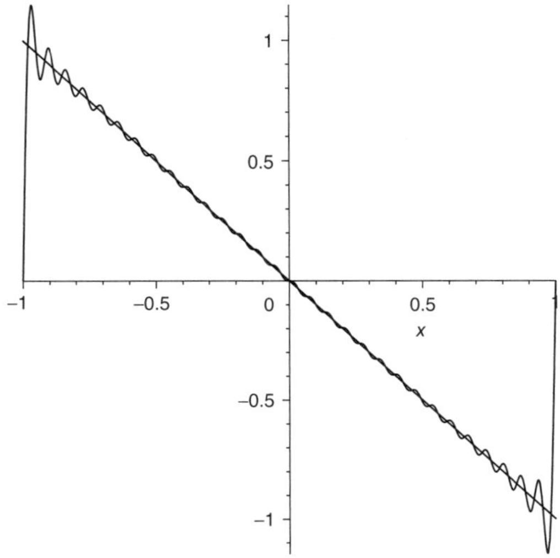

Figure 9.3: $f(x)$ and the 30th partial sum of the Fourier series in problem 1.

17. $B^2 - 4AC = -8 < 0$, so the equation is elliptic. Let $\xi = x + t/3$ and $\eta = \sqrt{2}t/3$ to obtain the canonical form

$$V_{\xi\xi} + V_{\eta\eta} + 18V^2 = 0.$$

1.2 Fourier Series

1. The Fourier series is

$$\sum_{n=1}^{\infty} \frac{2(-1)^n}{n\pi} \sin(n\pi x),$$

converging to $-x$ for $-1 < x < 1$ and to 0 at $x = \pm 1$. Figure 9.3 compares $f(x)$ with the 30th partial sum.

3. The series is

$$\sum_{n=1}^{\infty} \frac{-2n\pi \sin(4)(-1)^n}{n^2\pi^2 - 16} \sin(n\pi x/2),$$

converging to $\sin(2x)$ for $-2 < x < 2$ and to 0 at $x = \pm 2$. Figure 9.4 compares $f(x)$ to the 20th partial sum of the series.

5. The Fourier series converges to -4 for $-\pi < x < 0$, to 4 for $0 < x < \pi$, and to 0 at $x = 0, \pi, -\pi$. The series is

$$\sum_{n=1}^{\infty} \frac{8}{n\pi}(1 - (-1)^n) \sin(nx).$$

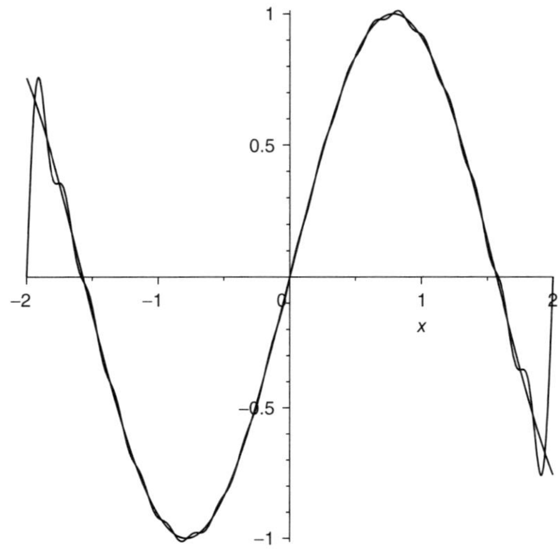

Figure 9.4: $f(x)$ and the 20th partial sum of the Fourier series in problem 3.

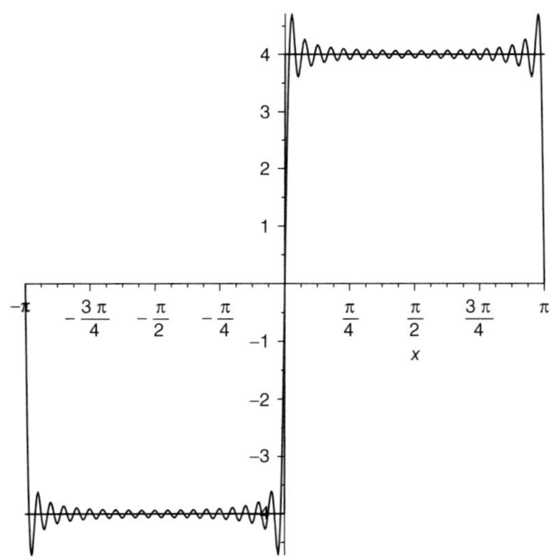

Figure 9.5: $f(x)$ and the 40th partial sum of the Fourier series in problem 5.

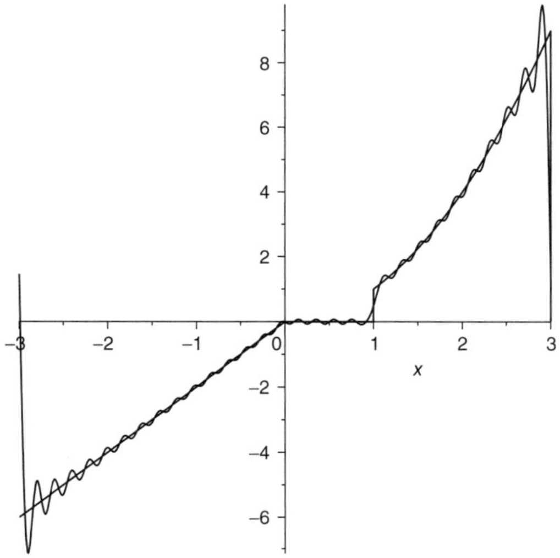

Figure 9.6: $f(x)$ and the 30th partial sum of the Fourier series in problem 7.

Figure 9.5 shows $f(x)$ and the 40th partial sum of this series.

For problems 7, 9 and 11, a graph of the function and a partial sum of the series expansion is given for comparison.

7. The Fourier series converges to

$$\begin{cases} 2x & \text{for } -3 < x < 0, \\ 0 & \text{for } 0 < x < 1, \\ x^2 & \text{for } 1 < x < 3, \\ 0 & \text{for } x = 0, \\ 1/2 & \text{for } x = 1, \\ 3/2 & \text{for } x = \pm 3. \end{cases}$$

Figure 9.6 compares $f(x)$ with the 30th partial sum of the Fourier series.

9. The Fourier series converges to

$$\begin{cases} -x & \text{for } -4 < x < 0, \\ 2 & \text{for } 0 < x < 4, \\ 1 & \text{at } x = 0, \\ 3 & \text{for } x = \pm 4. \end{cases}$$

Figure 9.7 compares $f(x)$ with the 25th partial sum of the series.

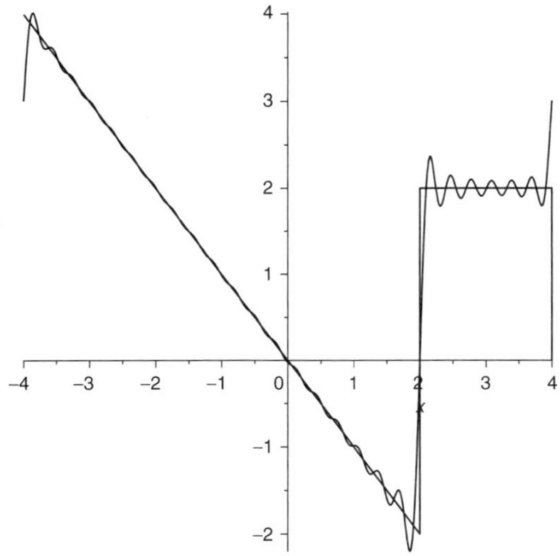

Figure 9.7: $f(x)$ and the 25th partial sum of the Fourier series in problem 9.

11. The Fourier series converges to

$$\begin{cases} \cos(\pi x) & \text{for } -2 < x < 0, \\ x & \text{for } 0 < x < 2, \\ 1/2 & \text{at } x = 0, \\ 3/2 & \text{at } x = \pm 2. \end{cases}$$

Figure 9.8 shows the function and the 50th partial sum of the Fourier series.

13. First show that

$$\frac{1}{2}x^2 = \frac{1}{6}\pi^2 + \sum_{n=1}^{\infty} \frac{2(-1)^n}{n^3} \cos(nx)$$

for $-\pi \leq x \leq \pi$. Now choose values of x so that the Fourier series becomes the requested series.

15. The cosine series is the function itself. The sine series on $[0, 3]$ is

$$\sum_{n=1}^{\infty} \frac{8}{n\pi}(1 - (-1)^n) \sin(n\pi x/3).$$

This sine series converges to 0 at $x = 0, 3$, and to 4 for $0 < x < 3$.

Figure 9.9 compares $f(x)$ with the 100th partial sum of this series.

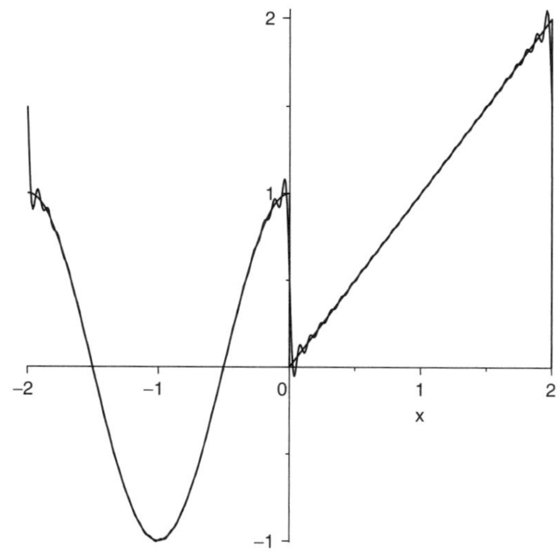

Figure 9.8: $f(x)$ and the 50th partial sum of the Fourier series in problem 11.

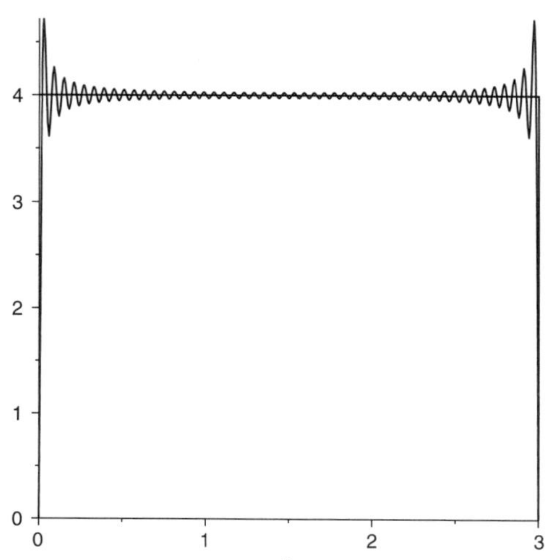

Figure 9.9: $f(x)$ and the 100th partial sum of the sine series of problem 15.

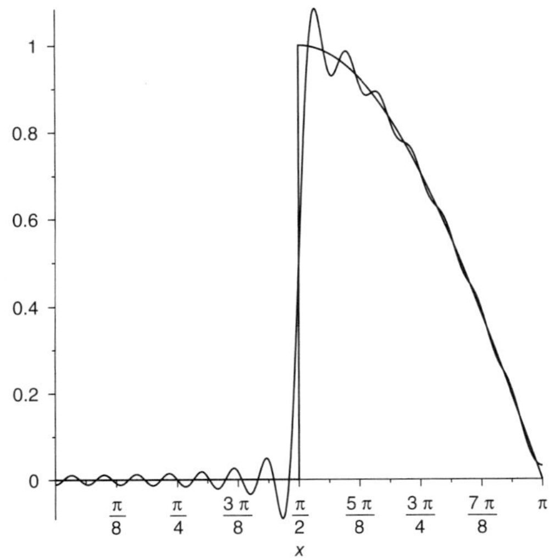

Figure 9.10: $f(x)$ and the 30th partial sum of the cosine series in problem 17.

17. The cosine series is

$$\frac{1}{\pi} - \frac{1}{\pi}\cos(x) + \sum_{n=2}^{\infty} \frac{2}{(-1)^{n+1}}\pi(n^2-1)\cos(nx),$$

converging to 0 for $0 < x < \pi/2$, $1/2$ for $x = \pi/2$, $\sin(x)$ for $\pi/2 < x < \pi$, and to 0 at $x = 0$ and $x = \pi$.

Figure 9.10 compares $f(x)$ with the 30th partial sum of this cosine series.

The sine series is

$$\frac{1}{2}\sin(x) + \sum_{n=2}^{\infty} \frac{2n\cos(n\pi/2)}{\pi(n^2-1)}\sin(nx)$$

converging to 0 for $0 \leq x < \pi/2$, $1/2$ at $x = \pi/2$, and 0 at $x = 0$ and $x = \pi$. Figure 9.11 compares $f(x)$ with the 50th partial sum of this sine series.

19. The cosine series is

$$\frac{4}{3} + \sum_{n=1}^{\infty} \frac{16}{n^2\pi^2}(-1)^n \cos(n\pi x/2),$$

converging to x^2 for $0 \leq x \leq 2$. Figure 9.12 compares $f(x)$with the 10th partial sum of this cosine series.

The sine series is

$$\sum_{n=1}^{\infty} \frac{16(-1+(-1)^n) - 8n^2\pi^2(-1)^n}{n^3\pi^3}\sin(n\pi x/2),$$

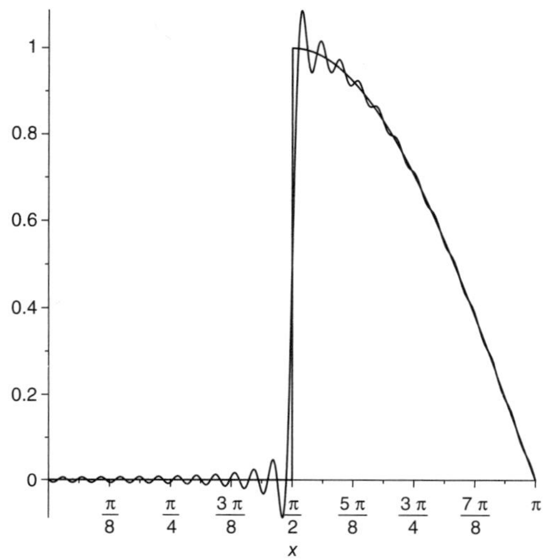

Figure 9.11: $f(x)$ and the 50th partial sum of the sine series in problem 17.

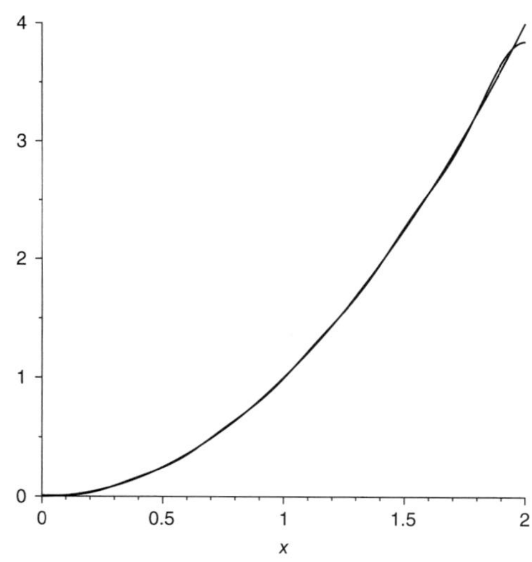

Figure 9.12: $f(x)$ and the 10th partial sum of the cosine series in problem 19.

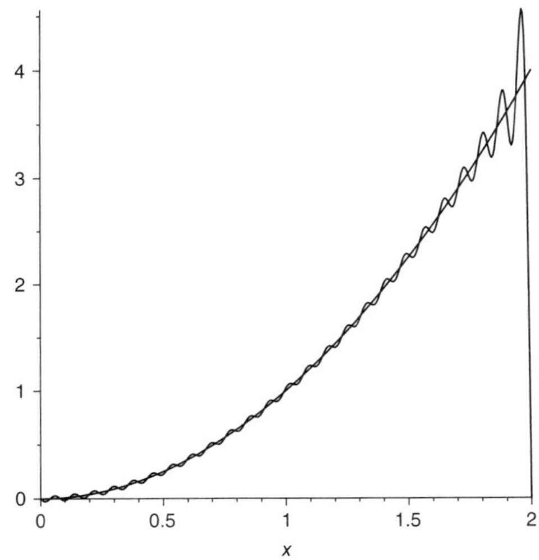

Figure 9.13: $f(x)$ and the 50th partial sum of the sine series in problem 19.

converging to x^2 for $0 < x < 2$ and to 0 for $x = 0$ and $x = 2$. Figure 9.13 compares $f(x)$ with the 50th partial sum of this series.

21. The cosine series is

$$\frac{2}{3\pi} + \frac{12}{5\pi}\cos(2x) + \sum_{n=4}^{\infty} \frac{-6(1 + (-1)^n)}{\pi(n^2 - 9)} \cos(nx).$$

This converges to $\sin(3x)$ for $0 \leq x \leq \pi$. Figure 9.14 compares the function with a partial sum of this series.

The sine series is $\sin(3x)$ on $[0, \pi]$.

1.3 Two Eigenvalue Problems

1.

$$\lambda_n = \left(\frac{(2n+1)\pi}{2L}\right)^2, X_n(x) = \sin\left(\frac{(2n+1)\pi x}{2L}\right)$$

for $n = 1, 2, \cdots$.

3.

$$\lambda_n = \frac{4n^2\pi^2}{L^2} \text{ for } n = 1, 2, \cdots$$

and

$$X_n(x) = a\cos\left(\frac{2n\pi x}{L}\right) + b\sin\left(\frac{2n\pi x}{L}\right).$$

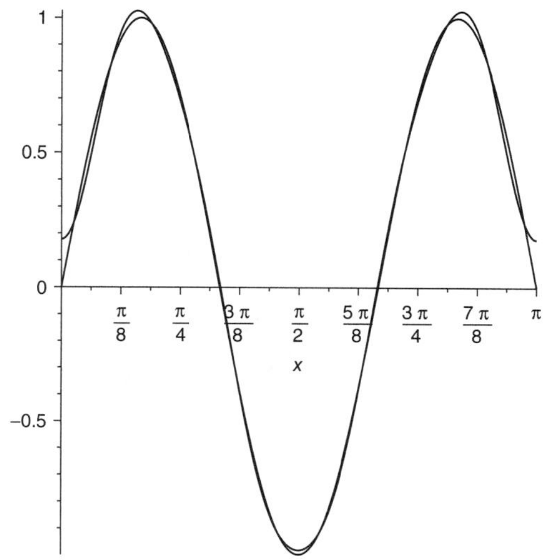

Figure 9.14: $f(x)$ and a partial sum of the cosine expansion of problem 21.

Chapter 2: Solutions of the Heat Equation

2.1 Solutions on an Interval $[0, L]$

For some problems graphics are shown to provide some insight into the behavior of the solution, even when this is not requested in the problem.

1. $u(x,t) = \sin(\pi x)e^{-k\pi^2 t}$

   Figure 9.15 displays the behavior of $u(x,t)$, as a surface in $x, t, u-$space.

3.
$$u(x,t) = \frac{16}{3} \sum_{n=1}^{\infty} \frac{64(-1)^n}{n^2\pi^2} \cos(n\pi x/4)e^{-n^2\pi^2 kt/16}.$$

   Figure 9.16 is a surface plot of $u(x,t)$.

5.
$$u(x,t) = \frac{1}{6}(1-e^{-6}) + \sum_{n=1}^{\infty} \frac{12(1-(-1)^n e^{-6})}{36+n^2\pi^2} cos(n\pi x/6)e^{-kn^2\pi^2 t/36}.$$

   Figure 9.17 shows $u(x,t)$ at times 0.1, 0.3, 0.9, 1.3, 2.5, 4.2, 9.2. The graphs move downward as t increases, appearing to approach the steady-state value $(1-e^{-6})/6$, which is about 0.1663.

7.
$$u(x,t) = L - \frac{1}{3}L^2 + \sum_{n=1}^{\infty} \frac{4L^2(-1)^{n+1}}{n^2\pi^2} \cos(n\pi x/L)e^{-n^2\pi^2 kt/L^2}.$$

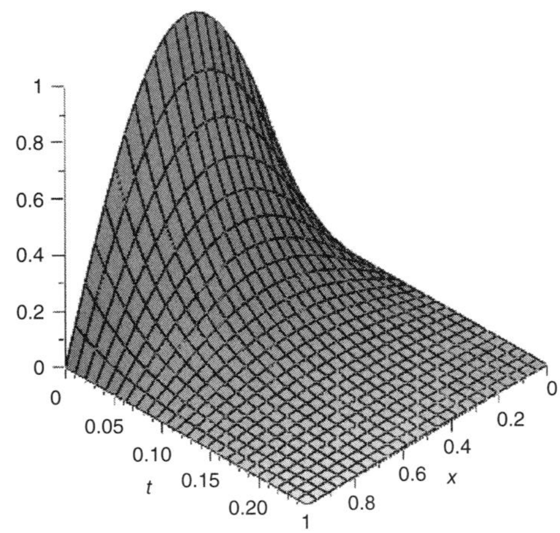

Figure 9.15: $u(x,t)$ in problem 1, for $k=2$ and $0 \le x \le 1, 0 \le t \le 1/4$.

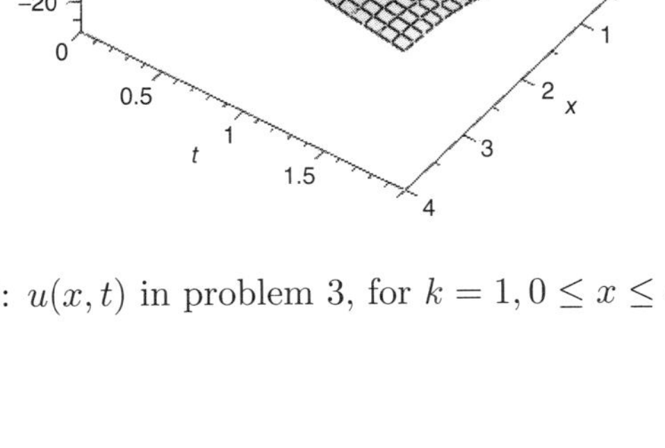

Figure 9.16: $u(x,t)$ in problem 3, for $k=1, 0 \le x \le 4, 0 \le t \le 6$.

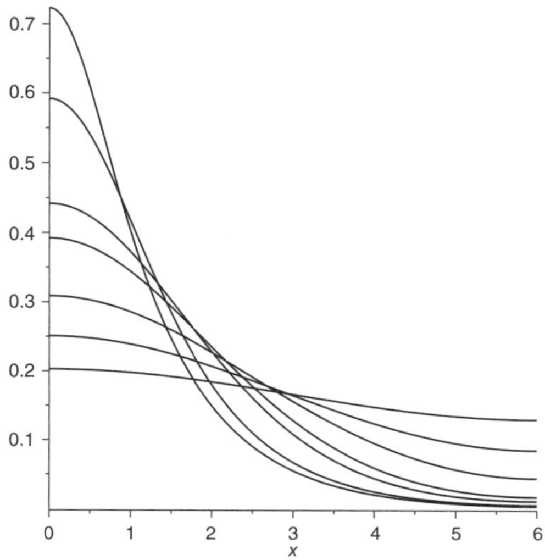

Figure 9.17: $u(x,t)$ at the requested times in problem 5.

9. $u(x,t) = v(x,t) + 3x$, where

$$v(x,t) =$$
$$\sum_{n=1}^{\infty} \frac{8}{n^3\pi^2}(-32 - 64(-1)^n + 3n^2\pi^2(-1)^n)\sin(n\pi x/4)e^{-n^2\pi^2 kt/16}.$$

11. Let $u(x,r) = e^{\alpha x+\beta t}v(x,t)$ and obtain $\alpha = 1/4$ and $\beta = -1/4$. The problem for $v(x,t)$ is

$$v_t = 4v_{xx} \text{ for } 0 < x < L, t > 0,$$
$$v(0,t) = v(L,t) = 0 \text{ for } t > 0,$$
$$v(x,0) = e^{-x/4}.$$

Then $u(x,t) = e^{(x-t)/4}v(x,t)$, where

$$v(x,t) = \sum_{n=1}^{\infty} \frac{-32n\pi}{L^2 + 16n^2\pi^2}\left(1 + e^{-L/4}(-1)^n\right)\sin(n\pi x/L)e^{-n^2\pi^2 kt/L^2}.$$

13. Let $u(x,t) = e^{\alpha x+\beta t}v(x,t)$. We find that $\alpha = 0$ and $\beta = -h$, and v satisfies:

$$v_t = kv_{xx} \text{ for } 0 < x < L, t > 0$$
$$v_x(0,t) = v_x(L,t) = 0 \text{ for } t > 0,$$
$$v(x,0) = x - \frac{x^2}{L} = f(x) \text{ for } 0 < x < L.$$

There is no exponential factor in $v(x,0)$ because $\alpha = 0$ for this problem. Then

$$v(x,t) = \frac{L}{6} + \sum_{n=1}^{\infty} \frac{-2L}{n^2\pi^2}(1+(-1)^n)\cos(n\pi x/L)e^{-n^2\pi^2 kt/L^2}$$

and

$$u(x,t) = e^{-ht}v(x,t).$$

15. Let $u(x,t) = e^{\alpha x+\beta t}v(x,t)$ and find that $\alpha = 3/2$ and $\beta = -k/4$. The problem for $v(x,t)$ is

$$\begin{aligned} v_t &= kv_{xx} \text{ for } 0 < x < L, t > 0, \\ v(0,t) &= v(L,t) = 0 \text{ for } t > 0, \\ v(x,0) &= e^{-3x/2}f(x). \end{aligned}$$

Then

$$v(x,t) = \sum_{n=1}^{\infty} b_n \sin(n\pi x/L)e^{-n^2\pi^2 kt/L^2},$$

where

$$b_n = \frac{2}{L}\int_0^L e^{-3\xi/2}f(\xi)\sin(n\pi\xi/L)\,d\xi.$$

19. $u(x,t)$ satisfies the initial-boundary value problem of problem 17, with $k = 9$ and $L = 2$. The solution is

$$u(x,t) = \sum_{n=1}^{\infty} b_n \sin((2n-1)\pi x/4)e^{-9n^2\pi^2 t/4},$$

in which

$$\begin{aligned} b_n &= \int_0^2 f(\xi)\sin((2n-1)\pi x/4)\,d\xi \\ &= \frac{-64}{(2n-1)^2\pi^2}(2+(2n-1)\pi(-1)^n). \end{aligned}$$

21. The solution has the form

$$u(x,t) = \frac{1}{2}a_0 + \sum_{n=1}^{\infty}(a_n\cos(n\pi x/L) + b_n\sin(n\pi x/L)),$$

where

$$a_n = \frac{1}{L}\int_{-L}^{L} f(\xi)\cos(n\pi\xi/L)\,d\xi \text{ for } n = 0,1,2,\cdots$$

and

$$b_n = \frac{1}{L}\int_{-L}^{L} f(\xi)\sin(n\pi\xi/L)\,d\xi \text{ for } n = 1,2,\cdots.$$

2.2 A Nonhomogeneous Problem

1.

$$u(x,t) =$$
$$\frac{2L^2}{k^2\pi^5}\sum_{n=1}^{\infty}\left[\frac{1-(-1)^n}{n^5}(kn^2\pi^2 t + e^{-kn^2\pi^2 t/L^2} - 1)\right]\sin(n\pi x/L)$$
$$+\sum_{n=1}^{\infty}\frac{4L^2}{n^3\pi^3}\sin(n\pi x/L)e^{-kn^2\pi^2 t/L^2}.$$

3. If $L < \pi$,

$$u(x,t) =$$
$$\sum_{n=1}^{\infty}\frac{e^{-kn^2\pi^2 t/L^2} - e^{-kn^2\pi^2(t-L)/L^2}}{kn^2\pi^2}c_n\sin(n\pi x/L)$$
$$+\sum_{n=1}^{\infty}\left(\frac{-4L^3}{n^3\pi^3}(2(-1)^n+1)\right)\sin(n\pi x/L)e^{-kn^2\pi^2 t/L^2},$$

where

$$c_n = \frac{2n\pi L^2}{L^2 - n^2\pi^2}\left(1-(-1)^n\cos(L)\right).$$

If $L = \pi$,

$$u(x,t) = \sum_{n=2}^{\infty}\left(\frac{2n(1+(-1)^n)}{\pi(n^2-1)}\frac{1-e^{-kn^2 t}}{kn^2}\right)\sin(nx)$$
$$+\sum_{n=1}^{\infty}\left(-4\frac{2(-1)^n+1}{n^3}\right)\sin(nx)e^{-kn^2 t}.$$

Figure 9.18 shows part of $u(x,t)$ for given values of L and k.

5.

$$u(x,t) =$$
$$\sum_{n=1}^{\infty}\frac{2L^3}{k^2n^5\pi^5}(-1)^{n+1}\left(kn^2\pi^2 t + L^2 e^{-kn^2\pi^2 t/L^2} - L^2\right)\sin(n\pi x/L)$$
$$+\frac{2K}{\pi}\sum_{n=1}^{\infty}\frac{1-(-1)^n}{n}\sin(n\pi x/L)e^{-kn^2\pi^2 t/L^2}.$$

Figure 9.19 shows part of the surface for selected values of L, k and K.

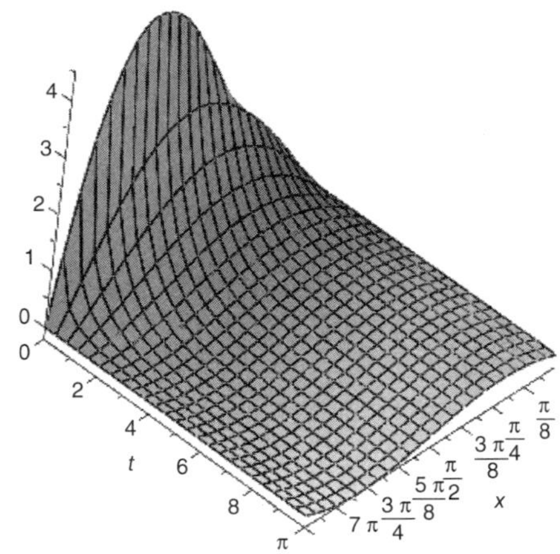

Figure 9.18: $u(x,t)$ in problem 3, for $L=\pi, k=1$, $0 \le t \le 10$.

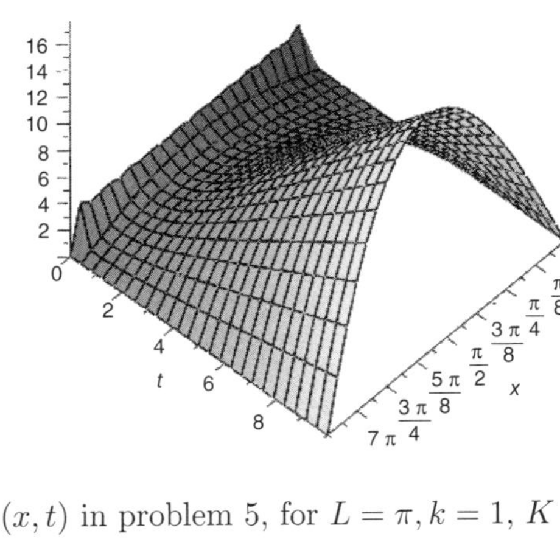

Figure 9.19: $u(x,t)$ in problem 5, for $L=\pi, k=1$, $K=3$, $0 \le t \le 10$.

7.

$$u(x,t) = \frac{1}{4}Lt^2 + 1 + \sum_{n=1}^{\infty} \frac{L^2}{k^2 n^4 \pi^4} c_n \left(kn^2\pi^2 t + L^2 e^{-kn^2\pi^2 t/L^2} - L^2\right) \cos(n\pi x/L),$$

where

$$c_n = \left(\frac{2L((-1)^n - 1)}{n\pi}\right).$$

9.

$$u(x,t) = \sum_{n=1}^{\infty} T_n(t) \sin((2n-1)\pi x/2L),$$

where

$$T_n(t) = \frac{8L(-1)^{n+1}}{k}\left(\frac{-4L^2 + 4kn^2\pi^2 t - 4k\pi^2 nt + k\pi^2 t}{(2n-1)^4\pi^4}\right) + \frac{8L(-1)^{n+1}}{k}\left(\frac{4L^2 e^{-k(2n-1)^2\pi^2 t/4L^2}}{(2n-1)^4\pi^4}\right) - 8\sqrt{\lambda_n}\left(\frac{-1 + e^{-k(2n-1)^2\pi^2 t/4L^2}}{(2n-1)^2\pi^2}\right).$$

2.3 The Heat Equation in Two Space Variables

1.

$$u(x,y,t) = \sum_{n=1}^{\infty} \frac{4}{\pi n^3}((-1)^n - 1)\sin(x)\sin(ny)e^{-k(1+n^2)t}.$$

3.

$$u(x,y,t) = \sum_{m=1}^{\infty} c_{0m} \sin(my) e^{-km^2 t} + \sum_{n=1}^{\infty}\sum_{m=1}^{\infty} c_{nm}\cos(nx)\sin(my)e^{-k(n^2+m^2)t},$$

where

$$c_{0m} = \frac{4\pi - 8}{\pi^2}\left(\frac{(1-(-1)^n)}{m^3}\right)$$

and

$$c_{nm} = \frac{16}{\pi^2}\left(\frac{-\pi(-1)^n + 4\pi(-1)^n n^2 + 2 + 8n^2}{(4n^2-1)^2}\right)\left(\frac{(-1)^m - 1}{m^3}\right).$$

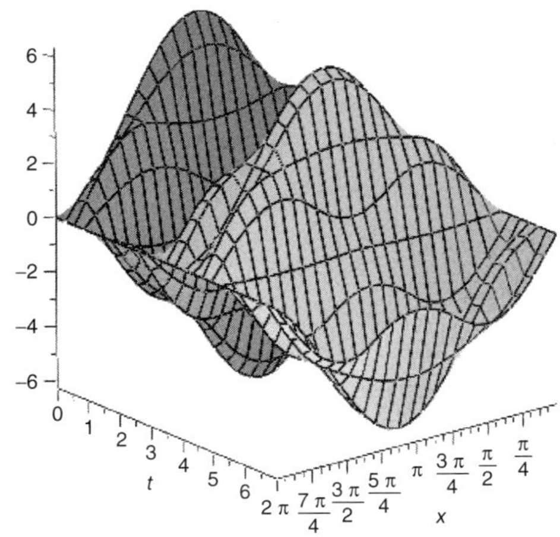

Figure 9.20: $u(x,t)$ in problem 1, for $0 \leq x \leq \pi, 0 \leq t \leq 7$.

Chapter 3: Solutions of the Wave Equation

3.1 Solutions on Bounded Intervals

1.

$$u(x,t) = \sum_{n=1,n\neq 2}^{\infty} \frac{8((-1)^n - 1)}{n(n^2-4)} \sin(nx/2)\cos(3nt/2).$$

Figure 9.20 shows part of the surface $u(x,t)$.

3.

$$u(x,t) = \sum_{n=1}^{\infty} \frac{8}{(2n-1)^3\pi^3} \sin((2n-1)\pi x)\cos(6(2n-1)\pi t).$$

Figure 9.21 shows part of the surface $u(x,t)$.

5.

$$u(x,t) = \sum_{n=1}^{\infty} \frac{4}{3}\frac{1-e^{-2}(-1)^n}{4+n^2\pi^2} \sin(n\pi x/2)\sin(3n\pi t/2).$$

Figure 9.22 is a graph of part of the surface $u(x,t)$.

7.

$$u(x,t) = \sin(x)\cos(3t) + \sum_{n=1}^{\infty} \frac{2}{3n^2} \sin(nx)\sin(3nt).$$

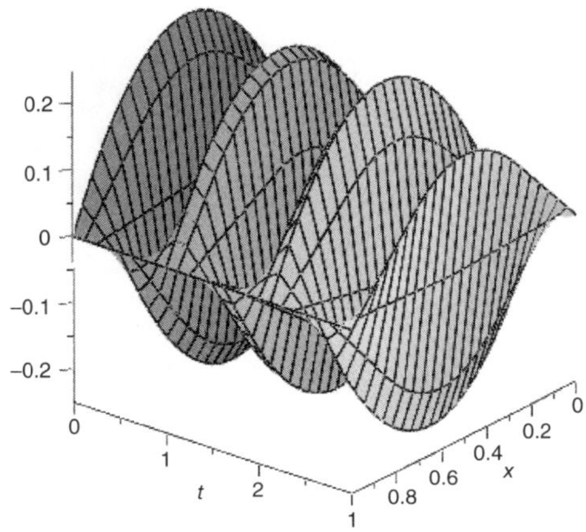

Figure 9.21: $u(x,t)$ in problem 3, for $0 \leq x \leq 1, 0 \leq t \leq 3$.

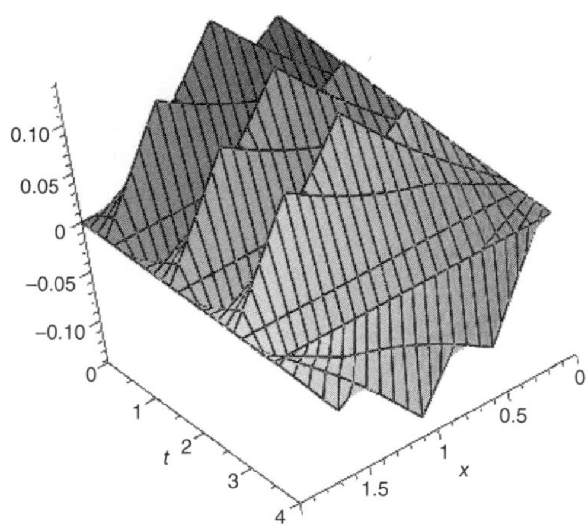

Figure 9.22: $u(x,t)$ in problem 5, for $0 \leq x \leq 2, 0 \leq t \leq 4$.

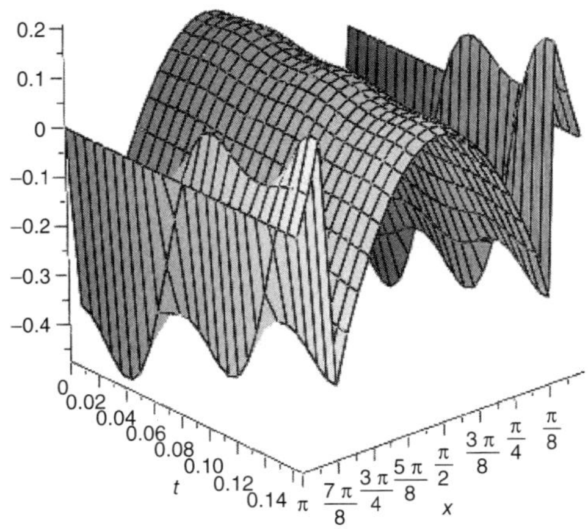

Figure 9.23: $u(x,t)$ in problem 19.

9.

$$u(x,t) = \sum_{n=1}^{\infty} c_n(t)\sin((2n-1)x),$$

where

$$c_n = \frac{-8}{(2n-1)\pi((2n-1)^2-4)}\cos((2n-1)t) + \frac{4}{(2n-1)^2\pi}\sin((2n-1)t).$$

15. *Hint*: Differentiate the first equation with respect to t and the second with respect to x. Then substitute v_{xt} from the first equation into the second.

19.

$$u(x,t) = v(x,t) + \frac{1}{4}(\cos(x)-1) + \frac{1}{2\pi}x,$$

where

$$v(x,t) = \sum_{n=2}^{\infty} c_n \sin(nx)\cos(2nt)$$

and

$$c_n = -\frac{1-(-1)^n}{2n\pi} - \frac{1}{2\pi}\left(\frac{n}{n^2-1}\right)(1+(-1)^n) + \frac{(-1)^n}{n\pi}.$$

Figure 9.23 is a graph of $u(x,t)$ for $0 \le x \le \pi$ and $0 \le t \le 1/6$.

21.

$$u(x,t) = v(x,t) + 3e^{-x} + \frac{3}{2}(1-e^{-2})x - 3,$$

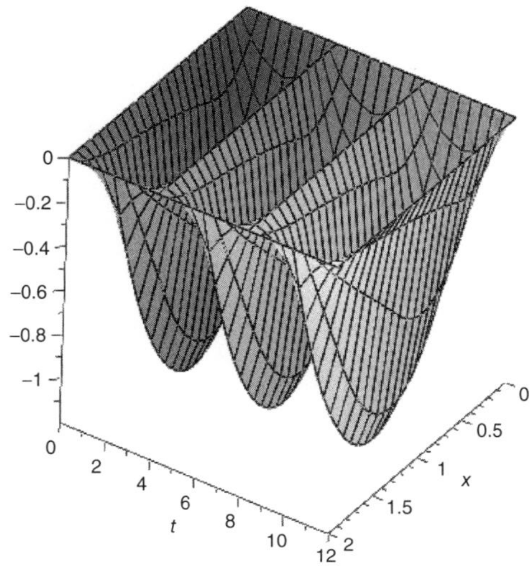

Figure 9.24: $u(x,t)$ in problem 21.

where

$$v(x,t) = \sum_{n=1}^{\infty} c_n \sin(n\pi x/2)\cos(n\pi t/2)$$

and

$$c_n = \frac{6}{n\pi}(1-(-1)^n e^{-2}) + \frac{6n\pi}{4+n^2\pi^2}((-1)^n e^{-2} - 1).$$

Figure 9.24 is a graph of $u(x,t)$ for $0 \le x \le 2$ and $0 \le t \le 10$.

23.

$$u(x,t) = v(x,t) + \frac{2}{27}x(1-x^2),$$

where

$$v(x,t) = \sum_{n=1}^{\infty}\left[\frac{8(-1)^n}{9n^3\pi^3}\cos(3n\pi t) + \frac{2}{3n^2\pi^2}(1-(-1)^n)\sin(3n\pi t)\right]\sin(n\pi x).$$

Figure 9.25 shows wave profiles moving downward over successive times $t = 0.037, 0.37, 0.52$, and 0.89, while Figure 9.26 is part of the wave surface $u(x,t)$.

25. The method does not adapt to a forcing function that is a function of t. If we try $u(x,t) = v(x,t) + g(t)$, then we need

$$u_t(x,0) = v_t(x,0) + g'(t) = 0,$$

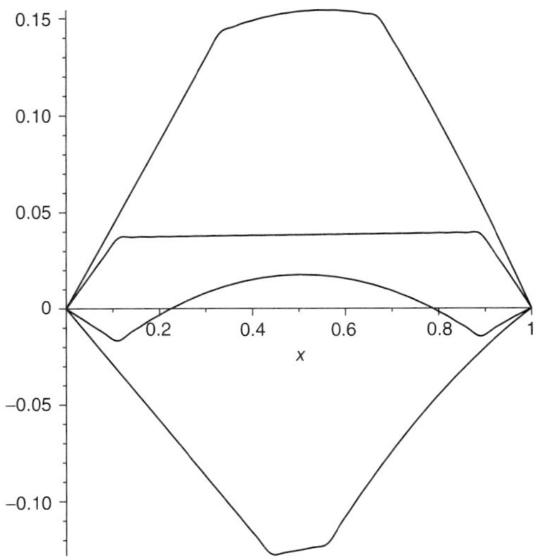

Figure 9.25: Wave profiles moving downward in problem 23.

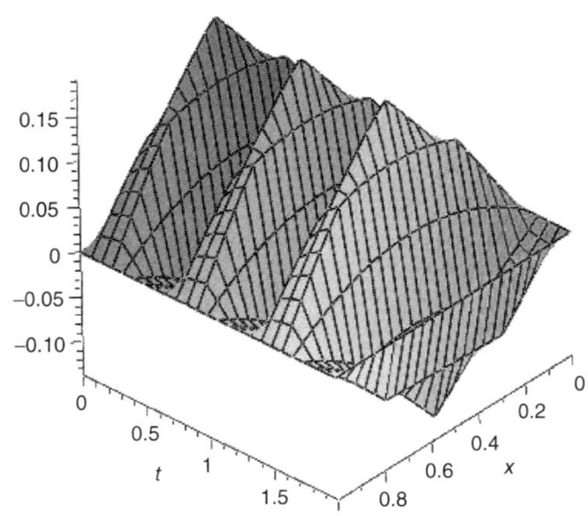

Figure 9.26: Part of the wave surface in problem 23.

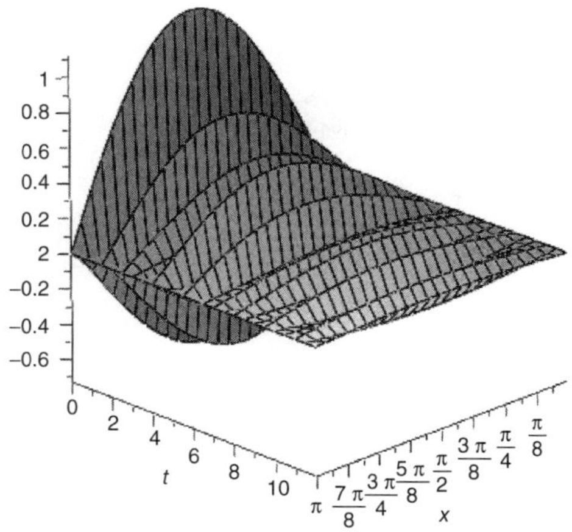

Figure 9.27: Part of the wave surface in problem 29.

which implies that $u_t(x,0) = -g'(t)$. But the left side is a function of x and the right side a function of t, so this makes no sense.

29. With

$$\alpha_n = \frac{1}{2}\sqrt{64n^2-1},$$

$$u(x,t) =$$

$$e^{-t/4}\sum_{n=1}^{\infty} b_n \left[\cos(\sqrt{64n^2-1}t) + \frac{1}{2\sqrt{64n^2-1}}\sin(\sqrt{64n^2-1}t)\right]\sin(nx).$$

in which

$$b_n = \frac{2}{\pi}\int_0^{\pi} \xi\cos(\xi/2)\sin(n\xi)\,d\xi = \frac{-32(-1)^n}{\pi(4n^2-1)^2}.$$

Figure 9.27 is a graph of part of the surface $u(x,t)$.

31. With

$$\alpha_n = 2\sqrt{16n^2-1},$$

$$u(x,t) =$$

$$e^{-t}\sum_{n=1}^{\infty} b_n \left[\cos(8\sqrt{16n^2-1}t) + \frac{1}{2\sqrt{16n^2-1}}\sin(8\sqrt{16n^2-1}t)\right]\sin(nx),$$

in which

$$b_n = \frac{2}{\pi}\int_0^{\pi} \xi^2(\pi-\xi)\sin(n\xi)\,d\xi = -\frac{4}{n^3}(1+2(-1)^n).$$

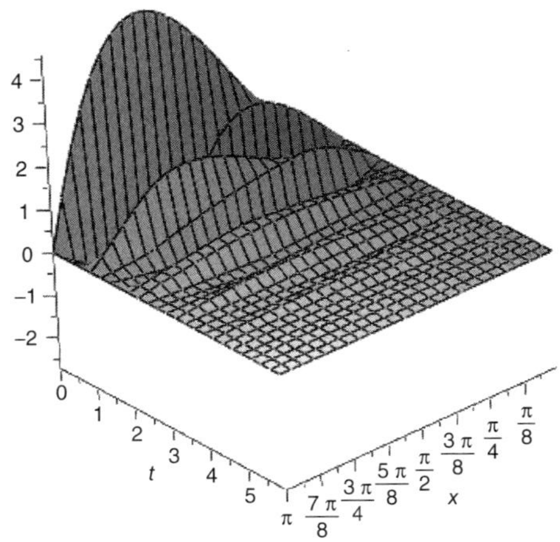

Figure 9.28: Part of the wave surface in problem 31.

Figure 9.28 shows part of the surface $u(x, t)$.

3.2.1 d'Alembert's Solution

1.

$$u(x,t) = \frac{1}{2}(\cos(3(x-7t)) + \cos(3(x+7t))) + \frac{1}{14}\int_{x-7t}^{x+7t} \xi \, d\xi$$
$$= \frac{1}{2}(\cos(3(x-7t)) + \cos(3(x+7t))) + xt,$$

which can be written

$$u(x,t) = \cos(3x)\cos(21t) + xt.$$

3.

$$u(x,t) = \frac{1}{2}\left(e^{-|x-3t|} + e^{-|x+3t|}\right)$$
$$+ \frac{1}{2}t - \frac{1}{24}\sin(2x+6t) + \frac{1}{24}\sin(2x-6t).$$

As with problem 1, using trigonometric identities, this solution can be written in various other ways.

5.

$$u(x,t) = \frac{1}{2}(\cos(x-2t) - \sin(x-2t) + \cos(x+2t) - \sin(x+2t))$$
$$+ \frac{1}{2}\sin(x)\sin(2t).$$

7.

$$u(x,t) = \cos(x)\cos(4t) + \frac{1}{8}x\left(e^{-x+4t} - e^{-x-4t}\right)$$
$$- \frac{1}{2}t\left(e^{-x+4t} + e^{-x-4t}\right) + \frac{1}{8}\left(e^{-x+4t} - e^{-x-4t}\right).$$

9.

$$u(x,t) = x^3 + 27xt^2 + \frac{1}{6}(\cos(x+3t) - \cos(x-3t))$$
$$+ \frac{1}{6}(x+3t)\sin(x+3t) - \frac{1}{6}(x-3t)\sin(x-3t).$$

11. Because $c = 1$ and $\psi(x) = 0$, the solution is

$$u(x,t) = \frac{1}{2}(\varphi(x-t) + \varphi(x+t))$$
$$= \frac{1}{2}\left(e^{-(x-t)^2} + e^{-(x+t)^2}\right).$$

15. With $c = 1$ and $\psi(x) = 0$, the sum of forward and backward waves is

$$u(x,t) = \frac{1}{2}(e^{-(x-t)^2} + \frac{1}{2}e^{-(x+t)^2}.$$

The wave is shown at times $t = 0, 1, 2, 3,$ and 5 in Figures 9.29–9.33. By time $t = 5$, the forward and backward waves have separated and move independently to the right and left along the $x-$axis. This occurs because $\psi(x) = 0$ and $\varphi(x)$ is nonzero only over a finite interval.

17. The sum of forward and backward waves is

$$u(x,t) = \frac{1}{2}(\varphi(x-t) + \varphi(x+t)).$$

and Figures 9.34–9.37 show the wave at times $t = 0, 1/2, 1,$ and 2. In this problem, the forward and backward waves separate by time $t = 2$.

19. The sum of forward and backward waves is

$$u(x,t) = \frac{1}{2}(\varphi(x-t) + \varphi(x+t)).$$

Figures 9.38–9.41 show the wave at times $t = 0, 1/2,$ 1, and 2.

3.2.2 The Cauchy Problem on a Half Line

1.

$$u(x,t) = \begin{cases} x^2 + t^2 + \sin(x)\sin(t) & \text{for } x - t \geq 0, \\ 2xt + \sin(x)\sin(t) & \text{for } x - t < 0. \end{cases}$$

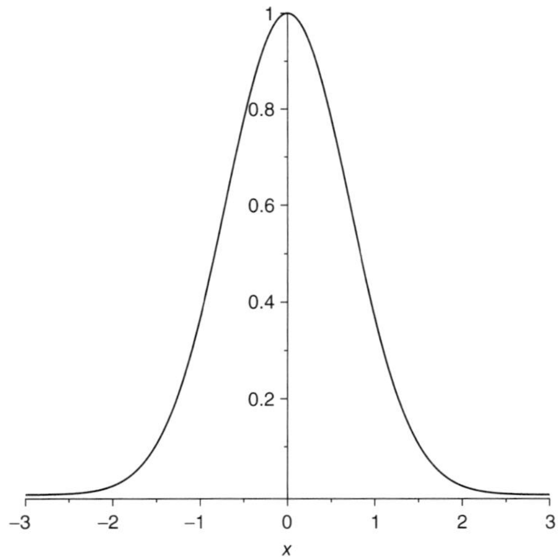

Figure 9.29: $u(x, 0)$ in problem 15.

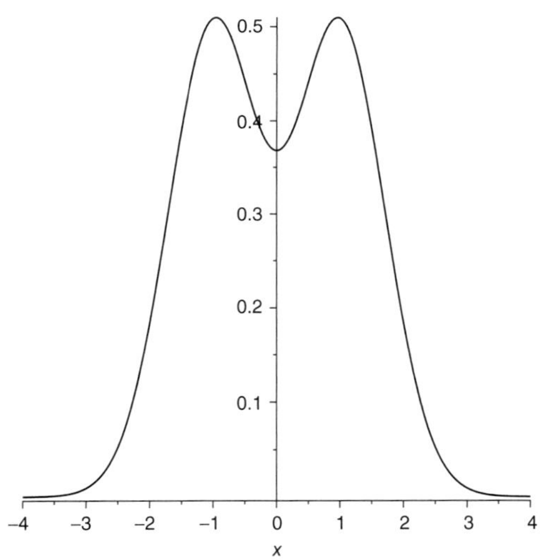

Figure 9.30: $u(x, 1)$ in problem 15.

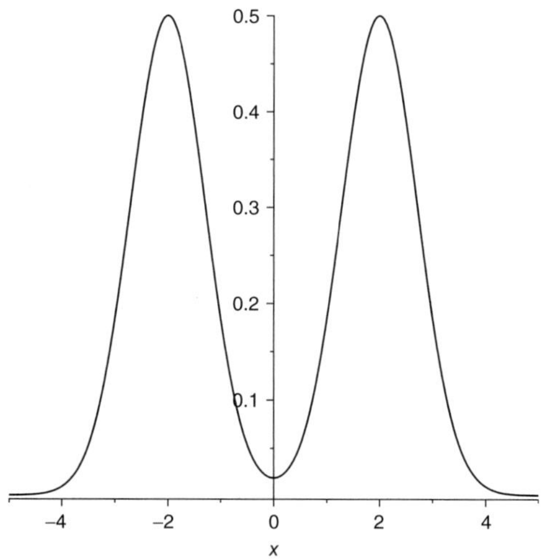

Figure 9.31: $u(x,2)$ in problem 15.

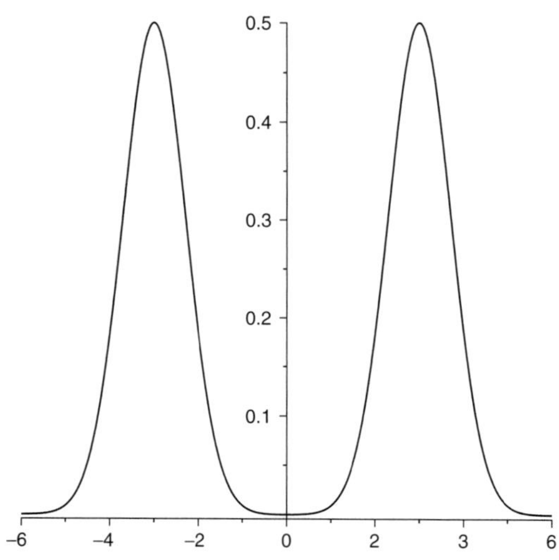

Figure 9.32: $u(x,3)$ in problem 15.

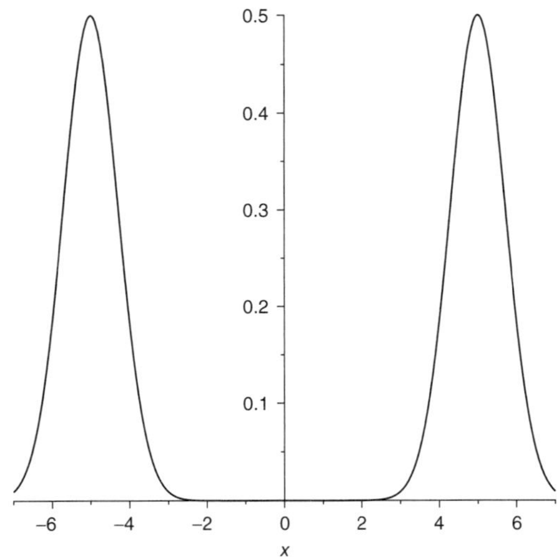

Figure 9.33: $u(x, 5)$ in problem 15.

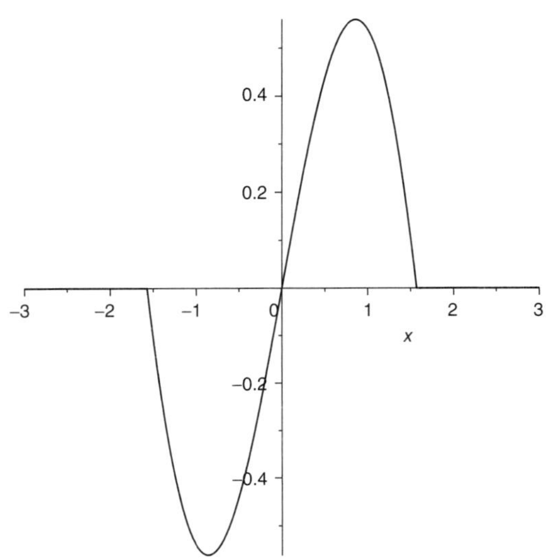

Figure 9.34: $u(x, 0)$ in problem 17.

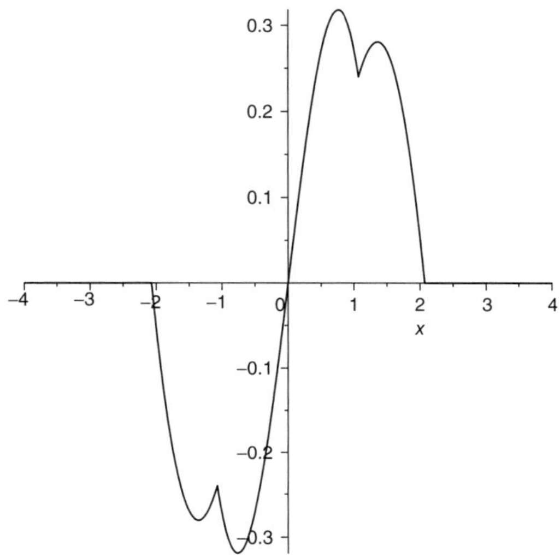

Figure 9.35: $u(x, 1/2)$ in problem 17.

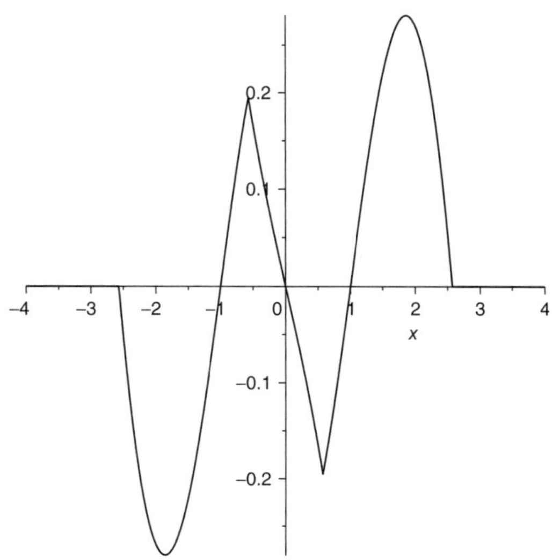

Figure 9.36: $u(x, 1)$ in problem 17.

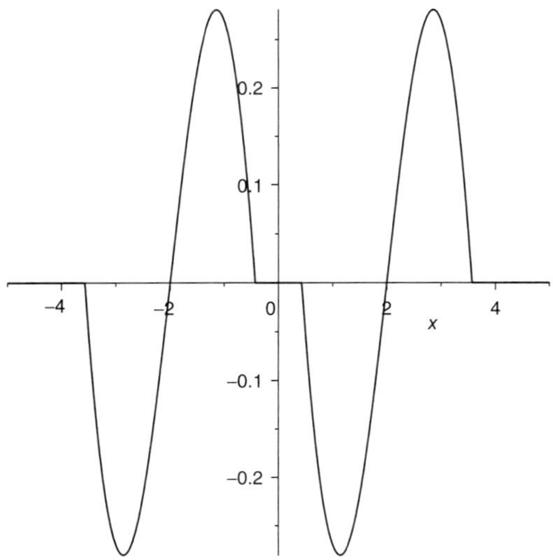

Figure 9.37: $u(x,2)$ in problem 17.

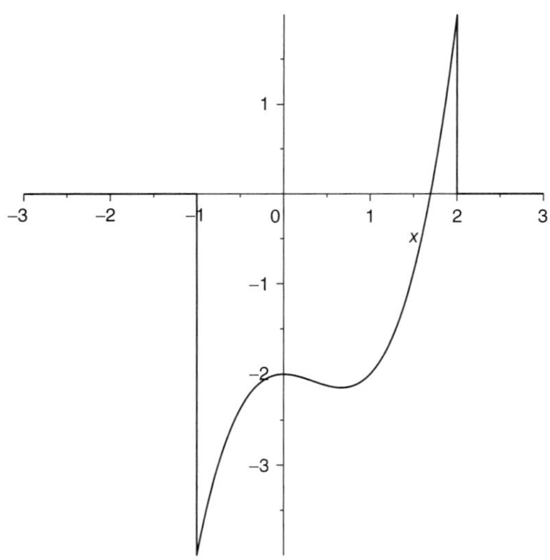

Figure 9.38: $u(x,0)$ in problem 19.

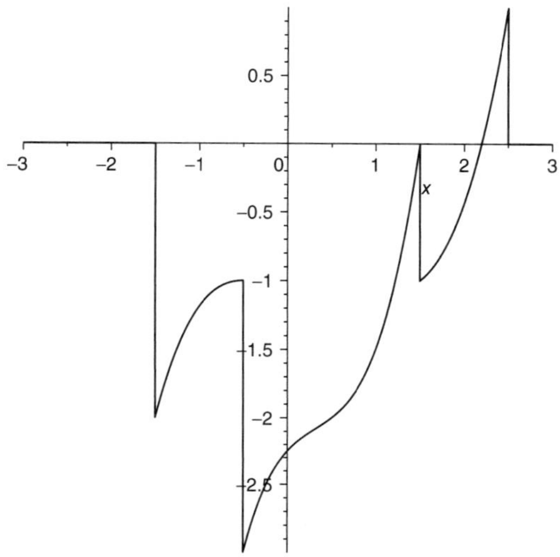

Figure 9.39: $u(x,1/2)$ in problem 19.

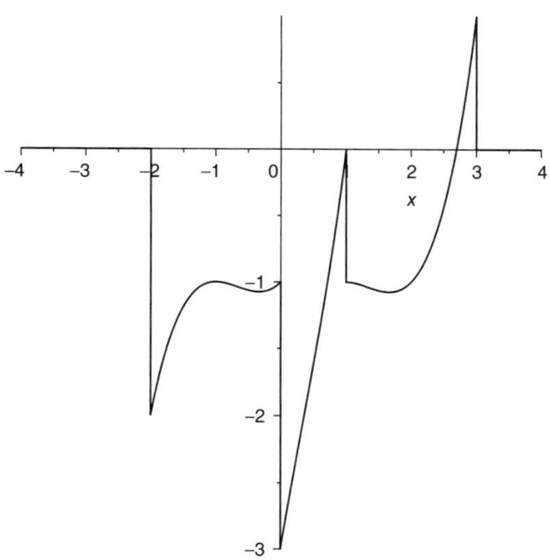

Figure 9.40: $u(x,1)$ in problem 19.

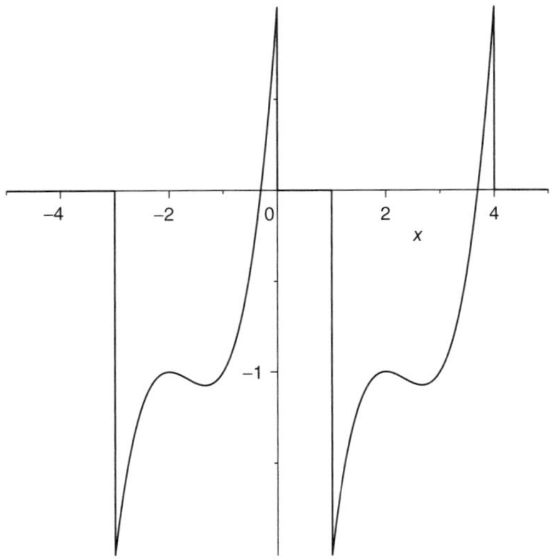

Figure 9.41: $u(x,2)$ in problem 19.

3.

$$u(x,t) = \begin{cases} 1 - e^x\cosh(2t) + x^2t + 4t^3/3 & \text{for } x - 2t \geq 0, \\ -e^{2t}\sinh(x) + 2xt^2 + x^3/6 & \text{for } x - 2t < 0. \end{cases}$$

5. If $x - 2t \geq 0$, then

$$u(x,t) = x\sin(x)\cos(2t) + 2t\cos(x)\sin(2t) + x^2t + 4t^3/3,$$

and if $x - 2t < 0$,

$$u(x,t) = x\cos(x)\sin(2t) + 2t\sin(x)\cos(2t) + x^3/6 + 2xt^2.$$

7.

$$u(x,t) = \begin{cases} x^3 + 27xt^2 + (1/3)e^{-x}\sinh(3t) & \text{for } x - 3t \geq 0, \\ x^3 + 27xt^2 + (1/3)e^{-3t}\sinh(x) & \text{for } x - 3t < 0. \end{cases}$$

9. If $x - 5t \geq 0$, then

$$u(x,t) = -1 + \frac{1}{2}(\cosh(x+5t) + \cosh(x-5t)) + \frac{1}{5}\sin(x)\sin(5t)$$

and if $x - 5t < 0$,

$$u(x,t) = \frac{1}{2}(\cosh(x+5t) - \cosh(x-5t)) + \frac{1}{5}\sin(x)\sin(5t).$$

3.2.3 Characteristic Triangles and Quadrilaterals

1. Begin with the d'Alembert solution

$$u(x,t) = \frac{1}{2}(\varphi(x-ct) + \varphi(x+ct)) + \frac{1}{2c}\int_{x-ct}^{x+ct} \psi(s)\,ds.$$

If $t > 0$ and $x > x + ct$, then for some positive h, $x = a + ct + h$. Show that $\varphi(x+ct) = \varphi(x+2ct+h) = 0$, and $\varphi(a+h) = 0$. Finally, check that

$$\int_{x-ct}^{x+ct} \psi(s)\,ds = \int_{a+h}^{a+h+2ct} \psi(s))\,ds = 0,$$

and hence conclude that $u(x,t) = 0$ if $x > a + ct$. A similar argument shows that $u(x,t) = 0$ if $x < -a - ct$.

3.2.4 A Cauchy Problem with a Forcing Term

1.

$$\begin{aligned} u(x,t) &= \frac{1}{2}((x+4t) + (x-4t)) + \frac{1}{8}\int_{x-4t}^{x+4t} e^s\,ds \\ &+ \frac{1}{8}\int_0^t \int_{x-4t+4T}^{x+4t-4T} (X+T)\,dX\,dT \\ &= x + \frac{1}{4}e^{-x}\sinh(4t) + \frac{1}{2}xt^2 + \frac{1}{6}t^3. \end{aligned}$$

3.

$$\begin{aligned} u(x,t) &= \frac{1}{2}((x+8t)^2 - (x+8t) + (x-8t)^2 - (x-8t)) \\ &+ \frac{1}{16}\int_{x-8t}^{x+8t} \cos(2s)\,ds + \frac{1}{16}\int_0^t \int_{x-8t+8T}^{x+8t-8T} T\cos(X)\,dX\,dT \\ &= x^2 + 64t^2 - x + \frac{1}{32}(\sin(2(x+8t)) - \sin(2(x-8t))) \\ &- \frac{1}{1024}(\sin(-x+8t) + \sin(x+8t)) + \frac{1}{64}t\cos(x). \end{aligned}$$

5.

$$\begin{aligned} u(x,t) &= \frac{1}{2}(\cosh(x+3t) + \cosh(x-3t)) \\ &+ \frac{1}{6}\int_{x-3t}^{x+3t} ds + \frac{1}{6}\int_0^t \int_{x-3t+3T}^{x+3t-3T} 3XT^2\,dX\,dT \\ &= \frac{1}{2}(\cosh(x+3t) + \cosh(x-3t)) + t + \frac{1}{4}xt^4. \end{aligned}$$

7.

$$\begin{aligned} u(x,t) &= \frac{1}{2}(\cos(2(x+2t)) + \cos(2(x-2t))) \\ &+ \frac{1}{4}\int_{x-2t}^{x+2t} (1-\cos(s))\,ds + \frac{1}{4}\int_0^t \int_{x-2t+2T}^{x+2t-2T} T^2\,dT \\ &= \cos(2x)\cos(4t) + t - \cos(x)\sin(t)\cos(t) + \frac{1}{12}t^4. \end{aligned}$$

9.

$$\begin{aligned} u(x,t) &= \frac{1}{2}((x+2t)\sin(x+2t) + (x-2t)\sin(x-2t)) \\ &+ \frac{1}{4}\int_{x-2t}^{x+2t} e^{-s}\,ds + \frac{1}{4}\int_0^t \int_{x-2t+2T}^{x+2t-2T} XT\,dX\,dT \\ &= \frac{1}{2}((x+2t)\sin(x+2t) + (x-2t)\sin(x-2t)) \\ &+ \frac{1}{2}e^{-x}\sinh(2t) + \frac{1}{6}xt^3. \end{aligned}$$

3.2.5 String with Moving Ends

1. In all three regions we obtain (by separate calculations)

$$u(x,t) = 2x - x^2 - t^2.$$

3. In all regions, we obtain (by separate calculations),

$$u(x,t) = 2x^2 - x^3 - 27xt^2 + 18t^2.$$

5.

$$u(x,t) = \begin{cases} 4x - 4x^2 + x^3 + 3xt^2 - 4t^2 + x^2t + \frac{1}{3}t^3 & \text{in I,} \\ -4t^2 - 4x^2 + 4xt^2 + \frac{4}{3}x^3 + 4x & \text{in II,} \\ -\frac{28}{3} + 10x + 8t - 5x^2 - 5t^2 + \frac{7}{6}x^3 + \frac{7}{2}xt^2 - 2xt & \text{in III.} \end{cases}$$

It is routine to check that the expressions in regions I and III agree on their common boundary segments.

3.3 The Wave Equation in Higher Dimensions

1.

$$u(x,y,t) = \sum_{m=1}^{\infty} 72c_m \sin(\pi x/3)\sin(m\pi y/6)\cos(\alpha_{1m}t),$$

where

$$\alpha_{1m} = \sqrt{\frac{\pi^2}{9} + \frac{m^2\pi^2}{36}}$$

and

$$c_m = \frac{(-1)^{m+1}}{m\pi} + \frac{m^2\pi^2(-1)^m - 2(-1)^m + 1}{m^3\pi^3}.$$

5.

$$u(x, y, t) = \sum_{n=1}^{\infty}\sum_{m=1}^{\infty}(a_{nm}\cos(\alpha_{nm}t) + b_{nm}\sin(\alpha_{nm}t))\sin(n\pi x)\sin(m\pi y),$$

where

$$\alpha_{nm} = \pi\sqrt{n^2 + m^2},$$

$$a_{nm} = \frac{16}{n^3m^3\pi^6}(2 + (-1)^n)((-1)^m - 1),$$

and

$$b_{nm} = \frac{4}{\alpha_{nm}}\frac{(-1)^{n+1}}{n\pi}\left(\frac{m^2\pi^2(-1)^{m+1} + 2(-1)^m - 2}{m^3\pi^3}\right).$$

Chapter 4: Dirichlet and Neumann Problems

4.1 Laplace's Equation and Harmonic Functions

1. $u(x, y) = x^3 - 3xy^2 - 4x^2 + 4y^2 + x - 1$ and $v(x, y) = 3x^2y - y^3 - 8xy + y$. Check that $u_{xx} + u_{yy} = 0$ and $v_{xx} + v_{yy} = 0$.

3. From the equations of the transformation, we find that

$$x = x^*\cos(\theta) - y^*\sin(\theta), y = x^*\sin(\theta) + y^*\cos(\theta).$$

Now use the chain rule to compute the partial derivatives of w, beginning with

$$\begin{aligned}\frac{\partial w}{\partial x^*} &= \frac{\partial u}{\partial x}\frac{\partial x}{\partial x^*} + \frac{\partial u}{\partial y}\frac{\partial y}{\partial x^*}\\ &= u_x\cos(\theta) - u_y\sin(\theta),\end{aligned}$$

and so on for the other derivatives.

4.2 The Dirichlet Problem for a Rectangle

1.

$$u(x, y) = \frac{1}{1 - e^{2\pi^2}}\sin(nx)\left(e^{\pi y} - e^{2\pi^2}e^{-\pi y}\right).$$

3.

$$u(x,y)=\sum_{n=1}^{\infty}\frac{32}{\pi^2\sinh(4n\pi)}\frac{n(-1)^{n+1}}{(4n^2-1)^2}\sin(n\pi x)\sinh(n\pi y).$$

5. Write $u(x,y)=u_1(x,y)+u_2(x,y)$, where $u_1(x,y)$ is zero on three sides of the rectangle, and $u_1(x,\pi)=x\sin(\pi x)$, while $u_2(x,y)$ is zero on three sides of the rectangle, and $u_2(2,y)=\sin(y)$. We find that

$$u_1(x,y)=\sum_{n=1}^{\infty}b_n\sin(n\pi x/2)\sinh(n\pi y/2),$$

where

$$b_n=\begin{cases}\frac{16n}{\pi^2\sinh(n\pi^2/2)}\frac{(-1)^n-1}{(n^2-4)^2} & \text{for } n\neq 2,\\ \frac{1}{\sinh(\pi^2)} & \text{for } n=2,\end{cases}$$

and

$$u_2(x,y)=\frac{1}{\sinh(2)}\sin(y)\sinh(x).$$

7.

$$u(x,y)=\sum_{n=1}^{\infty}b_n\sin\left(\frac{(2n-1)\pi}{2L}x\right)\sinh\left(\frac{(2n-1)\pi}{2L}y\right),$$

where

$$b_n=\frac{2}{L\sinh((2n-1)\pi K/2L)}\int_0^L f(\xi)\sin\left(\frac{(2n-1)\pi}{2L}\xi\right)d\xi.$$

4.3 The Dirichlet Problem for a Disk

1. $u(r,\theta)=\frac{1}{2}+\frac{1}{2}\left(\frac{r}{\rho}\right)^2\cos(2\theta)$.

3. $u(r,\theta)=\frac{1}{8}\left[1-\left(\frac{r}{4}\right)\right]\cos(4\theta)$.

5. $u(x,y)=32+x^2y^2-\frac{1}{8}(x^2+y^2)^2$.

7. The solution is

$$u(r,\theta)=\frac{1}{2}a_0+\sum_{n=1}^{\infty}(a_nr^n\cos(n\theta)+b_nr^n\sin(n\theta))$$
$$+\sum_{n=1}^{\infty}(c_nr^{-n}\cos(n\theta)+d_nr^{-n}\sin(n\theta))+k\ln(r),$$

where

$$a_0=\frac{1}{\pi\ln(\rho_2/\rho_1)}\left[\ln(\rho_2)\int_{-\pi}^{\pi}g(\theta)\,d\theta-\ln(\rho_1)\int_{-\pi}^{\pi}f(\theta)\,d\theta\right],$$

$$k = \frac{1}{2\pi \ln(\rho_2/\rho_1)} \int_{-\pi}^{\pi} (f(\theta) - g(\theta))\, d\theta,$$

$$a_n = \frac{1}{\pi(\rho_1^{2n} - \rho_2^{2n})} \left[\rho_1^n \int_{-\pi}^{\pi} g(\theta)\cos(n\theta)\, d\theta - \rho_2^n \int_{-\pi}^{\pi} f(\theta)\cos(n\theta)\, d\theta \right],$$

$$c_n = \frac{1}{\pi(\rho_2^{-2n} - \rho_1^{-2n})} \left[\rho_2^{-n} \int_{-\pi}^{\pi} f(\theta)\cos(n\theta)\, d\theta - \rho_1^{-n} \int_{-\pi}^{\pi} g(\theta)\cos(n\theta)\, d\theta \right],$$

$$b_n = \frac{1}{\pi(\rho_1^{2n} - \rho_2^{2n})} \left[\rho_1^n \int_{-\pi}^{\pi} g(\theta)\sin(n\theta)\, d\theta - \rho_2^n \int_{-\pi}^{\pi} f(\theta)\sin(n\theta)\, d\theta \right],$$

and

$$d_n = \frac{1}{\pi(\rho_2^{-2n} - \rho_1^{-2n})} \left[\rho_2^{-n} \int_{-\pi}^{\pi} f(\theta)\sin(n\theta)\, d\theta - \rho_1^{-n} \int_{-\pi}^{\pi} g(\theta)\sin(n\theta)\, d\theta \right].$$

9. $u(r,\theta) = 1 - \frac{1}{\ln(2)}\ln(r) + \frac{2}{3}r\cos(\theta) - \frac{2}{3}r^{-1}\cos(\theta)$.

11. $u(r,\theta) = 1 - \frac{1}{2\ln(2)}\ln(r) + \frac{2}{15}r^2\cos(2\theta) - \frac{2}{15}r^{-2}\cos(2\theta)$.

13. $u(r,\theta) = \frac{1}{2} + \frac{1}{24}r^2\cos(2\theta) - \frac{8}{3}r^{-2}\cos(2\theta)$.

17. $\int_{-\pi}^{\pi} \frac{\cos(n\xi)}{5-4\cos(\xi)}\, d\xi = \frac{\pi}{3(2^{n-1})}$.

4.4.1 Topology of R^n

1. Interior points are points (x,y) with $0 < x < 1, 0 < y < 1$. Boundary points are points $(x,0)$ and $(x,1)$, with $0 \le x \le 1$ and points $(0,y)$ and $(1,y)$ with $0 \le y \le 1$. The set is not open, because it contains some of its boundary points (for example, $(0,1/2)$). The set is not closed because it does not contain all of its boundary points (for example $(0,0)$. S is bounded and connected, but is not compact (because S is not closed).

3. Q is open, because it consists only of interior points. The boundary of Q consists of all (x,y) with $x^2+y^2 = 4$ and all (x,y) with $x^2+y^2 = 1$. The set is not closed because it does not contain its boundary points. Q is bounded and connected, but not compact.

5. Every point in the plane is a boundary point of M (because any rational number has irrational numbers arbitrarily close). Since M contains some of its boundary points, M is not open. But M does not contain all of its boundary points, so M is not closed. M is not bounded, and is not connected (any line between two points of M contains points with an irrational coordinate). M is not compact.

7. Boundary points are all points $(-3,y)$ and points $(-2,y)$. Because P contains some boundary points, P is not open. But P does not contain all of its boundary points, so P is not closed. P is not connected, not bounded, and not compact.

9. The boundary points of B are all points (x, y, z) with $x^2 + y^2 + z^2 = 4$. Since B contains its boundary points, B is closed, and also not open. B is connected, but not bounded and not compact.

11. Suppose x is a boundary point of A. Then every neighborhood of x contains a point of A and a point of A^C (that is, a point not in A). But this makes x a boundary point of A^C because $(A^C)^C = A$.

13. R^n is open because every point of R^n is an interior point. R^n is also closed because R^n has no boundary points, hence there is no boundary point of R^n that is not in R^n.

    The empty set $\varnothing$ is therefore both open and closed, by the conclusion of problem 11, because $(R^n)^C = \varnothing$.

4.4.2 Representation Theorems

1. If $u(\mathbf{x}) = 1$ for all $\mathbf{x} \in \Omega$, then

$$\iint_{S(\mathbf{x},\epsilon)} u(\mathbf{y})\, d\sigma_y = \iint_{S(\mathbf{x},\epsilon)} d\sigma_y,$$

   and this is the surface area of $S(\mathbf{x}, \epsilon)$, which is $4\pi\epsilon^2$. Thus in this case Theorem 4.3 simply reaffirms that $u(\mathbf{x}, \epsilon) = 1$.

3. Parametrize $\partial\Omega$ by arc length along the curve, with $x = x(s), y = y(s)$. The unit normal to $\partial\Omega$ is

$$\mathbf{n} = y'(s)\mathbf{i} - x'(s)\mathbf{j}.$$

   Show that

$$\begin{aligned} &\int_{\partial\Omega} \left(u\frac{\partial u}{\partial n} - v\frac{\partial u}{\partial n} \right) ds \\ &= \int_{\partial\Omega} \left(v\frac{\partial u}{\partial y} - u\frac{\partial v}{\partial y} \right) dx + \left(u\frac{\partial v}{\partial x} - v\frac{\partial u}{\partial x} \right) dy. \end{aligned}$$

   Now apply Green's theorem to the last line integral.

4.4.3 The Mean Value Property and the Maximum Principle

1. Parametrize $\partial\Omega$ by arc length, $x = x(s), y = y(s)$, so the unit outer normal is

$$\frac{dy}{ds}\mathbf{i} - \frac{dx}{ds}\mathbf{j}.$$

   Check that

$$\oint_{\partial\Omega} f\, ds = \oint_{\partial\Omega} u_x\, dy - u_y\, dx$$

   and use Green's theorem to show that

$$\oint_{\partial\Omega} f\, ds = \iint_{\Omega} \nabla^2 u\, dA.$$

3. Follow the hint to problem 2, using surface and triple integrals and Gauss's theorem for the 3−space scenario.

5. Suppose u and v are solutions, let $w = u - v$ and use the conditions of the problem and integral theorems to show that w is identically zero.

4.5.2 Neumann's Problem for a Rectangle

In these solutions, α_0 and β_0 are arbitrary constants reflecting the non-uniqueness of solutions of Neumann problems.

1.
$$u(x,y) = \alpha_0 + \frac{4}{\pi(e^{2\pi}-1)}\left(e^{\pi y} + e^{\pi(2-y)}\right)\cos(\pi x).$$

3. Write $u = u_1 + u_2$, where all boundary conditions are zero in problem 1 except for $u_y(x,0) = \cos(3x)$, and all boundary conditions are zero in problem 2 except for $u_y(x,\pi) = 6x - 3\pi$. Obtain
$$u_1(x,y) = \alpha_0 + \frac{1}{3(1-e^6)}\left(e^{3y} - e^{3(2\pi-y)}\right)\cos(3x)$$
and
$$u_2(x,y) = \beta_0 + \sum_{n=1}^{\infty}\frac{12}{\pi n^3\sinh(n\pi)}((-1)^n - 1)\cosh(ny)\cos(nx).$$

5.
$$u(x,y) = \sum_{n=1}^{\infty} a_n\left(e^{n\pi x} - e^{n\pi(2-x)}\right)\sin(n\pi y),$$
where
$$a_n = \frac{2}{n\pi(1-e^{2n\pi})}\left(\frac{-n^2\pi^2(-1)^n + 6((-1)^n-1)}{n^3\pi^3}\right).$$

4.5.3 Neumann Problem for a Disk

1. By equation 4.16, obtain
$$u(r,\theta) = \alpha_0 + \frac{1}{3\rho^2}r^3\sin(3\theta).$$
Using equation 3.18, obtain
$$u(r,\theta) = \alpha_0 - \frac{\rho}{2\pi}\int_{-\pi}^{\pi}\ln\left(1 + \frac{r^2}{\rho^2} - 2\left(\frac{r}{\rho}\right)\cos(\theta-\xi)\right)\sin(3\xi)\,d\xi.$$

3. By inspection, $u(x,y) = \alpha_0 + 4xy$.

5. $u(x,y) = \alpha_0 + \frac{1}{4}(x - x^2 + 3xy^2)$.

4.6 Poisson's Equation

1. $u(x,y) = Y_5(y)\sin(5x)$, where

$$Y_5(y) = c_1 e^{5y} + c_2 e^{-5y} - \frac{1}{25}$$

and

$$c_1 = \frac{e^{-15} - 1}{25(e^{-15} - e^{15})},\, c_2 = \frac{1 - e^{15}}{25(e^{-15} - e^{15})}.$$

3.

$$u(x,y) = \sum_{n=1}^{\infty} Y_n(y)\sin(nx),$$

where

$$Y_n(y) = \frac{2(-1)^{n+1}}{n^3 \sinh(n)}\sinh(ny) + \frac{2(-1)^n y}{n^3}.$$

4.7 An Existence Theorem for the Dirichlet Problem

1. In one direction, observe that, if v is harmonic on Ω, then $f_{K,v} = v$ for every open disk $K \subset \Omega$.

3. By the conclusion of problem 3 of section 4.1, the rotation, applied to the solution $u(x,y)$ for Ω, produces a function $u^*(x^*, y^*)$ that is harmonic on Ω^*, the rotated domain. Finally, show that, on the boundary of Ω^*, the rotated function takes on the rotated data function from the problem for Ω.

Chapter 5: Fourier Integral Methods of Solution

5.1 The Fourier Integral of a Function

1. The Fourier integral is

$$\int_0^{\infty} \frac{2}{\pi} \frac{\sin(\pi\omega) - \pi\omega\cos(\pi\omega)}{\omega^2} \sin(\omega x)\, d\omega,$$

converging to

$$\begin{cases} x & \text{for } -\pi < x < \pi, \\ 0 & \text{for } |x| > \pi, \\ -\pi/2 & \text{at } x = -\pi, \\ \pi/2 & \text{at } x = \pi. \end{cases}$$

3. The Fourier integral is

$$\int_0^\infty \frac{2}{\pi(1+\omega^2)} \cos(\omega x)\, d\omega,$$

converging to $e^{-|x|}$ for all x.

5. The Fourier integral is

$$\int_0^\infty \frac{2}{\pi\omega^2}(\cos(\alpha\omega) + \alpha\omega\sin(\alpha\omega))\cos(\omega x)\, d\omega,$$

converging to

$$\begin{cases} |x| & \text{for } -\alpha < x < \alpha, \\ \alpha/2 & \text{for } x = \pm\alpha, \\ 0 & \text{for } |x| > \alpha. \end{cases}$$

9. The sine integral is

$$\int_0^\infty B_\omega \sin(\omega x)\, d\omega,$$

where

$$B_\omega = \frac{2}{\pi(1+\omega^2)}(-\omega\cos(k\omega)\sinh(k) + \sin(k\omega)\sinh(k)).$$

This sine integral converges to

$$\begin{cases} \sinh(x) & \text{for } 0 < x < k, \\ \sinh(k)/2 & \text{at } x = k, \\ 0 & \text{for } x = 0 \text{ and } x > k. \end{cases}$$

The cosine integral is

$$\int_0^\infty A_\omega \cos(\omega x)\, d\omega,$$

where

$$A_\omega = \frac{1}{\pi(1+\omega^2)}(2\omega + \sin(k\omega)\sinh(k) - 2\omega\cos(k\omega)\cosh(k)).$$

This cosine integral converges to

$$\begin{cases} \sinh(x) & \text{for } 0 \le x < k, \\ \sinh(k)/2 & \text{at } x = k, \\ 0 & \text{for } x > k. \end{cases}$$

11. The sine integral is

$$\frac{2}{\pi}\int_0^\infty \frac{\omega^3}{(2-2\omega+\omega^2)(2+2\omega+\omega^2)}\sin(\omega x)\,d\omega,$$

converging to $e^{-x}\cos(x)$ for $x > 0$ and to 0 at $x = 0$.

The cosine integral is

$$\frac{2}{\pi}\int_0^\infty \frac{2+\omega^2}{(2-2\omega+\omega^2)(2+2\omega+\omega^2)}\cos(\omega x)\,d\omega,$$

converging to $e^{-x}\cos(x)$ for $x \geq 0$.

13. The sine integral is

$$\frac{2k}{\pi}\int_0^\infty \frac{1-\cos(\alpha\omega)}{\omega}\sin(\omega x)\,d\omega,$$

converging to

$$\begin{cases} k & \text{for } 0 < x < \alpha, \\ k/2 & \text{at } x = \alpha, \\ 0 & \text{at } x = 0 \text{ and for } x > \alpha. \end{cases}$$

The cosine integral is

$$\frac{2k}{\pi}\int_0^\infty \frac{\sin(\alpha\omega)}{\omega}\cos(\alpha x)\,d\omega,$$

converging to

$$\begin{cases} k & \text{for } 0 \leq x < \alpha, \\ k/2 & \text{for } x = \alpha, \\ 0 & \text{for } x > \alpha. \end{cases}$$

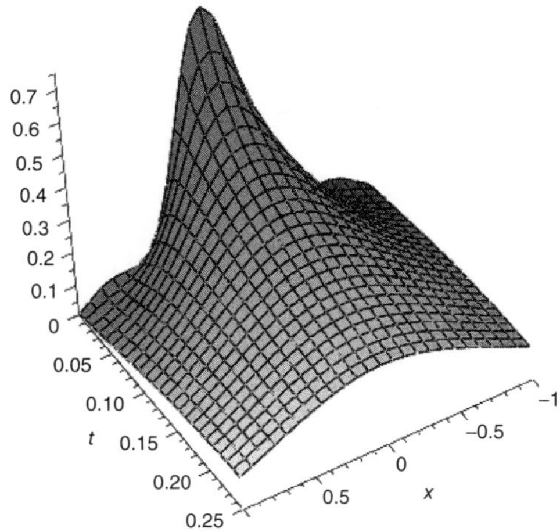

Figure 9.42: $u(x,t)$ in problem 3, with $k = 1$ and $-1 \leq x \leq 1, 0 \leq t \leq 1/4$.

5.2 The Heat Equation on the Real Line

1. If $f(x)$ is an odd function, then $f(x)\cos(\omega x)$ is odd, so $a_\omega = 0$ and

$$u(x,t) = \int_0^\infty b_\omega \sin(\omega x) e^{-k\omega^2 t}\, d\omega$$

is odd in the space variable. Similar reasoning applies if $f(x)$ is even.

3. By using the Fourier integral, obtain

$$u(x,t) = \frac{8}{\pi}\int_0^\infty \frac{1}{16+\omega^2}\cos(\omega x) e^{-\omega^2 kt}\, d\omega.$$

Using equation 5.9, we can also write

$$u(x,t) = \frac{1}{2\sqrt{\pi k t}}\int_{-\infty}^\infty e^{-(x-\xi)^2/4kt} e^{-4|\xi|}\, d\xi.$$

Figure 9.42 shows part of this solution surface.

5. By Fourier integral, we obtain

$$u(x,t) = \int_0^\infty (a_\omega \cos(\omega x) + b_\omega \sin(\omega x)) e^{-\omega^2 kt}\, d\omega,$$

where

$$a_\omega = \frac{2}{\pi\omega^2}(\cos(4\omega) + 4\omega\sin(4\omega) - 1)$$

and

$$b_\omega = \frac{2}{\pi\omega^2}(\sin(4\omega) - 4\omega\cos(4\omega)).$$

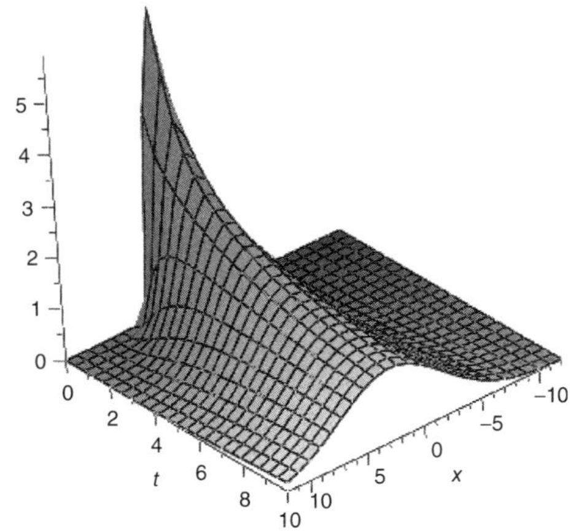

Figure 9.43: $u(x,t)$ in problem 5.

Another expression for the solution is obtained using equation 4.9:

$$u(x,t) = \frac{1}{2\sqrt{\pi k t}} \int_0^4 e^{-(x-\xi)^2/4kt} \xi \, d\xi.$$

Figure 9.43 shows part of the surface $u(x,t)$ with $k = 1$, $-12 \leq x \leq 12$ and $0 \leq t \leq 10$.

7. Using the Fourier integral, obtain

$$u(x,t) = \frac{2}{\pi} \int_0^\infty \frac{1-\cos(\omega)}{\omega} \sin(\omega x) e^{-\omega^2 kt} \, d\omega.$$

Using equation 4.9, we can also write the solution as

$$\begin{aligned} u(x,t) = &\frac{1}{2\sqrt{\pi k t}} \int_0^1 e^{-(x-\xi)^2/4kt} \, d\xi \\ &- \frac{1}{2\sqrt{\pi k t}} \int_{-1}^0 e^{-(x-\xi)^2/4kt} \, d\xi. \end{aligned}$$

13. By Fourier integral, the solution is

$$u(x,t) = \frac{4\alpha}{\pi} \int_0^\infty \frac{\omega}{(\alpha^2+\omega^2)^2} \sin(\omega x) e^{-\omega^2 kt} \, d\omega.$$

We can also write the solution in the form

$$u(x,t) = \frac{1}{2\sqrt{\pi k t}} \int_0^\infty \left(e^{-(x-\xi)^2/4kt} - e^{-(x+\xi)^2/4kt} \right) \xi e^{-\alpha\xi} \, d\xi.$$

15. By Fourier integral, the solution is

$$u(x,t) = \frac{4}{\pi}\int_0^\infty \frac{\sin(\omega)\cos(\omega) - 2\omega\cos^2(\omega) + \omega}{\omega^2}\sin(\omega x)e^{-\omega^2 kt}\,d\omega.$$

We can also write

$$u(x,t) = \frac{1}{2\sqrt{\pi kt}}\int_0^2 \left(e^{-(x-\xi)^2/4kt} - e^{-(x+\xi)^2/4kt}\right)\xi\,d\xi.$$

17.

$$u(x,t) = \frac{A}{\pi}\int_0^\infty \frac{1}{\omega^2}((\cos(h\omega)-1)\cos(\omega x) + \sin(h\omega)\sin(\omega x))e^{-\omega^2 kt}\,d\omega.$$

5.4 Burger's Equation

1. All that is needed is to determine $g(x)$ to insert into the solution 5.15. Given $f(x)$, then

$$g(x) = \begin{cases} e^{-x(1+x/2)/2k} & \text{for } -1 < x < 0, \\ e^{-x(1-x/2)/2k} & \text{for } 0 < x < 1, \\ e^{-1/4k} & \text{for } x < -1 \text{ and } x > 1. \end{cases}$$

3. Use equation 5.15, with

$$g(x) = \begin{cases} e^{-3/k} & \text{for } x < -3, \\ e^{x/k} & \text{for } -3 < x < 0, \\ e^{-5x/2k} & \text{for } 0 < x < 2, \\ e^{-5/k} & \text{for } x > 2. \end{cases}$$

7.

$$u(x,t) = \frac{10 + 5e^{5(x-15t/2)/6}}{1 + e^{5(x-15t/2)/6}}.$$

9.

$$u(x,t) = \frac{2 + e^{(x-3t/2)/16}}{1 + e^{(x-3t/2)/16}}.$$

5.5 The Cauchy Problem for the Wave Equation

1.
$$u(x,t) = \frac{2}{\pi}\int_0^\infty \frac{1}{1+\omega^2} cos(\omega x)\cos(\omega ct)\, d\omega.$$

3.
$$u(x,t) = \frac{2}{\pi}\int_0^\infty \frac{\sin(\pi\omega)}{1-\omega^2}\sin(\omega x)\cos(\omega ct)\, d\omega.$$

5.
$$u(x,t) = \frac{1}{\pi}\int_0^\infty \left[\frac{1}{1+\omega^2}\cos(\omega x) + \frac{\omega}{1+\omega^2}\sin(\omega x)\right]\cos(\omega ct)\, d\omega.$$

7.
$$u(x,t) = \int_0^\infty [a_\omega \cos(\omega x) + b_\omega \sin(\omega x)]\sin(\omega ct)\, d\omega,$$

where
$$a_\omega = \frac{1}{\pi\omega c}\int_{-\infty}^\infty \psi(s)\cos(\omega s)\, ds$$

and
$$b_\omega = \frac{1}{\pi\omega c}\int_{-\infty}^\infty \psi(s)\sin(\omega s)\, ds.$$

9.
$$u(x,t) = \frac{2k}{\pi c}\int_0^\infty \frac{\sin(\alpha\omega)}{\omega^2}\cos(\omega x)\sin(\omega ct)\, d\omega.$$

11.
$$u(x,t) = \int_0^\infty \left[\frac{2}{\pi\omega^3 c}\sin(5\omega) - \frac{10}{\pi\omega^2 c}\cos(5\omega)\right]\sin(\omega x)\sin(\omega ct)\, d\omega.$$

13.
$$u(x,t) = \int_0^\infty (a_\omega \cos(\omega x) + b_\omega \sin(\omega x))\sin(\omega ct)\, d\omega,$$

where
$$a_\omega = \frac{1}{\pi\omega c}\int_0^\pi \sin(s)\cos(\omega s)\, ds = \begin{cases} \frac{1+\cos(\pi\omega)}{\pi\omega c(\omega^2-1)} & \text{if } \omega \neq 1, \\ 0 & \text{if } \omega = 1 \end{cases}$$

and
$$b_\omega = \frac{1}{\pi\omega c}\int_0^\pi \sin(s)\sin(\omega s)\, ds = \begin{cases} \frac{\sin(\pi\omega)}{\pi\omega c(1-\omega^2)} & \text{if } \omega \neq 1, \\ \frac{1}{2\pi c} & \text{if } \omega = 1. \end{cases}$$

5.6 Laplace's Equation on Unbounded Domains

1.
$$u(x,y) = \frac{y}{\pi}\int_{-\infty}^{\infty} \frac{e^{-6\xi}}{y^2+(\xi-x)^2}\,d\xi.$$

3.
$$u(x,y) = \frac{y}{\pi}\int_{-\pi/2}^{\pi/2} \frac{\cos(\xi)}{y^2+(\xi-x)^2}\,d\xi.$$

5.
$$u(x,y) = \frac{y}{\pi}\int_{0}^{\infty} \left(\frac{1}{y^2+(\xi-x)^2} - \frac{1}{y^2+(\xi+x)^2}\right)\xi\cos(\xi)\,d\xi.$$

7.
$$u(x,y) = \frac{y}{\pi}\int_{-\infty}^{0} \left[\frac{1}{y^2+(\xi-x)^2} - \frac{1}{y^2+(\xi+x)^2}\right] f(\xi)\,d\xi.$$

9. Split the problem into two problems. In problem 1, $u(x,0) = 0$ and $u(0,y) = k$, whereas in problem 2, $u(x,0) = f(x)$ and $u(0,y) = 0$. Then

$$\begin{aligned} u(x,y) &= u_1(x,y) + u_2(x,y) \\ &= \frac{2k}{\pi}\int_0^{\infty} \frac{1}{\omega}\sin(\omega y)e^{\omega x}\,d\omega + \int_0^{\infty} c_\omega \sin(\omega x)e^{-\omega y}\,d\omega, \end{aligned}$$

where

$$c_\omega = \frac{2}{\pi}\int_0^{\infty} f(-\xi)\sin(\omega\xi)\,d\xi.$$

11.
$$u(x,y) = -\frac{1}{2\pi}\int_{-\infty}^{\infty} \ln(y^2+(\xi-x)^2)f(\xi)\,d\xi + c.$$

13.
$$u(x,y) = \int_0^{\infty} [a_\omega\cos(\omega y) + b_\omega\sin(\omega y)]e^{-\omega x}\,d\omega + c,$$

with

$$a_\omega = -\frac{1}{\pi\omega}\int_{-\infty}^{\infty} g(\xi)\cos(\omega\xi)\,d\xi$$

and

$$b_\omega = -\frac{1}{\pi\omega}\int_{-\infty}^{\infty} g(\xi)\sin(\omega\xi)\,d\xi.$$

15.
$$u(x,y) = \int_0^{\infty} b_\omega \sin(\omega x)e^{-\omega y}\,d\omega,$$

with

$$b_\omega = -\frac{2}{\pi\omega}\int_0^{\infty} f(\xi)\sin(\omega\xi)\,d\xi.$$

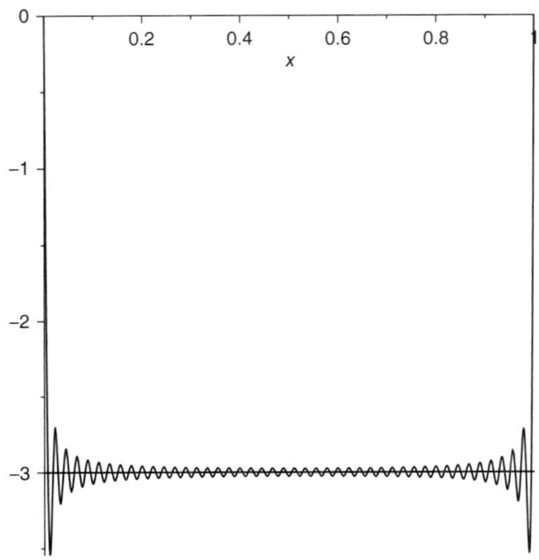

Figure 9.44: 90th partial sum of the eigenfunction expansion of problem 1.

Chapter 6: Solutions Using Eigenfunction Expansions

6.1 A Theory of Eigenfunction Expansions

1. Eigenvalues and eigenfunctions are

$$\lambda_n = -2 + n^2\pi^2, y_n(x) = e^x \sin(n\pi x).$$

The eigenfunction expansion of $f(x) = -3$ on $[0, 1]$ converges to -3 for $-1 < x < 1$. This expansion is

$$\sum_{n=1}^{\infty} b_n e^x \sin(n\pi x).$$

For the coefficients, write the differential equation in standard form to read the weight function $p(x) = e^{-2x}$. Then

$$\begin{aligned} b_n &= \frac{\int_0^1 -3e^{-2x} e^x \sin(n\pi x)\, dx}{\int_0^1 e^{-2x}(e^x \sin(n\pi x))^2\, dx} \\ &= \frac{6(-n\pi + e^{-1} n\pi(-1)^n)}{1 + n^2\pi^2}. \end{aligned}$$

Figure 9.44 compares $f(x)$ and the 90th partial sum of this expansion.

3. Eigenvalues and eigenfunctions are

$$\lambda_n = \frac{(2n-1)^2\pi^2}{4}, y_n(x) = \cos((2n-1)\pi x/2).$$

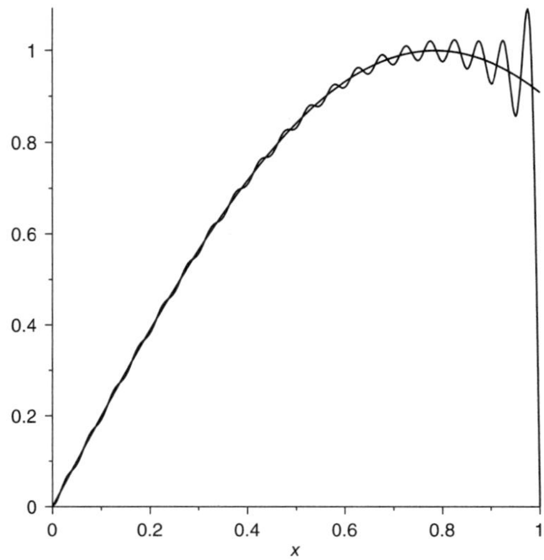

Figure 9.45: 40th partial sum of the eigenfunction expansion of problem 3.

The eigenfunction expansion is

$$\sum_{n=1}^{\infty} c_n \cos((2n-1)\pi x/2),$$

with

$$c_n = \frac{-16 - 8\pi n \sin(2)(-1)^n + 4n\pi(-1)^n}{-16 + 4n^2\pi^2 - 4n\pi^2 + \pi^2}.$$

Figure 9.45 compares $\sin(x)$ with the 40th partial sum of this expansion on $[0, 1]$.

5. Eigenvalues and eigenfunctions are

$$\lambda_n = \frac{n^2\pi^2}{4}, y_n = e^{3x/2}\sin(n\pi x/2).$$

For $0 < x < 2$, the expansion is

$$x = \sum_{n=1}^{\infty} c_n e^{3x/2}\sin(n\pi x/2),$$

where

$$c_n = \frac{-4(-6n\pi + e^{-3}n^3\pi^3(-1)^n + 15e^{-3}n\pi(-1)^n)}{(n^2\pi^2+9)^2}.$$

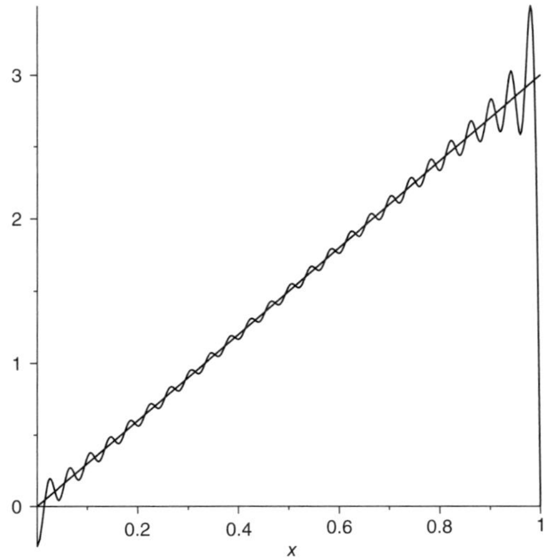

Figure 9.46: 50th partial sum of the Fourier-Bessel expansion of $3x$ on $(0, 1)$.

6.2 Bessel Functions

1. $y = c_1 x J_1(x) + c_2 x Y_1(x)$.

3.
$$y = c_1 x^6 J_{\sqrt{10/3}}(x^3) + c_2 x^6 J_{-\sqrt{10/3}}(x^3).$$

5.
$$y = c_1 x^{12} J_{10/3}\left(\frac{2}{3}x^3\right) + c_2 x^{12} J_{-10/3}\left(\frac{2}{3}x^3\right).$$

7.
$$y = c_1 x^2 J_{1/3}(x^2) + c_2 x^2 J_{-1/3}(x^2).$$

9. *Hint*: use a recurrence relation.

13. Figure 9.46 compares the function with this 50th partial sum.

The expansion is

$$\sum_{n=1}^{\infty} c_n J_0(j_n x),$$

where j_n is the nth positive zero of $J_0(x)$ (labeled in increasing order). To compute (approximately) these coefficients using MAPLE, first form a sequence of the zeros:

$$j : n \to \text{BesselJZeros}(0, j(n))$$

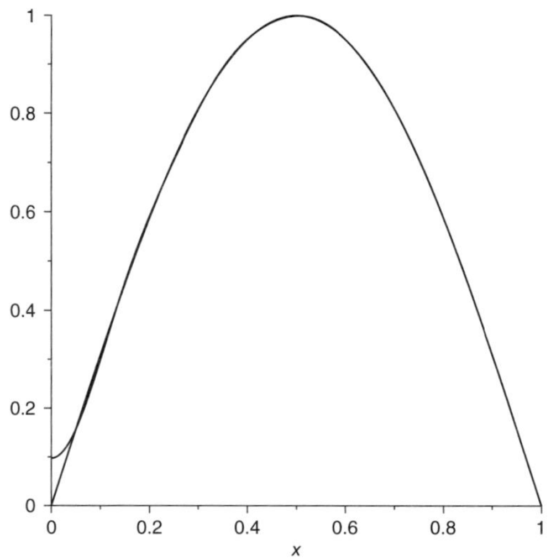

Figure 9.47: 10th partial sum of the Fourier-Bessel expansion of $\sin(\pi x)$ on $(0, 1)$.

and then define the sequence of coefficients

$$c := n \to \frac{2}{\text{BesselJ}(1, j(n)) \wedge 2} * \text{int}(3 * x \wedge 2 * \text{BesselJ}(0, j(n) * x), x = 0..1)$$

in MAPLE code. The fiftieth partial sum is the function

$$S := x - > \ \text{sum}(c(n) * \text{BesselJ}(0, j(n) * x), n = 1..50)$$

again in MAPLE code.

15. Figure 9.47 shows the 10th partial sum of this eigenfunction expansion, compared with $\sin(\pi x)$ on $(0, 1)$.

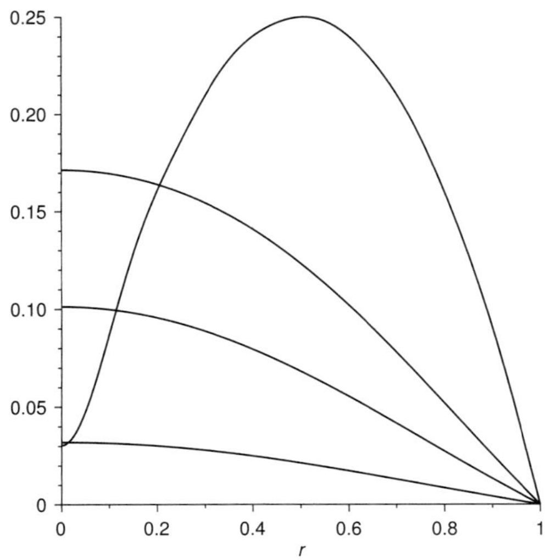

Figure 9.48: $U(r,t)$ decreasing through selected times in problem 1.

6.3.1 Temperature Distribution in a Solid Cylinder

1.

$$U(r,t) = \sum_{n=1}^{\infty} c_n J_0(j_n r) e^{-2j_n^2 t}$$

with

$$c_n = \frac{2}{(J_1(j_n))^2} \int_0^1 \xi^2(1-\xi) J_0(j_n \xi)\, d\xi.$$

Figure 9.48 shows $U(r,t)$ decreasing through times $t = 0, 1/20,\ 1/10,\ 1/5$.

3.

$$U(r,t) = \sum_{n=1}^{\infty} c_n J_0(j_n r/3) e^{-2j_n^2 t/9},$$

where

$$c_n = \frac{2}{(J_1(j_n))^2} \int_0^1 27\xi^3(1-\xi) J_0(j_n \xi)\, d\xi.$$

Figure 9.49 shows $U(r,t)$ for $t = 0, 1/10, 1/2, 1$.

5.

$$U(r,t) = \sum_{n=1}^{\infty} c_n J_0(j_n r) e^{-3j_n^2 t},$$

where

$$c_n = \frac{2}{(J_1(j_n))^2} \int_0^1 \xi e^{\xi}(1-\xi) J_0(j_n \xi)\, d\xi.$$

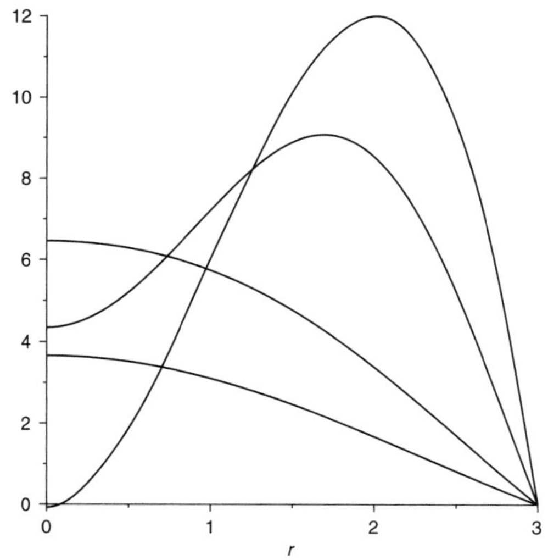

Figure 9.49: $U(r,t)$ at selected times in problem 3.

Figure 9.50 shows $U(r,t)$ for $t = 0, 1/10, 1/5$.

6.3.2 Vibrations of a Circular Drum

1.

$$z(r,t) = \sum_{n=1}^{\infty} c_n J_0(j_n r)\cos(4j_n t),$$

where

$$c_n = \frac{2}{J_1^2(j_n)}\int_0^1 \xi^3(1-x)J_0(j_n\xi)\,d\xi.$$

Figure 9.51 shows wave profiles $z(r,0)$ for $t = 0, 1/20, 1/15, 1/10$, with the waves moving upward from the initial position over this time frame. Figure 9.52 shows the surface $z(t,r)$ for $0 \le r \le 1$ and $0 \le t \le 2$.

3.

$$z(r,t) = \sum_{n=1}^{\infty} c_n J_0(j_n r/5)\cos(2j_n t/5),$$

where

$$c_n = \frac{2}{j_1^2(j_n)}\int_0^1 5\xi^2 J_0(j_n\xi)\,d\xi.$$

Figure 9.53 shows waves moving upward through times $t = 0, 1/2, 1$, and Figure 9.54 is the surface $z(r,t)$ for $0 \le r \le 5$ and $0 \le t \le 1$.

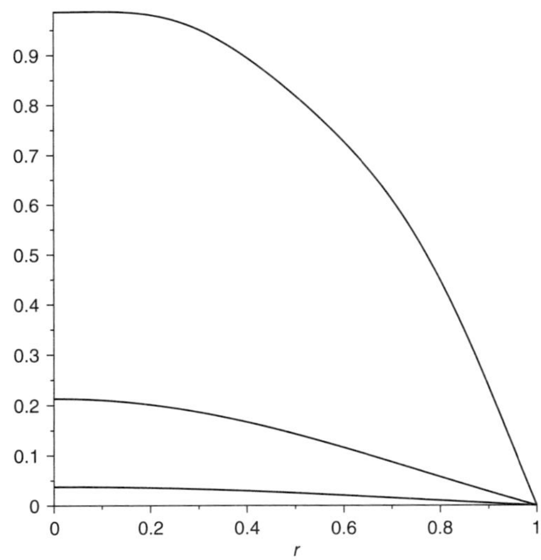

Figure 9.50: $U(r,t)$ at selected times in problem 5.

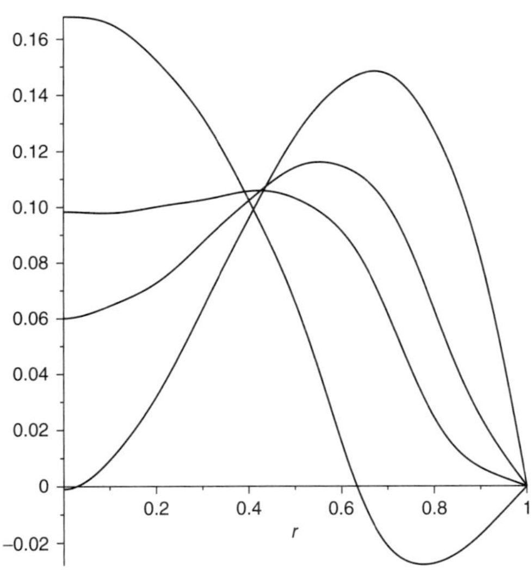

Figure 9.51: $z(r,t)$ at selected times in problem 1.

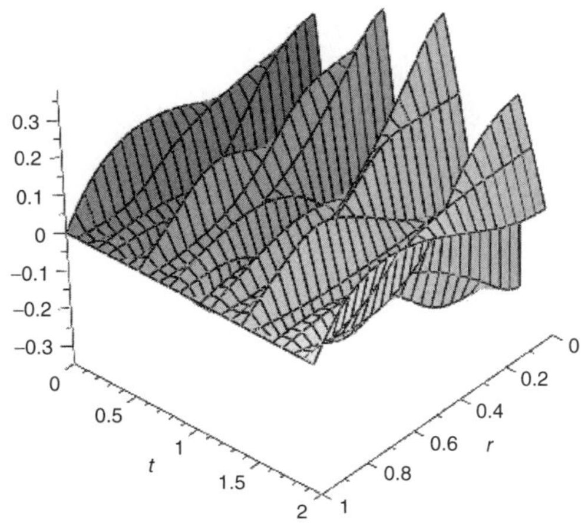

Figure 9.52: $z(r,t)$ in problem 1.

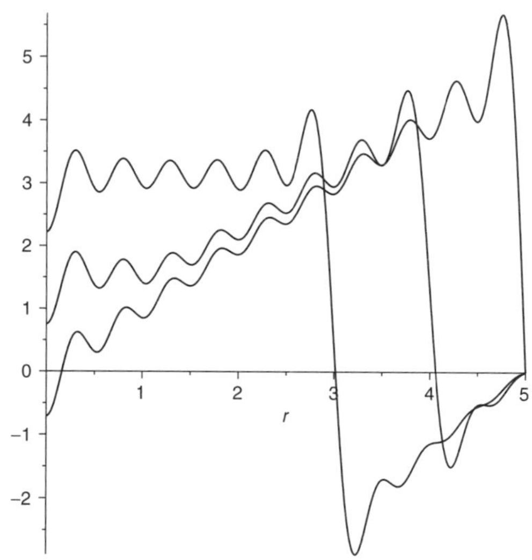

Figure 9.53: $z(r,t)$ at selected times in problem 3.

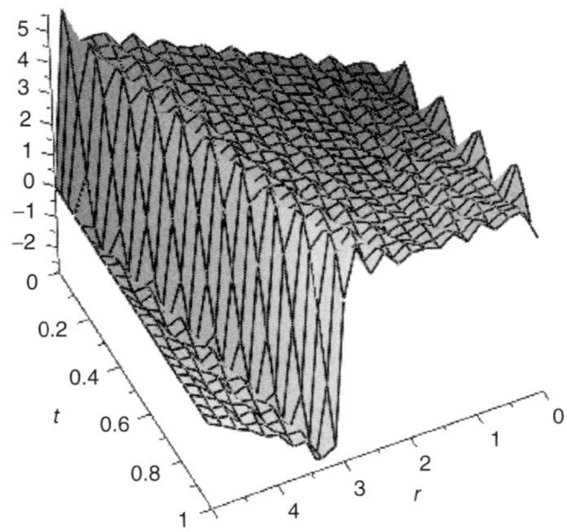

Figure 9.54: $z(r,t)$ in problem 3.

5.

$$z(r,t) = \sum_{n=1}^{\infty} c_n J_0(j_n r)\cos(3j_n t),$$

with

$$c_n = \frac{2}{J_1^2(j_n)} \int_0^1 \xi \cos(\pi\xi/2) J_0(j_n\xi)\, d\xi.$$

Figure 9.55 shows wave profiles $z(r,t)$ moving downward for times $t = 0, 1/10$ and $1/4$.

6.4 Legendre Polynomials and Applications

5. $x^2 - x + 1 = \frac{4}{3}P_0(x) - P_1(x) + \frac{2}{3}P_2(x)$.
7. $1 - x^3 = 1 - \frac{3}{5}P_1(x) - \frac{2}{5}P_3(x)$.
9. See Figure 9.56.
11. See Figure 9.57. With e^{-x}, the 11th partial sum is not a good fit to the function on $[-1, 1]$.
13. See Figure 9.58.
15.

$$u(\rho,\varphi) = \sum_{n=0}^{\infty} c_n \left(\frac{\rho}{2}\right)^n P_n(\cos(\varphi)),$$

where

$$c_n = \frac{2n+1}{2} \int_{-1}^1 (2 + \arccos^2(\xi)) P_n(\xi)\, d\xi.$$

Figure 9.59 is a graph of the surface $u(\rho,\varphi)$.

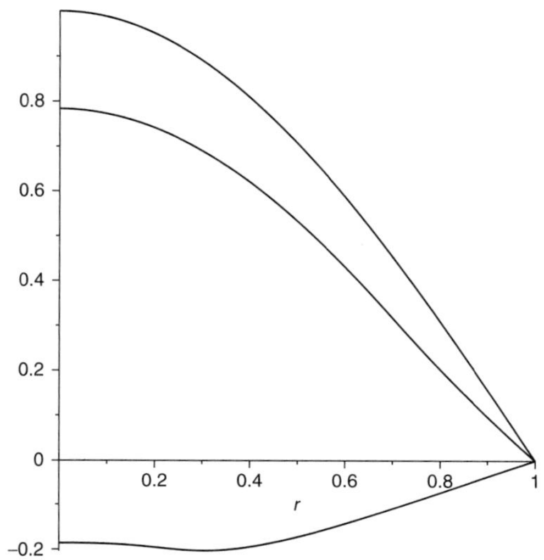

Figure 9.55: $z(r,t)$ at selected times in problem 5.

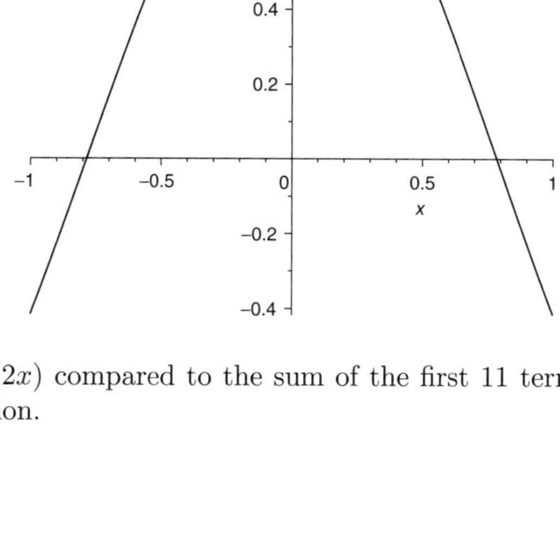

Figure 9.56: $\cos(2x)$ compared to the sum of the first 11 terms of its Fourier-Legendre expansion.

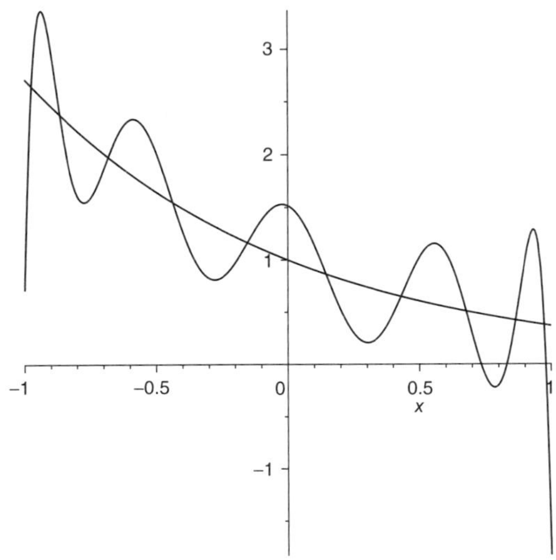

Figure 9.57: e^{-x} compared to the sum of the first 11 terms of its Fourier-Bessel expansion.

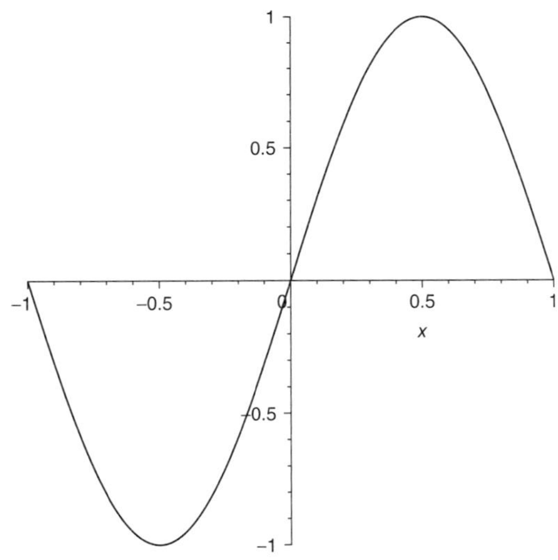

Figure 9.58: $\sin(\pi x)$ compared to the sum of the first 11 terms of its Fourier-Bessel expansion.

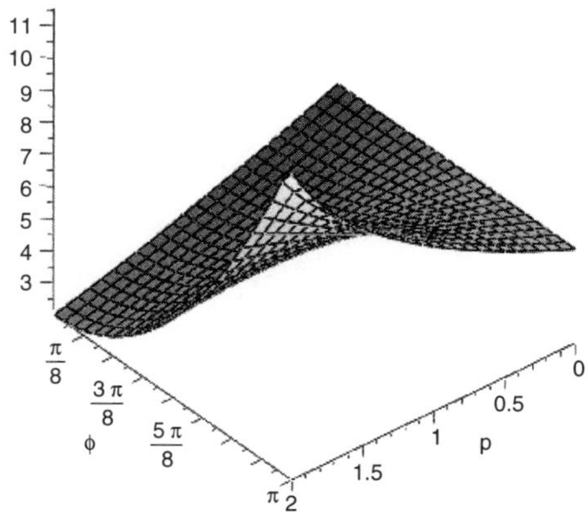

Figure 9.59: $u(\rho, \varphi)$ in problem 15.

17.

$$u(\rho, \varphi) = \sum_{n=0}^{\infty} c_n \left(\frac{\rho}{4}\right)^n P_n(\cos(\varphi)),$$

where

$$c_n = \frac{2n+1}{2} \int_{-1}^{1} (\cos(2\arccos(\xi)))P_n(\xi)\, d\xi.$$

Figure 9.60 is a graph of the surface $u(\rho, \varphi)$.

19.

$$u(\rho, \varphi) = \sum_{n=0}^{\infty} c_n \left(\frac{\rho}{5}\right)^n P_n(\cos(\varphi)),$$

where

$$c_n = \frac{2n+1}{2} \int_{-1}^{1} (\arccos(\xi) + (\arccos^2(\xi)))P_n(\xi)\, d\xi.$$

Figure 9.61 is a graph of $u(\rho, \varphi)$.

Chapter 7: Integral Transform Methods of Solution

7.1 The Fourier Transform

1.

$$\widehat{f}(\omega) = \frac{2}{\omega^2 - 1}(-\sin(a)\cos(a\omega) + \omega \sin(a\omega)\cos(a)).$$

3.

$$\widehat{f}(\omega) = \frac{-4i\omega}{(\omega^2+1)^2}.$$

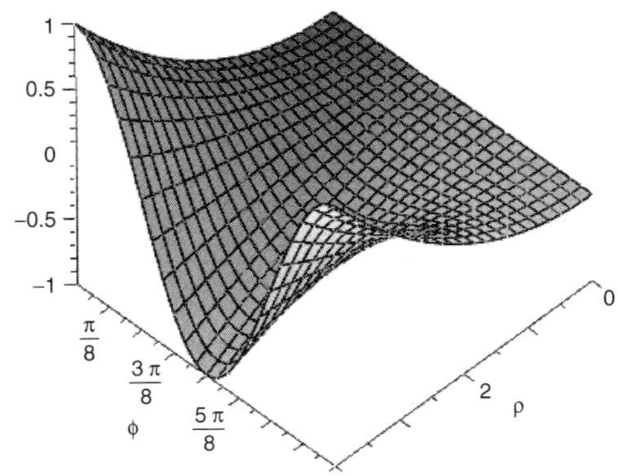

Figure 9.60: $u(\rho, \varphi)$ in problem 17.

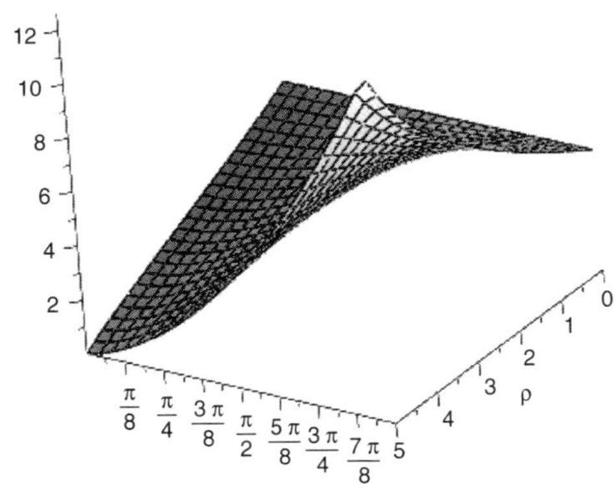

Figure 9.61: $u(\rho, \varphi)$ in problem 19.

5.

$$\widehat{f}(\omega) = -\frac{4ik}{\omega}\sin^2(\omega/2).$$

7.

$$\widehat{f}(\omega) = \frac{ik}{\omega}[\cos(\omega b) - \cos(\omega a) - i(\sin(\omega b) - \sin(\omega a))].$$

11.

$$\widehat{f}(\omega) = \frac{1}{k + i\omega}.$$

21. Carry out the integrations directly from the definition of the convolution, considering the cases $-1 < t < 0$ and $0 \leq t \leq 1$, to obtain

$$(f * g)(t) = \frac{2}{k^2 - 1}e^{kt}.$$

27.

$$\begin{aligned}\widehat{f}_S(\omega) &= \frac{2a\omega}{(a^2 + \omega^2)^2}\\ \widehat{f}_C(\omega) &= \frac{a^2 - \omega^2}{(a^2 + \omega^2)^2}.\end{aligned}$$

29.

$$\begin{aligned}&\widehat{f}_S(\omega)\\ &= \frac{k}{\omega}(1 - \cos(a\omega)) + \frac{k}{\omega}(2\cos^2(a\omega) - \cos(a\omega) - 1),\\ &\widehat{f}_C(\omega)\\ &= \frac{k}{\omega}\frac{\sin(a\omega)}{\omega} + \frac{k\sin(a\omega)}{\omega}(1 - 2\cos(a\omega)).\end{aligned}$$

7.2 Heat and Wave Equations by Fourier Transform

1. Write the solution as

$$u(x,t) = \int_{-\infty}^{\infty} \left(\frac{1}{2\pi} \int_{-\infty}^{\infty} \cos(\omega(x-\xi))e^{-k\omega^2 t}\, d\omega \right) f(\xi)\, d\xi.$$

From the term in large parentheses, define $G(x,t)$.

3.

$$u(x,t) =$$
$$\frac{2}{\pi}\int_0^{\infty} \left(\frac{1 - e^{-1}\cos(\omega) + e^{-1}\omega\sin(\omega)}{1+\omega^2} \right) \cos(\omega x)e^{-k\omega^2 t}\, d\omega.$$

5.

$$u(x,t) = -\frac{2k}{\pi}e^{-t}\int_0^{\infty}\int_0^t f(\xi)e^{(1+k\omega^2)\xi}\, d\xi \cos(\omega x)e^{-k\omega^2 t}\, d\omega.$$

7.

$$u(x,t) = \frac{2}{\pi}\int_0^{\infty} \frac{\sin(\pi\omega)}{1-\omega^2}\sin(\omega x)\cos(\omega ct)\, d\omega.$$

9.

$$u(x,t) =$$
$$\frac{1}{\pi}\int_0^{\infty}\left[\frac{1}{1+\omega^2}\cos(\omega x) + \frac{\omega}{1+\omega^2}\sin(\omega x)\right]\cos(\omega ct)\, d\omega.$$

11.

$$u(x,t) = \int_0^{\infty} \frac{2}{\pi\omega c}\frac{\cos(\pi\omega/2)}{1-\omega^2}\cos(\omega x)\sin(\omega ct)\, d\omega.$$

7.3 The Telegraph Equation

1. $b^2 - a = 3 > 0$, so case 3 of the discussion applies and

$$u(x,t) = \frac{1}{2\pi}\int_{|\omega|<1/\sqrt{3}} U_1(\omega,t)e^{i\omega x}\, d\omega$$
$$+ \frac{1}{2\pi}\int_{|\omega|\geq 1/\sqrt{3}} U_2(\omega,t)e^{i\omega x}\, d\omega.$$

Here

$$U_1(\omega,t) = e^{-2t}\left[c_1 e^{\sqrt{3-9\omega^2}t} + c_2 e^{-\sqrt{3-9\omega^2}t}\right],$$

with

$$c_1 = \frac{1-\cos(\omega)}{\omega^2}\left(1 + \frac{2}{\sqrt{3-9\omega^2}}\right),$$

and

$$c_2 = \frac{1-\cos(\omega)}{\omega^2}\left(1 - \frac{2}{\sqrt{3-9\omega^2}}\right).$$

And

$$\begin{aligned} U_2(\omega,t) &= c_3 e^{-2t}\cos(\sqrt{-3+9\omega^2}t) \\ &\quad + c_4 e^{-2t}\sin(\sqrt{-3+9\omega^2}t), \end{aligned}$$

with

$$c_3 = \frac{2}{\omega^2}(1-\cos(\omega))$$

and

$$c_4 = \frac{1-\cos(\omega)}{\omega^2}\frac{1}{\sqrt{-3+9\omega^2}}.$$

3. $b^2 - a = 0$, so case 2 applies. The solution is

$$u(x,t) = \frac{1}{2\pi}\int_{-\infty}^{\infty} U(\omega,t)e^{i\omega x}\,d\omega,$$

where

$$U(\omega,t) = e^{-3t}\left[f(\omega)\cos(9\omega t) + \frac{3f(\omega)}{4\omega}\sin(9\omega t)\right].$$

5. Because $b^2 - a = 13 > 0$, case 3 applies. Here $F(\omega) = 0$ and

$$G(\omega) = \frac{4}{\omega^2}(\sin(\omega) - \omega\cos(\omega)).$$

Let

$$c_1 = \frac{g(\omega)}{2\sqrt{13-\omega^2}}$$

and $c_2 = -c_1$. Define

$$U_1(\omega,t) = e^{-4t}\left[c_1 e^{\sqrt{13-\omega^2}t} + c_2 e^{-\sqrt{13-\omega^2}t}\right].$$

Let $c_3 = 0$. Let

$$c_4 = \frac{G(\omega)}{\sqrt{\omega^2-13}},$$

and

$$U_2(\omega,t) = e^{-4t}c_4\sin(\sqrt{\omega^2-13}t).$$

The solution is

$$\begin{aligned} u(x,t) &= \frac{1}{2\pi}\int_{|\omega|<\sqrt{13}} U_1(\omega,t)e^{i\omega x}\,d\omega \\ &\quad + \frac{1}{2\pi}\int_{|\omega|\geq\sqrt{13}} U_2(\omega,t)e^{i\omega x}\,d\omega. \end{aligned}$$

7.4 The Laplace Transform

1.

$$\begin{aligned} u(x,t) &= e^{kt-x} - e^{kt} * \mathcal{L}^{-1}\left(e^{-\sqrt{s/k}x}\right) \\ &= e^{kt-x} - e^{kt} * \frac{x}{2\sqrt{\pi t^3}} e^{-x^2/4kt}. \end{aligned}$$

3.

$$u(x,t) = T_0 \sum_{n=0}^{\infty} \left(\operatorname{erfc}\left(\frac{(2n+1)L - x}{2\sqrt{kt}}\right) - \operatorname{erfc}\left(\frac{(2n+1)L + x}{2\sqrt{kt}}\right)\right).$$

5.

$$u(x,t) = \frac{A}{6}\left(t - \frac{x}{c}\right)^3 H\left(t - \frac{x}{c}\right) - \frac{A}{6}t^3,$$

where $H(t)$ is the Heaviside function.

Chapter 8: First-Order Equations

8.1 Linear First-Order Equations

1. $u(x,y) = g(5x - 3y)e^{(yx^2 - 5x^3/9)/6}$.
3. $u(x,y) = -1 + e^{x^2/2}g(4x - y)$.
5. $u(x,y) = \frac{1}{2}x + \frac{1}{x}g(xy)$.
7. $u(x,y) = 4x + g(y + x^2/2)$.
9. $u(x,y) = -1 - y + e^x g(y - x)$.
11. $u(x,y) = e^{-x^2/2}g(x - \ln(y))$.
13. Obtain the general solution

$$u(x,y) = \left(1 + e^{-\alpha x + \ln(y-1)}\right)^n e^{\frac{1}{2}\beta e^{-\alpha x + \ln(y-1)} \int_0^x f(s)\,ds}.$$

Now set $u(0,y) = y^n$ to obtain the solution satisfying this condition.

8.2 The Significance of Characteristics

1. Characteristics are the ellipses $x^2 + 3y^2/2 = k$.
   (a) $u(x,y) = \frac{2}{5}\left(x^2 + \frac{3}{2}y^2\right)$.
   (b) $u(x,y) = 1 - \frac{2}{5}\left(x^2 + \frac{3}{2}y^2\right)$.
   (c) There is no solution with data specified along $2x^2 + 3y^2 = 4$, which is a characteristic.

3. Characteristics are lines $y = 2x + k$.

   (a) $u(x, y) = -1 + (1 + \cos(y - 2x))e^{-y/4}e^{3x/4}$.

   (b) There is no solution with the data given on this characteristic.

   (c) If $x = 1 + \sqrt{1+t}$, then

$$u(x, y) = -1 + (1 - \sqrt{1 + y - 2x})e^{x/4}e^{-(1+\sqrt{1+y-2x})/4}.$$

   If $x = 1 - \sqrt{1+t}$, then

$$u(x, y) = -1 + (1 + \sqrt{1 + y - 2x})e^{x/4}e^{-(1-\sqrt{1+y-2x})/4}.$$

5. The characteristics are curves $2x^3 - 3y^2 = k$.

   (a)

$$u(x, y) = \frac{1}{2}x^2 + 4\left(\frac{3}{5}(2x^3 - 3y^2)\right)^{1/3} - \frac{1}{2}\left(\frac{3}{5}(2x^3 - 3y^2)\right)^{2/3}.$$

   (b) There is no solution satisfying this condition.

   (c)

$$u(x, y) = \frac{1}{2}x^2 + \sin\left(\left(\frac{2x^3 - 3y^2}{2}\right)^{1/3}\right) - \frac{1}{2}\left(\frac{2x^3 - 3y^2}{2}\right)^{2/3}.$$

8.3 The Quasi-Linear Equation

1. $u(x, y) = \arcsin(\frac{1}{2}\ln(x^3/y))$.

3. $u(x, y)$ is implicitly defined by

$$y = \frac{1}{u + 1/((x - u)^2 + 2)}.$$

5. $u(x, y)$ is implicitly defined by

$$uy = 1 - x + \ln(u).$$

7.

$$u(x, y) = \frac{3}{y - 2x + \frac{7}{4}}.$$

Index

PURE AND APPLIED MATHEMATICS

A Wiley Series of Texts, Monographs, and Tracts

Founded by RICHARD COURANT

Editors Emeriti: MYRON B. ALLEN III, DAVID A. COX, PETER HILTON, HARRY HOCHSTADT, PETER LAX, JOHN TOLAND

ADÁMEK, HERRLICH, and STRECKER—Abstract and Concrete Catetories
ADAMOWICZ and ZBIERSKI—Logic of Mathematics
AINSWORTH and ODEN—A Posteriori Error Estimation in Finite Element Analysis
AKIVIS and GOLDBERG—Conformal Differential Geometry and Its Generalizations
ALLEN and ISAACSON—Numerical Analysis for Applied Science
*ARTIN—Geometric Algebra
ATKINSON, HAN, and STEWART—Numerical Solution of Ordinary Differential Equations
AUBIN—Applied Functional Analysis, Second Edition
AZIZOV and IOKHVIDOV—Linear Operators in Spaces with an Indefinite Metric
BASENER—Topology and Its Applications
BERG—The Fourier-Analytic Proof of Quadratic Reciprocity
BERKOVITZ—Convexity and Optimization in $\mathbb{R}^n$
BERMAN, NEUMANN, and STERN—Nonnegative Matrices in Dynamic Systems
BOYARINTSEV—Methods of Solving Singular Systems of Ordinary Differential Equations
BRIDGER—Real Analysis: A Constructive Approach
BURK—Lebesgue Measure and Integration: An Introduction
*CARTER—Finite Groups of Lie Type
CASTILLO, COBO, JUBETE, and PRUNEDA—Orthogonal Sets and Polar Methods in Linear Algebra: Applications to Matrix Calculations, Systems of Equations, Inequalities, and Linear Programming
CASTILLO, CONEJO, PEDREGAL, GARCIÁ, and ALGUACIL—Building and Solving Mathematical Programming Models in Engineering and Science
CHATELIN—Eigenvalues of Matrices
CLARK—Mathematical Bioeconomics: The Mathematics of Conservation, Third Edition
COX—Galois Theory, Second Edition
COX—Primes of the Form $x^2 + ny^2$: Fermat, Class Field Theory, and Complex Multiplication, Second Edition
*CURTIS and REINER—Representation Theory of Finite Groups and Associative Algebras
*CURTIS and REINER—Methods of Representation Theory: With Applications to Finite Groups and Orders, Volume I
CURTIS and REINER—Methods of Representation Theory: With Applications to Finite Groups and Orders, Volume II
DINCULEANU—Vector Integration and Stochastic Integration in Banach Spaces
*DUNFORD and SCHWARTZ—Linear Operators
Part 1—General Theory
Part 2—Spectral Theory, Self Adjoint Operators in Hilbert Space
Part 3—Spectral Operators
FARINA and RINALDI—Positive Linear Systems: Theory and Applications

*Now available in a lower priced paperback edition in the Wiley Classics Library.
†Now available in paperback.

FATICONI—The Mathematics of Infinity: A Guide to Great Ideas, Second Edition
FOLLAND—Real Analysis: Modern Techniques and Their Applications
FRÖLICHER and KRIEGL—Linear Spaces and Differentiation Theory
GARDINER—Teichmüller Theory and Quadratic Differentials
GILBERT and NICHOLSON—Modern Algebra with Applications, Second Edition
*GRIFFITHS and HARRIS—Principles of Algebraic Geometry
GRILLET—Algebra
GROVE—Groups and Characters
GUSTAFSSON, KREISS, and OLIGER—Time-Dependent Problems and Difference Methods, Second Edition
HANNA and ROWLAND—Fourier Series, Transforms, and Boundary Value Problems, Second Edition
*HENRICI—Applied and Computational Complex Analysis
Volume 1, Power Series—Integration—Conformal Mapping—Location of Zeros
Volume 2, Special Functions—Integral Transforms—Asymptotics—Continued Fractions
Volume 3, Discrete Fourier Analysis, Cauchy Integrals, Construction of Conformal Maps, Univalent Functions
*HILTON and WU—A Course in Modern Algebra
*HOCHSTADT—Integral Equations
JOST—Two-Dimensional Geometric Variational Procedures
KHAMSI and KIRK—An Introduction to Metric Spaces and Fixed Point Theory
*KOBAYASHI and NOMIZU—Foundations of Differential Geometry, Volume I
*KOBAYASHI and NOMIZU—Foundations of Differential Geometry, Volume II
KOSHY—Fibonacci and Lucas Numbers with Applications
LAX—Functional Analysis
LAX—Linear Algebra and Its Applications, Second Edition
LOGAN—An Introduction to Nonlinear Partial Differential Equations, Second Edition
LOGAN and WOLESENSKY—Mathematical Methods in Biology
LUI—Numerical Analysis of Partial Differential Equations
MARKLEY—Principles of Differential Equations
MORRISON—Functional Analysis: An Introduction to Banach Space Theory
NAYFEH—Perturbation Methods
NAYFEH and MOOK—Nonlinear Oscillations
O'LEARY—Revolutions of Geometry
O'NEIL—Beginning Partial Differential Equations, Third Edition
PANDEY—The Hilbert Transform of Schwartz Distributions and Applications
PETKOV—Geometry of Reflecting Rays and Inverse Spectral Problems
*PRENTER—Splines and Variational Methods
PROMISLOW—A First Course in Functional Analysis
RAO—Measure Theory and Integration
RASSIAS and SIMSA—Finite Sums Decompositions in Mathematical Analysis
RENELT—Elliptic Systems and Quasiconformal Mappings
RIVLIN—Chebyshev Polynomials: From Approximation Theory to Algebra and Number Theory, Second Edition
ROCKAFELLAR—Network Flows and Monotropic Optimization
ROITMAN—Introduction to Modern Set Theory
ROSSI—Theorems, Corollaries, Lemmas, and Methods of Proof
*RUDIN—Fourier Analysis on Groups
SENDOV—The Averaged Moduli of Smoothness: Applications in Numerical Methods and Approximations

*Now available in a lower priced paperback edition in the Wiley Classics Library.
†Now available in paperback.

SENDOV and POPOV—The Averaged Moduli of Smoothness
SEWELL—The Numerical Solution of Ordinary and Partial Differential Equations, Second Edition
SEWELL—Computational Methods of Linear Algebra, Second Edition
SHICK—Topology: Point-Set and Geometric
SHISKOWSKI and FRINKLE—Principles of Linear Algebra With *Maple*™
SHISKOWSKI and FRINKLE—Principles of Linear Algebra With *Mathematica*®
*SIEGEL—Topics in Complex Function Theory
Volume 1—Elliptic Functions and Uniformization Theory
Volume 2—Automorphic Functions and Abelian Integrals
Volume 3—Abelian Functions and Modular Functions of Several Variables
SMITH and ROMANOWSKA—Post-Modern Algebra
ŠOLÍN–Partial Differential Equations and the Finite Element Method
STADE—Fourier Analysis
STAHL and STENSON—Introduction to Topology and Geometry, Second Edition
STAHL—Real Analysis, Second Edition
STAKGOLD and HOLST—Green's Functions and Boundary Value Problems, Third Edition
STANOYEVITCH—Introduction to Numerical Ordinary and Partial Differential Equations Using MATLAB®
*STOKER—Differential Geometry
*STOKER—Nonlinear Vibrations in Mechanical and Electrical Systems
*STOKER—Water Waves: The Mathematical Theory with Applications
WATKINS—Fundamentals of Matrix Computations, Third Edition
WESSELING—An Introduction to Multigrid Methods
†WHITHAM—Linear and Nonlinear Waves
ZAUDERER—Partial Differential Equations of Applied Mathematics, Third Edition

*Now available in a lower priced paperback edition in the Wiley Classics Library.
†Now available in paperback.